INTRAOVARIAN CONTROL MECHANISMS

ADVANCES IN EXPERIMENTAL MEDICINE AND BIOLOGY

Recent Volumes in this Series

Volume 140
COMPOSITION AND FUNCTION OF CELL MEMBRANES: Application to the Pathophysiology of Muscle Diseases
Edited by Stewart Wolf and Allen K. Murray

Volume 141
BIOCHEMISTRY AND FUNCTION OF PHAGOCYTES
Edited by F. Rossi and P. Patriarca

Volume 142
BIOCHEMISTRY AND BIOLOGY OF CORONAVIRUSES
Edited by V. ter Meulen, S. Siddell, and H. Wege

Volume 143
RELAXIN
Edited by Ralph R. Anderson

Volume 144
MUCUS IN HEALTH AND DISEASE – II
Edited by Eric N. Chantler, James B. Elder, and Max Elstein

Volume 145
TERMINAL TRANSFERASE IN IMMUNOBIOLOGY AND LEUKEMIA
Edited by Umberto Bertazzoni and F. J. Bollum

Volume 146
MECHANISMS OF CELL-MEDIATED CYTOTOXICITY
Edited by William R. Clark and Pierre Golstein

Volume 147
INTRAOVARIAN CONTROL MECHANISMS
Edited by Cornelia P. Channing and Sheldon J. Segal

Volume 148
STRUCTURE AND FUNCTION RELATIONSHIPS IN BIOCHEMICAL SYSTEMS
Edited by Francesco Bossa, Emilia Chiancone, Alessandro Finazzi-Agrò, and Roberto Strom

INTRAOVARIAN CONTROL MECHANISMS

Edited by

Cornelia P. Channing

Department of Physiology
School of Medicine
University of Maryland
Baltimore, Maryland

and

Sheldon J. Segal

Population Division
The Rockefeller Foundation
New York, New York

PLENUM PRESS • NEW YORK AND LONDON

Library of Congress Cataloging in Publication Data

Conference on Intraovarian Control Mechanisms (1981: Bellagio, Italy)
Intraovarian control mechanisms.

(Advances in experimental medicine and biology; v. 147)
Includes bibliographical references and index.
1. Ovaries—Congresses. 2. Hormones, Sex—Physiological effect—Congresses. 3. Cellular control mechanisms—Congresses. I. Channing, Cornelia P. II. Segal, Sheldon Jerome, Date- . III. Title. IV. Series. [DNLM: 1. Ovary—Physiology—Congresses. W1 AD559 v. 147 / WP 320 C747i 1981]
QP261.C67 1981 599.01'6 82-9849
ISBN-13: 978-1-4615-9280-8 e-ISBN-13: 978-1-4615-9278-5
DOI: 10.1007/978-1-4615-9278-5 AACR2

Proceedings of a Conference on Intraovarian Control Mechanisms, held April 28-30, 1981, at the Villa Serbelloni, in Bellagio, Italy

Softcover reprint of the hardcover 1st edition 1982

A Division of Plenum Publishing Corporation
233 Spring Street, New York, N.Y. 10013

PREFACE

Recently intraovarian control of steroidogenesis and oocyte maturation has become a subject of interest. Non steroidal substances secreted locally by the ovarian follicle and corpus luteum have been implicated in control of oocyte maturation follicular maturation and steroidogenesis as well as luteal function. In addition some ovarian polypeptides such as inhibin F (folliculostatin) leave the ovary and act as modulators of pituitary function which in turn regulates ovarian function by exhibiting changes in follicle stimulating hormone secretion in response to changes in inhibin. It was desired to see how these intraovarian regulators act to control ovarian function. In addition it was of interest to see what the chemical nature of these intraovarian regulators is. To achieve these ends a meeting was held at the Villa Serbelloni in Bellagio situated overlooking Lake Como. Leaders in this field were invited and asked to vigorously discuss the subject of intraovarian control mechanisms. Their manuscripts comprise this volume.

C.P. Channing
S.J. Segal

CONTENTS

MATURATION, MORPHOLOGY, AND ENDOCRINE

FUNCTION OF THE OVARIAN FOLLICLE

Lars Bjersing

Institute of Pathology
University of Umeå
S-901 87 Umeå, Sweden

Introduction

> He, who sees things grow from the beginning will have the finest view of them.
> (Aristotle 384-322 B.C.)

These words still hold true and in looking for intraovarian control mechanisms I will go far back in ovarian and follicular history. Then we will follow the follicle from its formation up to its grand finale, follicle rupture. During that time many local factors determine its maturation, morphology and function. A lot of such factors will be discussed in detail later in the symposium. To supplement these some other fascinating intraovarian mechanisms will be the subject of this presentation.

Prefollicular History, Histology, and Hormones

Exciting advances have recently been made regarding gonadal embryology and differentiation but many problems remain unsolved. Six weeks after conception the gonad is still undifferenlated but in week seven gonadal sex is seen; testicular cords are obvious in the XY embryo (cf. van Wagenen and Simpson, 1965). By contrast the ovary is unchanged as can be seen in Table 1.

Only two questions about local factors of importance for follicle formation and maturation will be posed.

First, what is the role of the oocyte in follicle formation? Without oocytes no follicles are formed. But in the testis, tubules

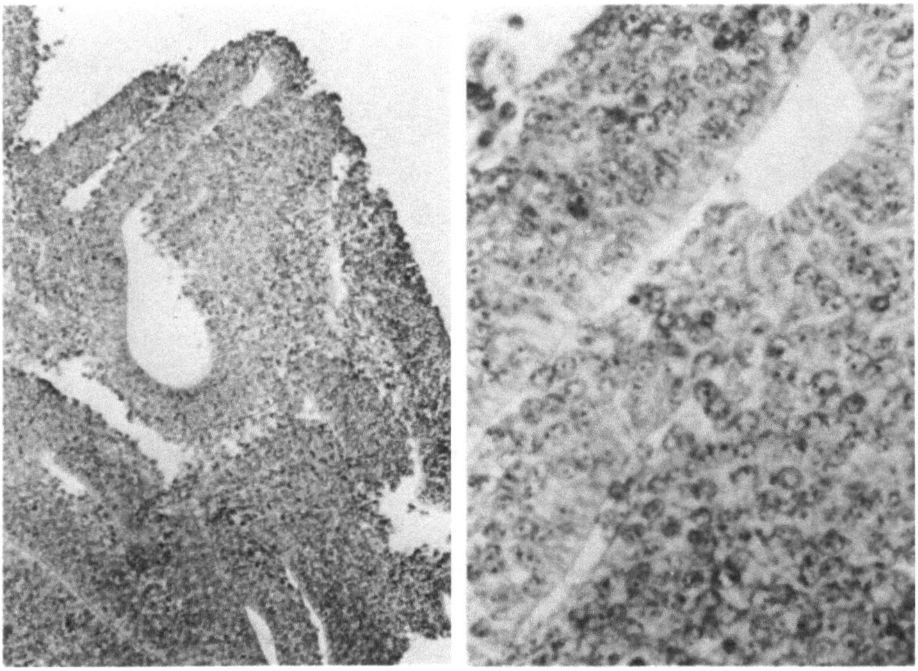

a b

Fig. 1. Reproduction of twenty-year-old, unpublished color micrographs of porcine granulosa cells, isolated according to the method of Bjersing (1962). The sharp contours denote the distinct separation of the avascular membrana granulosa from theca interna along the basement membrane. (a) ×150; (b) detailed view of (a), ×600.

are formed even in the absence of germ cells (cf. Gondos, 1980). The precise role of the oocyte is not known. In a recent paper Ohno and his co-workers (Ohno et al., 1979) suggested that its task was purely mechanic; they even planned to add variously charged Sephadex beads to see if follicles would form around the beads. So far, I have not seen any report of such experiments.

In another recent paper Ohno discussed and illustrated the potential role of ovary-organizing X-antigen (Ohno, 1980). The follicular cells become strongly attracted to individual oocytes as the dictytate oocyte begins to express both an ovary-organizing antigen and its specific receptor. The lack of both H-Y antigen and its receptor in male germ cells is probably important for the formation of the tubules in the male gonad. These ideas on gonadal morphogenesis are most interesting but they require more experimental support. But one thing seems clear, normal follicular differentiation depends on meiotic germ cells (cf. Jirásek, 1977; Jost, 1980).

Table 1. Major Events in Early Human Gonads

Weeks after conception	Events
5	Germ cell migration to gonadal primordia
6	Unisex
7	Sertoli cells form testicular cords
	Ovaries unchanged
8	Leydig cells appear
	Testes produce testosterone
	Ovaries produce estradiol

For references, see George and Wilson, 1978; Jirăsek, 1977; Peters and McNatty, 1980; van Wagenen and Simpson, 1965.

Second, what is the role of estrogen in germ cell differentiation and follicle formation? Only recently was it found that the fetal ovary produces and contains a lot of estrogen very early in its career, well before histologic differentiation and follicle formation in the ovary (George and Wilson, 1978, 1980). Almost at the same time the fetal testis starts to produce larger amounts of testosterone.

As pointed out by George, Milewich, and Wilson (1978) it is not known whether the early estrogen synthesis in the ovary has any functional significance or whether it is only a marker of differentiation. If it has any effect it is probably local — plasma levels are not elevated. And we do know that estrogen stimulates granulosa cell proliferation in the postnatal ovary (Bradbury, 1961).

But where in the ovary is the estrogen produced? In the adult ovary granulosa cells convert androgens to estrogens with great efficiency (cf. Bjersing, 1978). It then appears likely that it is the pregranulosa cells that produce (the main part of) the estrogen before follicle formation. If granulosa and pregranulosa cells stem from the ovarian rete, as proposed by Byskov (cf. Byskov et al., 1977; Byskov, 1978a) it is not unlikely that even the fetal rete cells can convert androgen to estrogen.

Byskov and others (Byskov, 1974; Byskov and Saxén, 1976; O and Baker, 1976; Rivelis et al., 1976) have shown that the fetal rete ovarii secretes a diffusible meiosis-inducing substance and that media from cultured testes prevent meiosis. Recently, large amounts of testosterone were demonstrated in such meiosis-inhibiting media (Byskov, 1978b). Could it simply be that estrogen is the meiosis-inducing substance and testosterone the meiosis-preventing substance?

If this is correct, estrogens should be obligatory for follicle formation, since no follicles form without meiotic oocytes (cf. Jirásek, 1977; Jost, 1980).

Steroids have been suggested as meiosis-inducers before. On the basis of studies in birds, Gregory Erickson (1974a) proposed that progestin may be involved. But, application of various steroids to early gonads of both sexes have failed to mimic the effects of meiosis-inducing and meiosis-preventing substances (cf. McCarrey and Abbott, 1979). One reason could be the inability to reach the proper concentration of the estrogen or androgen, another reason could be that a certain combination or sequence of steroids is necessary, just as estrogen followed by progestin is needed in creating secretory endometrium.

One way to study this would be to inhibit separate steps in steroidogenesis in culture experiments, another just to try differerent combinations of steroids on cultures.

Granulosa-Theca Cell Interaction and Control of Steroid Hormone Synthesis

Granulosa-thecal interaction is perhaps the oldest and best known example of an intraovarian control mechanism. Opinions on the endocrine role of granulosa and theca cells have differed from time to time. Only recently has the granulosa been generally accepted as an important partner.

One reason for the different opinions was the lack of direct evidence. But in 1959 Falck in Lund isolated single ovarian cell types from the rat ovary and cultured them in the anterior chamber of the eye. He studied their estrogenic activity both alone and in combination, and used vaginal autotransplants as indicators. From these experiments he concluded that both granulosa and theca cells were necessary for effective estrogen production.

Falck used elegant biological methods but the nature of the interaction between the two cell types was not evident from his studies. For detailed information and direct evidence of steroid hormone synthesis biochemical in vitro studies seemed necessary.

My own interest in the granulosa cells started more than twenty years ago and was aroused during work with feminizing and viriliz-

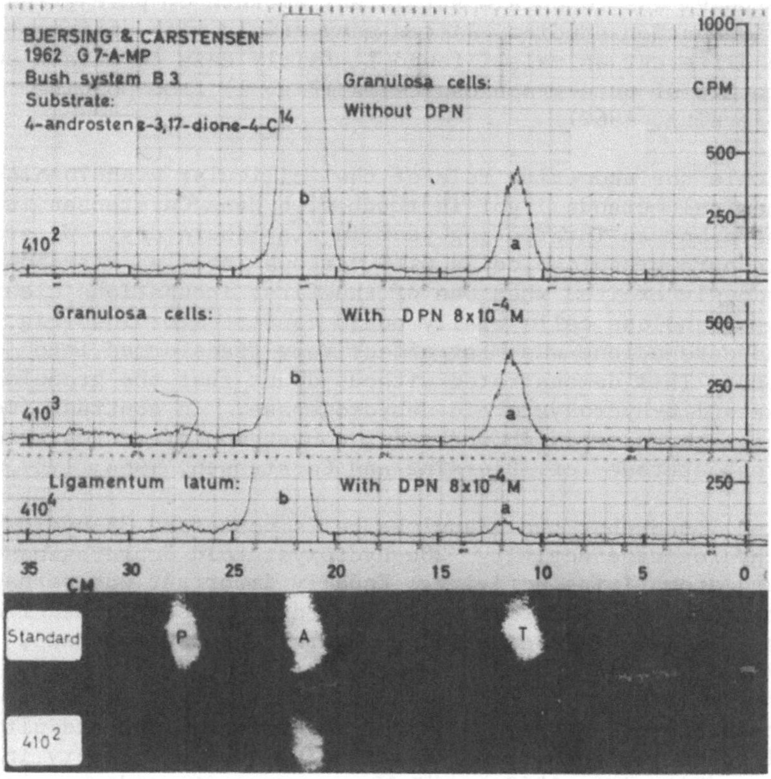

Fig. 2. Early recording of a paper chromatography strip from an incubation of pure granulosa cells (see Fig. 1) with 4-^{14}C-androstenedione. The labelled substances in this chromatography step were located by scanning in a Frieseke-Hoepfner gas flow counter. Below, standard steroids A (androstenedione) and T (testosterone), run on parallel strips and located by photography in ultraviolet light. In the top recording a marked conversion of androstenedione to testosterone can be seen.

ing ovarian tumors. During that work I realized that very little was definitely known about formation of steroid hormones in different types of normal ovarian cells. For detailed information and direct evidence of steroid hormone synthesis biochemical in vitro studies of single cell types seemed necessary. Much to my surprise I found no such biochemical studies on pure cells.

Since the follicular granulosa cells are completely separated from blood vessels and other cell types — by the basement membrane of the follicle — I considered these cells ideal for the purpose.

The only problem was to obtain sufficient amounts of pure granulosa cells for incubation studies. After many trials with different methods on different animals I found it fairly easy to obtain sufficient amounts of pure granulosa cells (Fig. 1) from porcine ovaries (Bjersing, 1962).

To tackle the endocrine role of the follicular granulosa cells with biochemical methods I got in touch with Hans Carstensen in Uppsala and spent an interesting year in his laboratory. We started to incubate pure granulosa cells with various labelled steroids and got tremendously excited when one of the first incubations clearly showed that granulosa cells really could convert steroids (Fig. 2). In fact androstenedione was most effectively transformed into testosterone. This demonstrated without doubt that the granulosa cells possess 17β-hydroxysteroid dehydrogenase. In contrast (more or less pure) theca cells showed a low conversion of androstenedione into testosterone (cf. Bjersing and Carstensen, 1964, 1967).

We also found that the granulosa cells possessed 3β-hydroxysteroid dehydrogenase activity, 20α-hydroxysteroid dehydrogenase activity and aromatizing activity. Equally important was perhaps the finding that the granulosa cells lacked or almost lacked 17α-hydroxylase and 17,20-lyase activity (Bjersing and Carstensen, 1964, 1967).

On the basis of these results and on histochemical and ultrastructural studies, I proposed the following two-cell theory in the late sixties (Bjersing, 1967a): "C-19 precursor steroids are elaborated by theca interna cells through 17α-hydroxylation and side-chain cleavage and transferred across the basement membrane of the follicle to the granulosa cells for the production of testosterone by these cells. Aromatization to oestrogens, particularly oestradiol-17β, may then be carried out by both granulosa and theca cells."

Kenneth Ryan has recently tabulated different theories on estrogen formation in the ovarian follicle (Ryan, 1980). Falck was the pioneer, 1959, and he was followed by Short (1962) who made early in vivo experiments in the mare and concluded that only the theca cell produced estrogen. Other important and ingenious in vivo experiments were performed by Channing in cooperation with Coudert (Channing and Coudert, 1976).

Some divergencies in the results may be due to species differences. More important, however, is probably time and location, when in the cycle an experiment was performed and from where the cells were taken, from the central or peripheral part of the granulosa layer (cf. Zoller and Weisz, 1979) and, of course, also if the isolated cells were free of contamination.

Opinions on the role of granulosa and theca cells still differ but it is becoming increasingly clear that the granulosa plays an important role for effective oestrogen production by the ovarian follicle; the two-cell theory or hypothesis proposed by Bjersing (1967) still seems to account for established facts (cf. Bjersing, 1978, 1981; Richards, 1980); Hillensjö, 1981; McNatty et al., 1980; Tsang et al., 1980).

A diagram of the hypothesis (Fig. 3) illustrates intraovarian control mechanisms with close cooperation between both cells and organelles.

Steroids and Some Other Factors Necessary for Normal Ovulation
(cf. Bjersing, 1979; Bjersing et al., 1980)

What is the mechanism of follicle rupture? What is the role of steroids?

Follicles below a certain size do not ovulate in response to LH. Estradiol is necessary for such follicle maturation. Estradiol causes an increase in FSH receptors, this in turn increases estradiol synthesis and in fact occasions the LH surge, which of course also is necessary for ovulation.

But are steroids necessary for the intraovarian mechanism of ovulation? What are their levels in the preovulatory follicle?

The pattern is essentially the same in both small and large mammals, including man. We can first take the hamster as an example. The LH peak occurs about 2 p.m. This peak is caused by estradiol. Progesterone is especially interesting.

Quite recently the fall in progesterone prior to ovulation has also been reported in man and in pigs.

Now, are all these sex steroids important or only one? Is perhaps progesterone the key substance?

Recently Mori and his coworkers (Mori et al., 1977a, b) found that antiserum to progesterone inhibits ovulation in the rat while progesterone restores ovulation. Antiserum to testosterone also inhibits ovulation in the rat. Again, testosterone or dihydrotestosterone restores ovulation after antiprogesterone has been given.

It has been found that cyanoketone that blocks steroidogenesis also blocks ovulation in rats if it is given within 4 hours of the LH peak (Lipner and Greep, 1971). We can then conclude that sex steroids are necessary for follicle rupture.

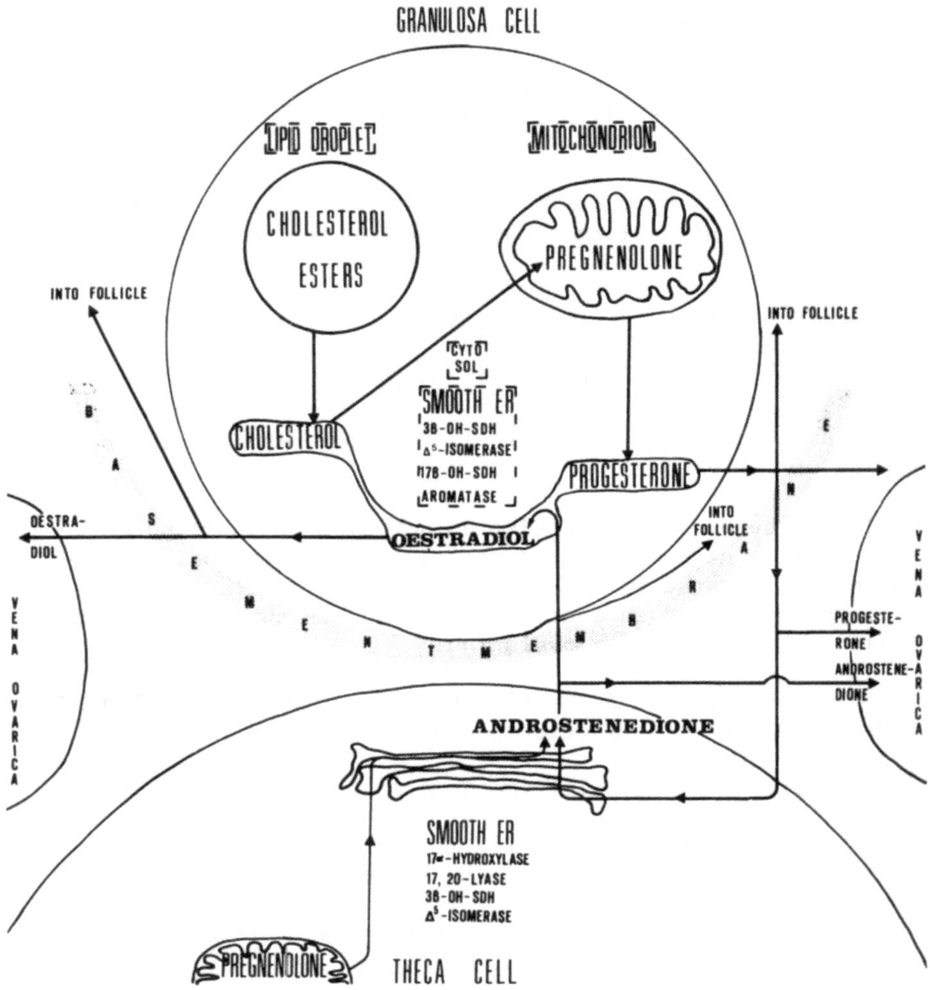

Fig. 3. Oestradiol synthesis in large Graafian follicle. Cholesterol esters are degraded to cholesterol which passes into the mitochondrion where it is converted into pregnenolone. This steroid is acted upon by 3β-hydroxysteroid-dehydrogenase (3β-OH-SDH) and Δ5-isomerase, resulting in progesterone which is produced in appreciable amounts during the preovulatory period. From C19-precursors supplied by the theca cells (and the granulosa cells) androstenedione is formed in the smooth endoplasmic reticulum of the theca interna cell. This is effected by enzymes present in the theca cell but practically lacking in the granulosa cell, namely 17α-hydroxylase and 17,20-lyase (Bjersing and Cartensen, 1967). The andro-

But how do they act? Estradiol and testosterone cause dose-dependent release of lysosomal enzymes. We also know that the steroid values are very high in the follicle. As an example we can mention that testosterone levels can be up to 3,000 times higher and estradiol levels up to 40,000 times higher in the human follicle than in peripheral plasma. More generally, steroids with two oxygenic functions per molecule — one at C-3 and the other at C-17 or C-20 — can labilize lysosomes at concentrations well below sex steroids in the preovulatory follicle.

The sex steroids probably work together, release lysosomal enzymes, help to weaken the follicle wall and to promote follicle rupture (cf. Bjersing, 1979; Bjersing et al., 1980). A number of other factors such as prostaglandins and plasminogen activators are also necessary or important for ovulation (Bjersing, 1979).

The resumption of the meiosis is also part of the preovulatory process (Tsafriri, 1978). Are perhaps some of the intraovarian factors involved in follicular rupture also involved in resumption of meiosis? Are steroids essential? Results by several groups indicate that steroids do take part in regulation of meiosis (cf. Baker, 1979; Moor and Warnes, 1979; Thibault, 1977). If steroids are essential, how do they interact with the oocyte maturation inhibitor, a factor that has been extensively studied by Channing, Hillensjö, Tsafriri, and others (cf. Channing et al., 1980; Hillensjö, 1976; Tsafriri, 1978).

Stimulation of cell membrane receptors, e.g., β-adrenergic receptors, leads to methylation of membrane phospholipids with formation of phosphatidylcholine and reduction in membrane viscosity with influx of Ca^{2+}. The increased fluidity in the membrane facilitates interaction of the receptor with the guanylnucleotide coupling factor and adenylate cyclase; as a result cyclic AMP is produced. In the presence of Ca^{2+} phospholipase A_2 is activated and this enzyme catalyzes hydrolysis of phosphatidylcholine to arachidonic acid and lysophosphatidylcholine (Hirata and Axelrod, 1980).

stenedione then passes into the follicular granulosa cells where 17β-hydroxysteroid-dehydrogenase converts androstenedione into testosterone. This steroid is in turn transformed into 17β-oestradiol by the aromatizing enzyme system present in the smooth endoplasmic reticulum (ER). (It is not indicated in the figure that the theca cell is able — mainly via the Δ5-pathway — to synthesize some oestrogen on its own).

For references regarding location of enzymes in cell organelles see Bjersing, 1967b; Christensen and Gillim, 1969; and Christensen, 1975.

According to Hirata and Axelrod (1980) desensitization occurs in cells exposed to excessive amounts of hormones and transmitters. This desensitization appears mediated by products of phospholipase A_2 activity, such as lysophosphatidylcholine, arachidonic acid or prostaglandin metabolites; treatment with phospholipase A_2 inhibitors has abolished refractoriness to cyclic AMP formation. Recently an endogenous inhibitor of phospholipase A_2 was identified in rabbit neutrophils. The synthesis of this inhibitor, a protein of a molecular weight of about 40,000, is stimulated by glucocorticoids and it is digestible by Pronase (Hirata et al., 1980).

It appears probable but remains to be established that ovarian cell membranes and their receptors react in a similar way to gonadotrophins and that glucocorticoids induce a phospholipase A_2 inhibitor protein in ovarian cells.

A massive stimulation of cell membrane receptors, e.g., the numerous LH receptors in preovulatory granulosa cells, might lead to marked structural alterations. In cells also exposed to very high levels of sex steroids that risk should be considerable since these hormones might incite membrane derangement by opposing the inhibitory effect of glucocorticoids on phospholipase A_2 activity — probably by competing with glucocorticoids for intracellular receptor binding (Hirata et al., 1980). In addition the steroids (and prostaglandin $F_{2\alpha}$) may labilize lysosomal membranes (cf. Weiner and Kaley, 1972; Bjersing and Cajander, 1975; Bjersing, 1979) with digestion of the phospholipase A_2 inhibitor as one possible result.

Follicular rupture should, perhaps, be seen as a consequence of extreme desensitization. Massive amounts of LH, sex steroids, and prostaglandins are all present in the preovulatory follicle, and according to or extrapolating from the results of Hirata and Axelrod (1980) and Hirata et al. (1980) they would all be expected to lead to desensitization. Could it also be that the combined effect of LH and sex steroids in the follicles triggers resumption of meiosis? The high levels of prostaglandin $F_{2\alpha}$ may have additional effect by labilizing lysosomes (cf. Weiner and Kaley, 1972).

To summarize and to be more precise I propose the following provocative hypothesis:

> "Preovulatory sex steroids and LH (and prostaglandins) increase phospholipase A_2 activity and trigger resumption of meiosis.
>
> Sex steroids counteract glucocorticoids and choke synthesis of phospholipase A_2 inhibitor (much the same as oocyte maturation inhibitor?)"

A seemingly important objection against the hypothesis comes from experiments with inhibition of steroidogenesis with cyanoketone or aminoglutethimide; LH induced oocyte maturation in spite of inhibition of steroidogenesis (Lieberman et al., 1976). However, the follicle destined to ovulate contains large amounts of steroid hormones already before the preovulatory LH surge (cf. Bjersing, 1978); estrogen rise precedes the increase in progestins, which may be important.

Concluding Remarks

We have now followed a follicle during its whole life time. After hard struggle in its early career it has managed to come in the forefront and win the big prize, nature's greatest reward, the possibility to create a new generation and pass on the genes.

Intraovarian control mechanisms are presenting themselves as more and more important factors and the old days are gone when FSH and LH were the only known regulators of ovarian morphology and function. The past was simple and peaceful, the present is complex and exciting, the future lies ahead and will be next year's writing.

References

Baker, T. G., 1979, The control of oogenesis in mammals, in: "Ovarian Follicular Development and Control," A. R. Midgley and W. A. Sadler, eds., Raven Press, New York, 353-364.

Bjersing, L., 1962, Method for isolating pig granulosa cell aggregates in amounts allowing biochemical investigation of steroid hormone synthesis in vitro, Acta Pathol. Microbiol. Scand., 55:127-128.

Bjersing, L., 1967a, On the morphology and endocrine function of granulosa cells in ovarian follicles and corpora lutea, Acta Endocrinol., 56:1-23.

Bjersing, L., 1967b. On the ultrastructure of granulosa lutein cells in porcine corpus luteum. With special reference to endoplasmic reticulum and steroid hormone synthesis. Z. Zellforsch., 82:187-211.

Bjersing, L., 1978. Maturation, morphology, and endocrine function of the follicular wall in mammals, in: "The Vertebrate Ovary," R. E. Jones, ed., Plenum Press, New York, pp. 181-214.

Bjersing, L., 1979, Intraovarian mechanisms of ovulation, in: "Human Ovulation," E. S. E. Hafez, ed., Elsevier, Amsterdam-New York, pp. 149-157.

Bjersing, L., 1981, Correlation of fine structure and endocrine function of the human corpus luteum, in: "Functional Morphology of the Human Ovary," J. R. T. Coutts, ed., MTP Press Ltd, Lancaster, England, pp. 119-136.

Bjersing, L. and Cajander, S., 1975, Ovulation and the role of the ovarian surface epithelium, Experientia, 31:605-608.
Bjersing, L., Cajander, S., and Bergdahl, E., 1980, The role of steroid and enzymes in follicle rupture, in: "Gynecology and Obstetrics," S. Sakamoto, S. Tojo, and T. Nakayama, eds., Excerpta Medica, Amsterdam, pp. 1225-1228.
Bjersing, L. and Carstensen, H., 1964, The role of the granulosa cell in the biosynthesis of ovarian steroid hormones, Biochim. Biophys. Acta, 86:639-640.
Bjersing, L. and Carstensen, H., 1967, Biosynthesis of steroids by granulosa cells of the porcine ovary in vitro, J. Reprod. Fert., 14:101-111.
Bradbury, J., 1961, Direct action of estrogen on the ovary of the immature rat, Endocrinology, 63:115-120.
Byskov, A. G., 1974, Does the rete ovarii act as a trigger for the onset of meiosis? Nature, 252:396-397.
Byskov, A. G., 1978, The anatomy and ultrastructure of the rete system in the fetal mouse ovary, Biol. Reprod., 19:720-735.
Byskov, A. G., 1978, Regulation of initiation of meiosis in fetal gonads, Int. J. Andrology, 2:29-38.
Byskov, A. G. and Saxén, L., 1976, Induction of meiosis in fetal mouse testis in vitro, Dev. Biol., 52:192-200.
Byskov, A. G., Skaggebaek, N. E. Stafanger, G., and Peters, H., 1977, Influence of ovarian surface epithelium and rete ovarii on follicle formation, J. Anat., 123:1, 77-86.
Channing, C. P. and Coudert, S. P., 1976, Contribution of granulosa cells and follicular fluid to ovarian estrogen secretion in the rhesus monkey in vivo, Endocinrology, 98:590-597.
Channing, C. P., Schaerf, F. W., Anderson, L. D., and Tsafriri, A., 1980, Ovarian follicular and luteal physiology, Int. Rev. Physiol., 22:117-154.
Christensen, A. K., 1975, Leydig cells, in: "Handbook of Physiology: Endocrinology," Vol. V, R. O. Greep and E. B. Ashwood, eds., American Physiological Society, Washington, D.C., pp. 57-94.
Christensen, A. K. and Gillim, S. V., 1969, The correlation of fine structure and function in steroid-secreting cells, with emphasis on those of the gonads, in: "The Gonads," K. W. McKerns, ed., North-Holland Publ. Co., Amsterdam, pp. 415-488.
Erikckson, G. F., 1974, The control of the differentiation of female embryonic germ cells in bird, Dev. Biol., 36:113-129.
Falck, B., 1959, Site of production of oestrogen in rat ovary as studied in microtransplants, Acta Physiol. Scand., 47:1-101 (Suppl. 163).
George, F. W. and Wilson, J. D., 1978, Conversion of androgen to estrogen by the human fetal ovary, J. Clin. Endocrinol. Metab., 47:550-555.
George, F. W. and Wilson, J. D., 1980, Endocrine differentiation of the fetal rabbit ovary in culture, Nature, 283:861-863.

George, F. W., Milewich, L., and Wilson, J. D., 1978, Oestrogen content of the embryonic rabbit ovary, Nature, 274:172-173.
Gondos, B., 1980, Development and differentiation of the testis and male reproductive tract, in: "Testicular Development, Structure, and Function," A. Steinberger and E. Steinberger, eds., Raven Press, New York, pp. 3-20.
Hillensjö, T., 1976, Oocyte maturation and glycolysis in isolated preovulatory follicles of PMS-injected immature rats, Acta Endocrinol., 82:809-830.
Hillensjö, T., 1981, Steroid biosynthesis by granulosa' thecal and stromal cells: Their interaction in Intragonadal regulation of reproduction or gonadal cybernines," P. Franshimont and C. P. Channing, eds., Academic Press, New York, pp. 33-60.
Hirata, F. and Axelrod, J., 1980, Phospholipid methylation and biological signal transmission, Science, 209:1082-1090.
Hirata, F., Schiffman, E., Krishnamoorthy, V., Salomon, D., and Axelrod, J., 1980, A phospholipase A_2 inhibitory protein in rabbit neutrophils induced by glucocorticoids, Proc. Natl. Acad. Sci. U.S.A. 77:2533-2536.
Jost, A., 1980, Comparative aspects of sexual differentiation, in: "Gynecology and Obstetrics," S. Sakamoto, S. Tojo, and T. Nakayama, eds., Excerpta Medica, Amsterdam, pp. 10-19.
Jirásek, J. E., 1977, Morphogenesis of the genital system in the human, Birth Defects: Original Article Series, XIII, 2:13-39.
Lieberman, M. E., Tsafriri, A., Bauminger, S., Collins, W. P., Ahrén, K., and Lindner, H. R., 1976, Oocytic meiosis in cultured rat follicles during inhibition of steroido-genesis, Acta Endocrinol., 83:151-157.
Lipner, H. and Greep, R. O., 1971, Inhibition of steroidogenesis at various sites in the biosynthetic pathway in relation to induced ovulation, Endocrinology, 88:602-607.
McCarrey, J. R. and Abbott, U. K., 1979, Mechanisms of genetic sex determination, gonadal sex differentiation, and germ-cell development in animals, Advances in Genetics, E. W. Caspari, ed., 20:217-290.
McNatty, K. P., Makris, A., DeGrazia, C., Osathanondh, R., and Ryan, K. J., 1980, Steroidogenesis by recombined follicular cells from human ovary in vitro, J. Clin Endocrinol. Metab., 51:1286-1292.
Moor, R. M. and Warnes, G. M., 1979, Regulation of meiosis in mammalian oocytes, Br. Med. Bull., 35:99-103.
Mori, T., Suzuki, A., Nishimura, T., and Kambegawa, A., 1977a, Inhibition of ovulation in immature rats by anti-progesterone antiserum, J. Endocr., 73:185-186.
Mori, T., Suzuki, A., Nishimura, T., and Kambegawa, A., 1977b, Evidence for androgen participation in induced ovulation in immature rats, Endocrinology, 101:623-626.
O, W. S. and Baker, T. G., 1976, Initiation and control of meiosis in hamster gonads in vitro, J. Reprod. Fert., 48:399-401.

Ohno, S., 1980, The identification of testis-organizing H-Y antigen of man as hydrophobic polymers of M. W. 18,000 subunit, Ann. Endocrinol. Paris, 41:263-274.

Ohno, S., Nagai, Y., Ciccarese, S., and Smith, R., 1979, In vitro studies of gonadal organogenesis in the presence and absence of H-Y antigen, In Vitro, 15:11-18.

Peters, H. and McNatty, K. P., 1980, A correlation of structure and function in mammals, The Ovary, Granada Publishing, London-New York.

Richards, J. S., 1980, Maturation of ovarian follicles: Actions and interactions of pituitary and ovarian hormones on follicular cell differentiation, Physiol. Rev., 60:51-89.

Rivelis, C., Prepin, J., Vigier, B., and Jost, A., 1976, Prophase méiotique dans les cellules germinales de l'ébauche ovarienne de rat cultivée in vitro en milieu anhormonal. C.R. Acad. Sci., Paris, Ser. D., 282:1429-1432.

Ryan, K. J., 1980, Human ovarian function, in: "Gynecology and Obstetrics," S. Sakamoto, S. Tojo, and T. Nakayama, eds., Excerpta Medica, Amsterdam, pp. 3-9.

Short, R. V., 1962, Steroids in the follicular fluids and the corpus luteum of the mare. A "two-cell type" theory of ovarian steroid synthesis, J. Endocrinol., 24:59-63.

Thibault, C., 1977, Hammond Memorial Lecture, Are follicular maturation and oocyte maturation independent processes? J. Reprod. Fert., 51:1-15.

Tsafriri, A., 1978, Oocyte maturation in mammals, in: "The Vertebrate Ovary," R. E. Jones, ed., Plenum, New York, pp. 409-442.

Tsang, B. K., Armstrong, D. T., and Whitfield, J. F., 1980, Steroid biosynthesis by isolated human ovarian follicular cells in vitro, J. Clin. Endocrinol. Metab., 51:1407-1411.

Wagenen van G. and Simpson, M. E., 1965, Embryology of the Ovary and Testis. Homo sapiens and Macaca Mulatta, Yale University Press, New Haven and London.

Weiner, R. and Kaley, G., 1972, Lysosomal fragility induced by prostaglandin $F_{2\alpha}$, Nature, New Biology, 236:46-47.

Zoller, L. C. and Weisz, J., 1979, A quantitative cytochemical study of glucose-6-phosphate dehydrogenase and Δ5-3β-hydroxysteroid dehydrogenase activity in the Membrana granulosa of the ovulable type of follicle of the rat, Histochemistry, 62:125-135.

ROLE OF OVARIAN INHIBIN (FOLLICULOSTATIN) IN REGULATING FSH SECRETION IN THE FEMALE RAT

Neena B. Schwartz

Department of Biological Sciences
Northwestern University
Evanston, Illinois 60201

Introduction

The purpose of this chapter is to review what we know at present about the role of ovarian inhibin, or as we have called it "folliculostatin" (Schwartz and Channing, 1977; Hoak and Schwartz, 1980), in the regulation of the secretion of FSH by the anterior pituitary gland. Most of the data to be presented are taken from studies in the rat, but some reference to the growing literature on the subject from other species is cited.

I begin by discussing the lability of serum FSH levels in situations where ovarian feedback is altered. I then discuss experimental evidence that this lability in FSH secretion is compensatory for changes in endogenous or exogenous folliculostatin levels. The next section of the chapter presents a method for quantifying total granulosa cell population in the ovary, and utilizes the technique for examining several experimental maneuvers which are associated with alterations in serum FSH. I than examine the direct evidence that there is a hormone secreted by the ovary which has the specfic ability to suppress FSH secretion — i.e., folliculostatin. The chapter ends with a model for displaying experimental observations relevant to LH and FSH secretion in the rat.

Lability of Serum FSH

Serum FSH values are often thought to be more stable than LH because brief challenges with GnRH are more likely to evoke LH than FSH secretion (see Savoy-Moore and Schwartz, 1980); treatment with antiserum to GnRH is more likely to inhibit LH than FSH (Kawakami

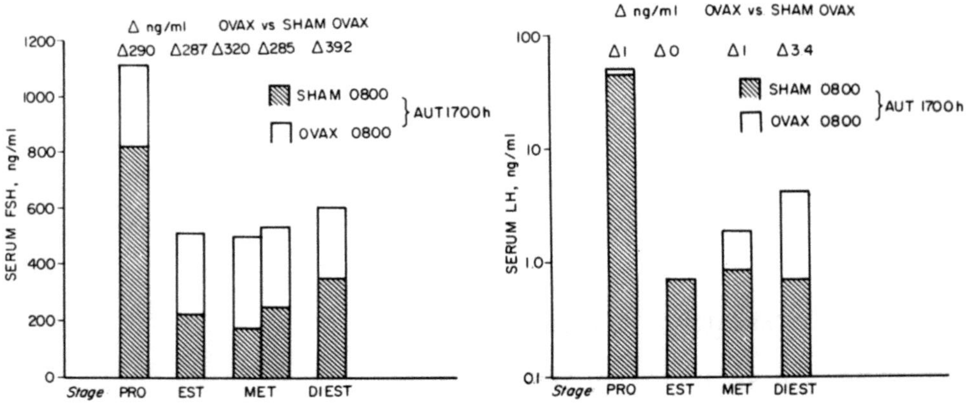

(a) (b)

Fig. 1. (a) Serum FSH (NIH-FSH-RP1 used as standard) in rats ovariectomized or sham ovariectomized at 0800h and autopsied at 1700h. Data on proestrus and one set of metestrous data are from Campbell et al., 1978. The estrous, metestrous, and diestrous days are from Grady et al., 1981. N = 4 or 5. (b) Serum LH (NIH-LH-S16 used as standard) in rats ovariectomized or sham ovariectomized at 0800h and autopsied at 1700h. Data on proestrus from Campbell et al., 1978. Estrous, metestrous, and diestrous data are from Grady et al., 1981. N = 4 or 5.

and Higuchi, 1979); LH has a much shorter half-life in serum than FSH (LaRochelle and Freeman, 1974), and abrupt non-specific stimuli such as stress are more likely to inhibit or enhance LH than FSH secretion (see Savoy-Moore and Schwartz, 1980). Furthermore, estradiol positive feedback is more likely to trigger a surge of LH than FSH (Henderson et al., 1977). In chronically ovariectomized rats, estradiol is less likely to suppress FSH to cyclic tonic levels then LH (Campbell and Schwartz, 1979).

From a different perspective, however, serum FSH is quite labile. Following ovariectomy on any day of the estrous cycle, there is a prompt increase in serum FSH by nine hours; the absolute increase is the same regardless of cycle day (Fig. 1a). By contrast, serum LH does not show any increase over sham treated rats by this time on proestrus or estrus, and shows only marginal increase on metestrus and diestrus (Fig. 1b). The latter increase is no greater if one waits for 24 hours postovariectomy (Lederhandler and Schwartz, 1977), nor for the next three days (Brown-Grant and Grieg, 1975).

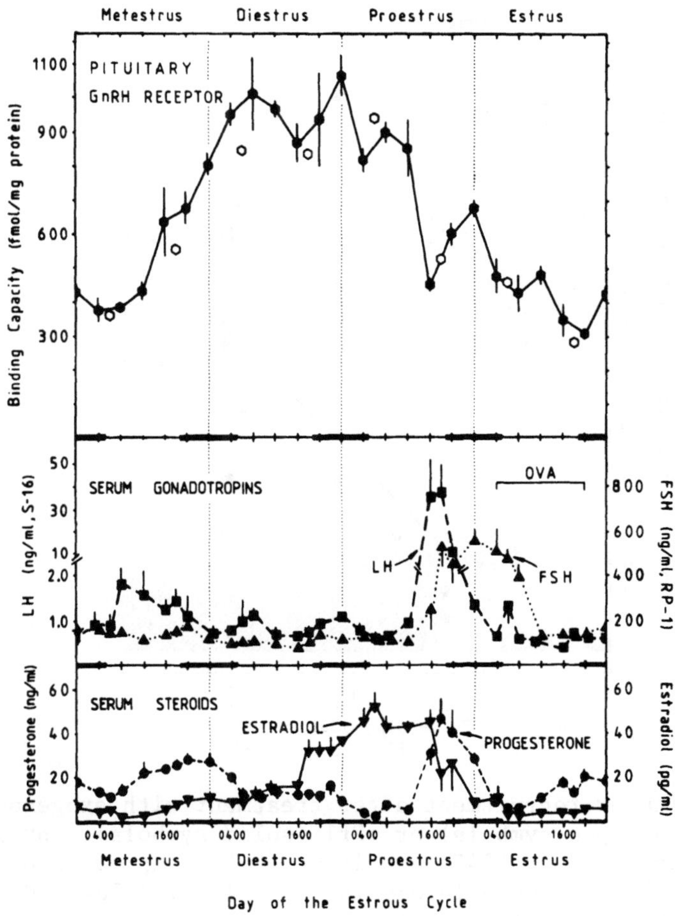

Fig. 2. (a) The binding capacity of d-Ala[6] analog to pituitary plasma membrane preparations from rats decapitated at 4-h intervals during the estrous cycle. Means ± S.E. for three to five triplicate determinations are shown at each time point. (b) Serum LH (■) measured in units of NIH-LH-S16 and serum FSH (▲) measured in units of NIH-FSH-RP-1. Note the scale break and scale change for LH values. Means ± S.E. are plotted, with N = 5 at each time point. Limit of detectability was 0.4 ng/ml for the LH assay and 58.5 ng/ml for the FSH assay. Presence of ova in the oviducts is indicated on estrus. Black bars on the time scale indicate periods of darkness, and the light vertical dotted lines divide the days of the cycle. (c) Serum estradiol (▼) and progesterone (●) are represented; N = 5 for each time point. This figure appeared in Savoy-Moore et al., 1980, and is used by permission of the publishers.

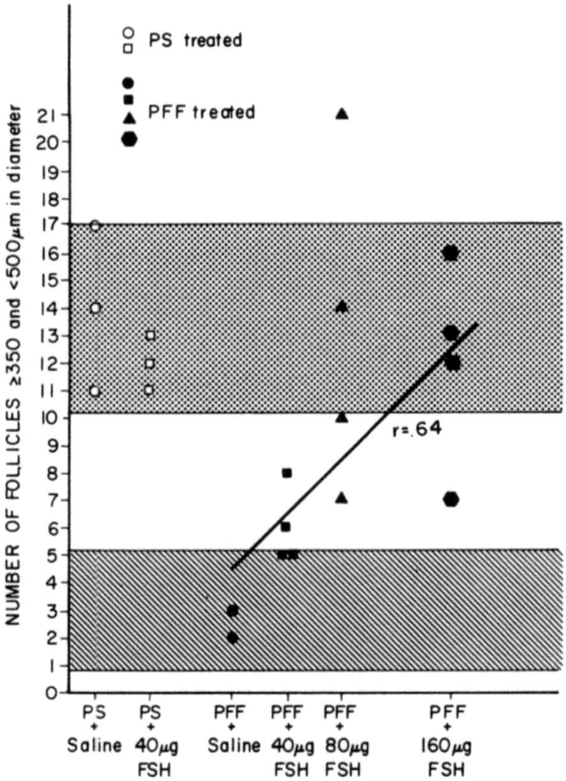

Fig. 3. Follicular recruitment after treatment with exogenous FSH and pS (open symbols) or pFfl (solid symbols). pS or pFfl was injected at 1545h and 1830h (0.5 ml ip at each time). The 95% confidence intervals for all follicles 350-500 μm (in one ovary at 1100h in estrus) is represented for pS-treated animals (stippled area) and for pFfl-treated animals (hatched area). The number of follicles 350-500 μm in diameter from one ovary of each individual animal in the treatment groups which follow is plotted: 0, pS + saline; □, pS + 40 μg of FSH; ●, pFfl + saline; ■, pFfl + 40 μg of FSH; ▲, pFfl + 80 μg of FSH; ⬣, pFfl + 160 μg of FSH. For line, r = 0.64. Reproduced from Hoak and Schwartz, 1980, with permission of the publishers.

A prolonged release of FSH follows the primary LH and FSH surge of proestrus (Schwartz and Talley, 1978; Fig. 2). It can be induced by exogenous GnRH, LH, FSH, or hCG (Schwartz and Talley, 1978; Ashiru and Blake, 1979; Sasamoto and Taya, 1980). Injection of FSH on any day of the estrous cycle causes a significant rise in FSH, although the response is smallest on estrus (Ashiru and Blake, 1980b).

An injection of pFfl on any day of the estrous cycle is very effective in suppressing serum FSH without suppressing serum LH whether the recipient is intact or acutely ovariectomized (Grady et al., 1981). When superimposed on estradiol, it can suppress serum FSH seven days after ovariectomy to the low values seen during the estrous cycle (Campbell and Schwartz, 1979).

We (Savoy-Moore et al., 1980a) have measured the latency of FSH suppression to pFl in female rats in the following experimental protocol. At 0800h, metestrous rats were ovariectomized. At 1130h, they were injected with 0.5 ml of charcoal extracted pFfl, or porcine serum intravenously, and were killed at 30 minute intervals thereafter. A significant suppression by the follicular fluid had occurred by 1.5 hours after injection. When it is remembered that the half-life of serum FSH is about 1.5h (Larochelle and Freeman, 1974), it can be appreciated that the latency of suppression is extremely short.

Compensatory FSH Secretion

Some years ago, we postulated that the FSH secretion which occurred between proestrus and estrus (Fig. 2) might be responsible for recruiting a cohort of follicles into the ovary for the next cycle (McClintock and Schwartz, 1968). Several investigators have tried to test this hypothesis with antiserum to FSH on the day of proestrus (Schwartz et al., 1973; Welschen and Dullaart, 1976). This treatment failed to prevent extensively the events of the next cycle in the rat; since it was not possible to measure serum FSH by RIA in the presence of circulating anti-rabbit-IgG, the failure to impact on the following cycle remained unexplained.

Having demonstrated that we could suppress either the primary surge of FSH (Hoffmann et al., 1979) or the secondary surge (Schwartz and Channing, 1977) with follicular fluid, we have examined whether such suppression can block follicular recruitment into the ovary. Blocking the primary surge did not prevent recruitment, but blocking the secondary surge did so (Fig. 3). Thus, there is not only a "folliculostatin" in the fluid, in the sense of serum FSH suppression, but also of follicle growth suppression. As can be seen in Fig. 3, exogenous FSH can overcome this pFfl blockade (Hoak and Schwartz, 1980).

Several studies have demonstrated a "rebound" in serum FSH levels following suppression by follicular fluid (Hoffmann et al., 1979; DePaolo et al., 1979). Presumably, this occurs as the exogenous fluid disappears in animals in which there is low endogenous secretion of folliculostatin activity. To see whether this rebound was occurring in our rats following pFfl suppression of the secondary surge of FSH, we followed serum hormone levels into the

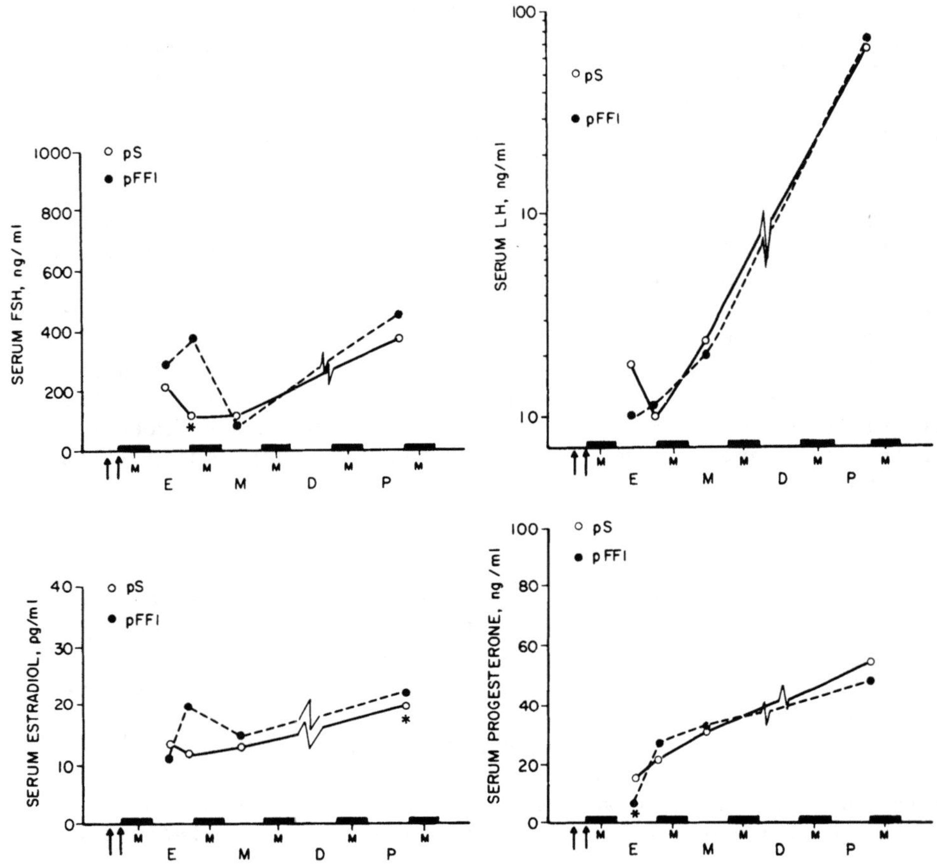

Fig. 4. Rats were treated on proestrus with pS or pFfl (0.5 ml ip at 1545 and 1830h) and were autopsied at 1100h estrus, 1100h metestrus, or 1830h proestrus. N = 4. "M indicates midnight = midpoint of dark period." (a) Serum FSH (NIH-FSH-RP1 is standard). (b) Serum LH (NIH-LH-S16 is standard. (c) Serum estradiol. (d) Serum progesterone. *$P < 0.05$ pS vs pFfl.

next cycle after treatment with fluid late in proestrus (see Schwartz and Channing, 1977; Hoak and Schwartz, 1980). At 1100h, estrous serum FSH was low regardless of treatment (Fig. 4) and follicular recruitment into the cohort size above 350 µm had not occurred, replicating our previous findings (Fig. 2). By the afternoon of estrus, serum FSH had rebounded and was elevated (Fig. 4). This elevation caused follicular recruitment by 1100 metestrus morning. (The data on follicle numbers on these data will be shown below, using a new method of data presentation.) The hormone

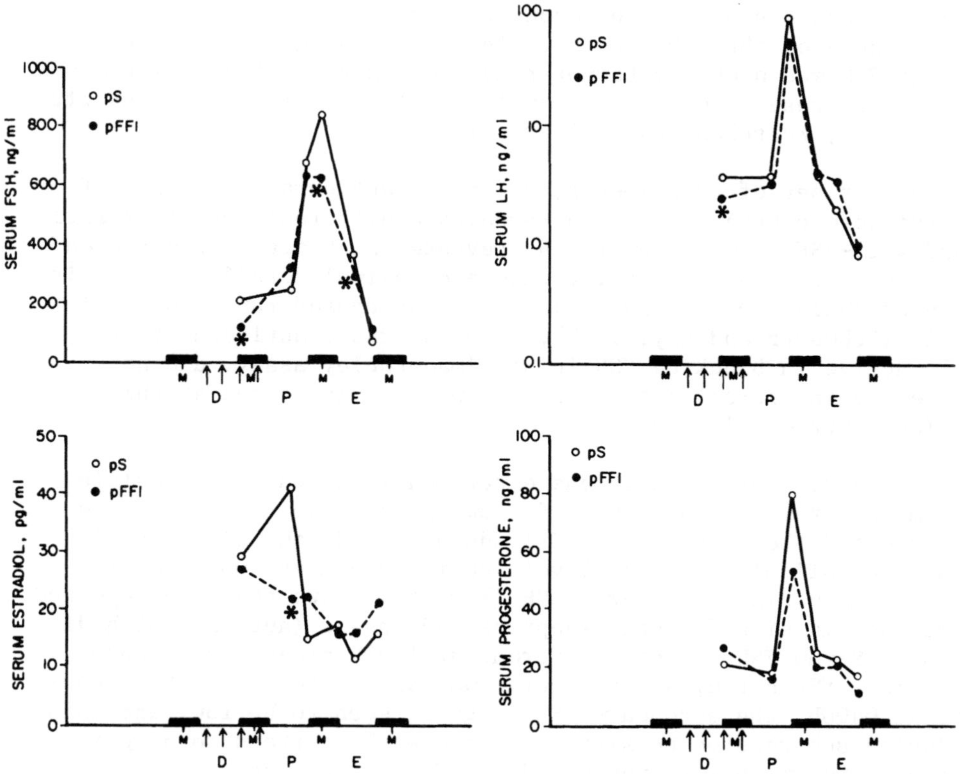

Fig. 5. Rats were treated on diestrus (day before proestrus in 4-day cycles) with 0.5 ml ip-pS or pFfl at 0800, 1400, 2000h diestrus, and 0200h proestrus. Autopsies were carried out at 2000h diestrus, 1300, 1830h proestrus and 0400, 1100, and 1830h estrus. (a) Serum FSH (NIH-FSH-RP1 as standard). (b) Serum LH (NIH-LH-S16 as standard). (c) Serum estradiol. (d) Serum progesterone. *P < .05 pS vs pFfl.

data (Fig. 4) seen during the cycle after delayed recruitment, are remarkably normal, with LH, FSH, and progesterone surges occurring at 1830h next proestrus. The "compensatory" FSH secretion which occurred late on estrus, can be due to the decrease of ovarian secretion of endogenous folliculostatin until the recruitment occurred (Welschen et al., 1980; vide infra).

The failure of suppression of the primary FSH surge on the afternoon of proestrus to block follicular recruitment by the next morning (Hoak and Schwartz, 1980), can be attributed to the high serum FSH values observed in such rats in the middle of the night (Hoffmann et al., 1979). Obviously, the data in Fig. 4 demonstrate that even successful blockade of follicular recruitment at estrus by suppression of the secondary surge, is overcome by 24 h later because of the FSH rebound. These data suggest that antiserum to FSH was probably temporarily successful, but that serum FSH levels later rose, compensating fully for the suppression (Schwartz et al., 1973; Welschen and Dullaart, 1976).

Another set of experiments, utilizing antisera on the day of diestrus before proestrus, also revealed a different impact of anti-LH and anti-FSH. Antiserum to LH prevented such signs of estrogen secretion as uterine weight increase and vaginal cornification: it prevented ovulation in very small doses and caused rapid follicular atresia (Schwartz and Ely, 1970). By contrast, antiserum to FSH, highly active in blocking FSH in Steelman-Pohley assay, did not block estrogen secretion or ovulation when administered at this time (Schwartz et al., 1973).

At the time of the previous experiments, it was not possible to examine serum hormone levels following anti-FSH treatment. We have repeated the study using pFfl administered four times at diestrus and following hormone levels during the next two days after treatment (see Fig. 5). Serum FSH (Fig. 5a) shows that the repeated injections did significantly suppress FSH levels (autopsy 2000h diestrus). Serum FSH was not different at 1300h proestrus, before the primary FSH surge, and the surge was also normal (1830h). However, at 0400h, the secondary FSH surge is seen to be low, and this continuing suppression persisted for the early estrous morning measurement. Serum LH was suppressed by pFfl during the treatment (Fig. 5b), but all subsequent LH measurements were normal. This includes a normal LH surge at 1830h proestrus. Serum estradiol showed a clear attenuation just before the onset of the proestrous critical period in the treated rats (Fig. 5c), but this attenuated level permitted normal primary surges of LH and FSH, at least at 1830h. Serum progesterone measurements were normal throughout (Fig. 5d).

We have not completed follicle counts in the ovaries of the animals depicted in Fig. 5. However, ovulation had occurred in only two of five treated rats autopsied at 1100 estrus (one and four ova apiece). All pFfl-treated rats did show large LH-stimulated follicles by that time. by 1830h estrus, all four pFfl treated rats had ovulated, but fewer ova appeared in the oviducts, and large LH-stimulated follicles were still present in the ovaries.

Thus, blockade of FSH secretion at diestrus can impact on the events of the next two days, but the blocking effects are not complete. Enough estrogen was secreted to trigger the proestrous surges. The follicles were definitely not atretic, although they appeared delayed in their responsiveness to the ovulatory surge of LH.

Table 1. Calculations of Follicle Volume in One Ovary on Each Day of the Estrous Cycle (4-day)

Diameter μm	Volume (x 10^5) μm^3		Estrus AM	Metestrus AM	Diestrus AM	Proestrus AM	Proestrus PM
150 - 249	41.8	N	31[a]	31[a]	31[a]	31[a]	31[a]
		V_t	1296	1296	1296	1296	1296
250 - 349	141	N	19.4[b]	20[b]	13[a]	10[b]	9.7[b]
		V_t	2735	2820	1833	1410	1368
350 - 449	335	N	9.8[b]	11.6[c]	7.8[c]	1.4[c]	2[c]
		V_t	3283	3886	2613	916	670
450 - 549	654	N	6.8[c]	6[c]	5.2[c]	4.1[c]	1.0[b]
		V_t	4447	3924	3401	2681	654
550 - 649	1131	N	1[b]	0.2[b]	3[c]	4.7[c]	(8.8)[b]
		V_t	1131	226	3393	5316	(9953)
	Total V_t($x10^5$)		12892	12152	12536	11619	3988 (13941)

$V = 1/6\ \pi D^3$

N = Number of healthy follicles in this class size/ovary

$V_t = N \times V = \mu m^3$ in each class size

References - (a) Mandl and Zuckerman, 1952
(b) Hoak and Schwartz, 1980 and unpublished observations
(c) Peppler and Greenwald, 1970

Follicle Cell Counts: A Technique for Data Presentation and a Hypothesis

The data presented and reviewed thus far, demonstrate that elevated serum FSH between proestrus and estrus (Fig. 2), recruits follicles into a size class >350 μm (Fig. 3). We have suggested: "It may be that the sudden growth or recruitment of follicles produces increased circulating folliculostatin levels which lead to the termination of the secondary FSH surge and low tonic levels of FSH during the next two days of the cycle" (Hoak and Schwartz, 1980). Similarly, Welschen and co-workers (1980) have hypothesized: "The inhibin-feedback mechanism may therefore be involved in the regulation of the number of follicles $\geq 200 \times 10^5$ μm^3 in volume which are to ovulate or to become atretic during each cycle."

Since Ericsson and Hsueh (1978) demonstrated that cultures of follicular cells can secrete an activity which suppresses FSH secretion by pituitary cells, it seems plausible to invoke a reciprocal feedback mechanism between follicles and FSH secretion in vivo. In seeking a less clumsy way than follicle size classifications to quantitate this feedback system, it occurred to me that it would be useful to have an expression for total follicle cell population to substitute for follicle size classes, whether these are expressed as diameter, volume, or number of cells per cross section (Pedersen, 1970). In order to do this, I started with the assumption that the follicle shape could be approximated by a sphere and thus volume would be approximated by a cubic exponent of diameter. Pedersen (1970) demonstrated that as follicles get larger, the number of follicle cells increased as a (modified) cubic function, so that this assumption does not seem outlandish. If one knows the average maximum diameter of each size class in the ovary, one can calculate an average volume of each follicle in that class; mulitplying this by the number of follicles in that class will yield the total average follicle volume in that size class in each ovary. Adding up all of these volumes yields the total "volume of follicle" in the ovary; this number will be more dependent on the few large follicles present then on the many nonproliferating follicles present because of the cubic exponent feature. Now, one more assumption: the number of granulosa cells within each follicles is proportional to the volume of the follicle — this means that the total number of follicle cells within an ovary can be represented by the total volume figure calculated (see Hage et al., 1978).

In order to carry out the calculation, I needed to find a differential count of follicles in normal rats at each stage of the estrous cycle. Mandl and Zuckerman (1952) had counted follicles from 100 μm to over 650 μm diameter. The class between 100 and 149 μm had an average of 54 follicles; the average volume of each follicle is 10 $\mu m^3 \times 10^5$. The next largest class was 150-249 μm; average number/ovary was 31. For this class the volume calculations

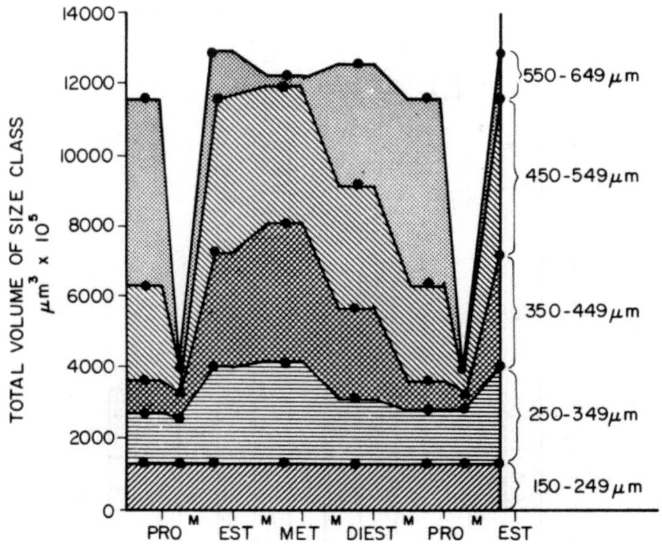

Fig. 6. Plot of V_t or total volume of follicle in each size class as a function of cycle stage. These data are from Table 1.

are presented in Table 1. Since no differences were found between stages in this class (and lower diameters), the same V_t is entered across stage. For larger follicles, where stage differences emerged, it is not possible to utilize the same source of data, since Mandl and Zuckerman used vaginal smear cytology as the sole stage criterion and it is not possible to translate these criteria into more modern staging (Schwartz, 1969). For larger size follicles, I used our own data, or data of Peppler and Greenwald (1970). When the same stage was measured in both of the latter studies, I consistently chose the data which had lower numbers.

When one looks at the grand totals of follicle volume at the bottom of the table, it becomes obvious why I did not bother recording the 54 × 10 $\mu m^3 \times 10^5$ of the 100-149 μm class, since 540 $\mu m^3 \times 10^5$ makes an insignificant contribution to the totals.

The bottom line of the table indicates a remarkably consistent total follicle volume (and presumably, total follicle cells) across the four mornings of the cycle. On the late afternoon of proestrus, the preovulatory swelling has commenced, inflating the volume of the largest follicle class. Here, I have taken a speculative leap. By 1830h, the LH surge has already initiated germinal vesicle breakdown (Ayalon et al., 1972), has terminated estradiol secretion and stimulated progesterone secretion (Figs. 2, 5). I believe that with these known shifts in steroidogenesis, there is a simultaneous cessation of folliculostatin secretion, so that this pool of

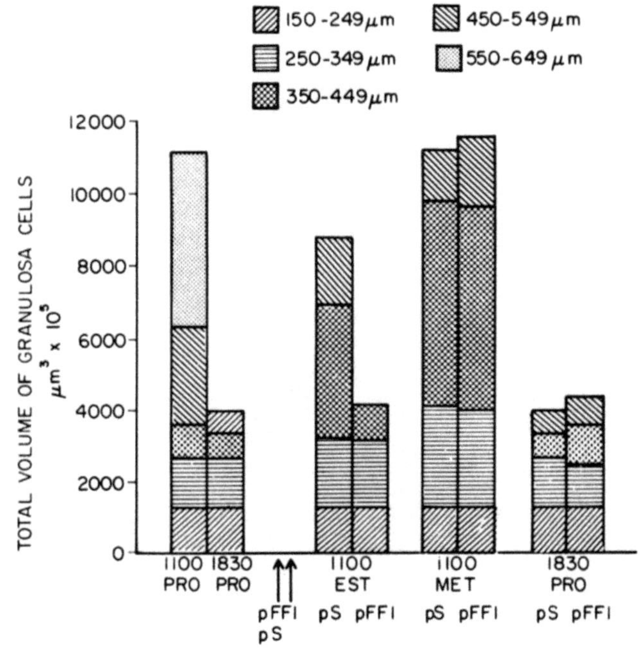

Fig. 7. Plot of V_t of follicle (granulosa cells) in control proestrous rats and in treated rats from the study in Fig. 4.

follicles ceases its contribution to the feedback signal for FSH. Thus, the total follicle volume, which we assume to be related to total follicle cells, is also proportional to the folliculostatin secretion levels, assuming that every follicle cell in the ovary secretes the same amount of folliculostatin. The data in Table 1 are plotted in Fig. 6, reinforcing the similarity across cycle stage of total follicle volume, except after the onset of the critical period of proestrus.

I have delayed until now showing the follicle counts of the experiment summarized in Fig. 4. You will recall that pFf1 was administered late in proestrus and blocked recruitment of follicles into the 350 μm class. (Data from a similar experiments is shown in Fig. 3.) The results from the new experiment are plotted in the new way in Fig. 7; at 1100h estrus, the total volume is much attenuated; it can be seen that this is indeed because the >350 μm diameter range is missing. By 24 h later, due to the delayed FSH release (Fig. 4a), the total volume of follicle has caught up (Fig. 7).

Is this new way of plotting follicle volume (in one ovary) useful under other circumstances? Unilateral ovariectomy is a case

Table 2. Calculations of Follicle Volume Following Unilateral Ovariectomy*

Diameter μm	Volume $(\times 10^5)$ μm^3		Sham - Time (h) Post-Surgery 0	4	12	16	20	24	28	32
400	335	N	5.0	4.5	3.8	5.0	3.0	0.3	0.5	1.0
		V_t	1675	1508	1273	1675	1005	101	168	335
550	871	N	5.6	4.8	7.3	4.3	6.0	8.8	4.7	6.3
		V_t	4878	4181	6358	3745	5226	7665	4094	5487
		Total V_t	6533	5689	7631	5420	6231	7766	4262	5822
		Total V_tx2	13,106	11,378	15,262	10,840	12,462	15,532	8524	11,644

Diameter μm	Volume $(\times 10^5)$ μm^3		Unilateral Ovariectomy - Time (h) Post-Surgery 0	4	12	16	20	24	28	32
400	335	N	5.0	5.3	5.0	8.3	4.3	6.0	6.0	7.0
		V_t	1675	1776	1675	2781	1441	2010	2010	2345
550	871	N	5.6	6.3	7.0	6.0	6.3	8.0	9.5	12.0
		V_t	4878	5487	6097	5226	5487	6968	8275	10,452
		Total V_t	6553	7263	7772	8007	6928	8978	10,285	12,797

*The data on numbers of follicles of the two diameter size classes come from Welschen, et al., 1977.

where serum FSH rises and then falls, co-incident with recruitment of follicles in the volume class >350 μm (Table 1; Welschen et al., 1980; Peppler and Greenwald, 1970). It has been demonstrated that estradiol levels do not fall following unilateral ovariectomy, LH does not change and progesterone decreases; injection of estradiol or progesterone does not prevent the rise in FSH (Welschen et al., 1978; Butcher, 1977). Injection of pFfl does block the rise in FSH (Welschen et al., 1977).

Welschen and colleagues (1977) have measured follicle numbers at frequent time intervals following unilateral ovariectomy or sham surgery on metestrous morning. They distinguished two sizes of follicles-200-500 × 10^5 $μm^3$ and >500 × 10^5 $μm^3$; no smaller follicles were counted. I have subjected their data (Table 2) to the same data analysis used on the cycle control data presented in Table 1. As can be seen in Table 2, the cell volume in the unilaterally ovariectomized rats is below the cell volume in controls (multiplied by two for both ovaries) until about 28 h, when the serum FSH started to fall toward normal. Using different reasoning about follicle growth (Welschen et al., 1978) conclude that these particular follicles are secreting inhibin or folliculostatin once compensatory follicle growth has occurred. Table 2 assumes only that every follicle cell always secretes the same folliculostatin until after the preovulatory LH surge.

Butcher (1977) has demonstrated that the periovulatory surges of FSH are elevated in the first cycle and the third cycle after unilateral ovariectomy (the only cycles studied). Greenwald has shown the same thing for the hamster (Bast and Greenwald, 1977). If the total follicle volume has compensated (Table 2), why does the FSH elevate more than normal between proestrus and estrus? In animals with one ovary only instead of two, there is probably an exaggerated drop of follicle volume after the primary LH surge reaches the preovulatory follicles (Fig. 6). Also, since there is only one ovary, the follicle classes below 250 μm can "buffer" less well, since they are present only in one ovary. Unfortunately, the studies of follicle groups after unilateral ovariectomy have counted only follicles above 250 μm in diameter (Peppler and Greenwald, 1970; Welschen et al., 1978).

Hormonal Status of Ovarian Inhibin (Folliculostatin)

Ovarian steroid appears to be an inadeuqate feedback explanation for the regulation of serum FSH levels in the rat, as well as in the human (Sherman and Korenman, 1975); rhesus monkey (Goodman et al., 1979); rabbit (Dufy-Barbe et al., 1978); sheep (Goodman et al., 1981); bovine (Roche and Ireland, 1981); and mouse (Bronson and Channing, 1978). Specifically, in the rat it is not necessary for estradiol levels to fall for FSH to rise after bilateral ovariectomy (Campbell et al., 1977), or unilateral ovariectomy (Butcher, 1977; Welschen et al., 1978). Estradiol does not suppress serum FSH to cyclic levels in ovariectomized rats seven days after ovariectomy (Campbell and Schwartz, 1979). Exogenous progesterone (Schwartz and Talley, 1978), estradiol (Butcher, 1977) or testosterone (Welschen et al., 1980) do not alter the magnitude of the secondary FSH surge which follows the primary surges. Estradiol injections do not suppress the singular rise in FSH which follows unilateral ovariectomy (Butcher, 1977), although it can prevent compensatory ovarian hypertropy (Ramirez and Sawyer, 1974). The latter is probably a reflection of a need for background LH to permit follicular growth (Schwartz and Ely, 1970).

In contrast to the relative ineffectiveness of ovarien steroids, ovarian follicular fluid is remarkably effective in suppressing FSH specifically in a number of in vivo models as well as in vitro on a pituitary cell culture (DeJong et al., 1978). The FSH suppressing activity is present even after steroid extraction, and is associated with a peptide greater than 10,000 MW (Lorenzen et al., 1978). The injection of pFfl can prevent the primary (Hoffmann et al., 1979) and the secondary proestrous FSH surges (Schwartz and Channing, 1977; Welschen et al., 1980); the elevated serum FSH which follows bilateral ovariectomy acutely (Marder et al., 1977) or chronically (Campbell and Schwartz, 1979), as well as unilateral ovariectomy (Welschen et al., 1978). Reliable suppression of serum FSH can also be achieved in male rats (Lorenzen et al., 1981).

In other species, follicular fluid also can suppress FSH secretion. Follicular fluid (or ovarian extracts) of porcine, bovine, equine, rat, hamster, mouse, and human origin can suppress FSH secretion in mouse (Bronson and Channing, 1978), hamster (Chappel et al., 1980; Chappel, 1979), horse (Miller et al., 1979), and rhesus monkey (Stouffer et al., 1981).

More direct evidence, other than effectiveness of ovarian nonsteroidal extracts in suppressing FSH, is also available. Rat granulosa cells secrete a substance which can suppress FSH (Ericsson and Hsueh, 1978). Ovarian vein plasma contains a substance which can supppress FSH secretion specifically in pituitary cell cultures; during a normal estrous cycle, the concentration of the "FSH-IA" decreases during the afternoon of proestrus (DePaolo et al., 1979b); this can be blocked by barbiturate treatment and induced by exogenous gonadotrophic hormone injection (Shander et al., 1980). This confirms our hypothesis that preovulatory LH/FSH primary surges suppress endogenous folliculostatin secretion and thus induce secondary FSH release, or permit it (Schwartz and Talley, 1978). Thus, the follicle cell volume changes as depicted in Fig. 6, coincide not only with the serum FSH (Fig. 2 — FSH high only when follicular volume low — Fig. 6), but with ovarian secretion of folliculostatin (DePaolo et al., 1979; Shander et al., 1980).

Concentration of FSH-suppressing activity has also been shown to vary across follicle size (Lorenzen et al., 1978; Welschen et al., 1978), as well as to change during the cycle in human follicles as well as hamster ovaries (Chappel et al., 1980; Chappel, 1979). While change in follicular content as a function of follicle size or stage cannot be used to predict secretion rate of this hormone, changes across time in content are suggestive of changes in secretion rate.

Uilenbroek et al. (1978 has demonstrated that ovaries transplanted into the spleen of gonadectomized male or female rats suppress FSH secretion quite well, without suppressing LH. Since hepatic metabolism inactivates steroids quite effectively, these observations are strongly suggestive that the peptide folliculostatin, demonstrated to be in follicular fluid, is indeed secreted by ovarian tissue in vivo.

Temporal Sequence of Follicular Growth during the Rat Cycle: Relation to LH and FSH Secretion

Figure 8 provides a model to summarize the ideas in this chapter. From left to right horizontally, a cohort of follicles is traced as it is recruited from the total pool of <250 μm in diameter. As one moves down vertically, the drawing moves from the secondary FSH surge of the last cycle, to the present cycle, and then the recruitment of follicles is shown for the next cycle. Folliculostatin is hypothesized to be secreted by all granulosa cells, with a size cohort secreting in proportion to its total number of cells, which is proportional to its volume (Fig. 6). In animals with both ovaries intact, there appears to be no variation in numbers or volume across the cycle, in the pool of follicles under <349 μm in diameter, except for the primordial follicles which, of course, decline slowly in numbers with age. This slow decline is not reflected in elevated serum FSH values until very late in reproductive life (Sherman and Korenman, 1975) since most contribution of folliculostatin comes from large follicles (Fig. 6).

The secondary FSH surge of the previous cycle recruits the cohort of follicles (diameter above 350 μm) which will provide the major contribution to estradiol and folliculostatin secretion for the next cycle. Once estradiol has reached a sufficient level to trigger the preovulatory surges of LH and FSH, germinal vesicle breakdown occurs, estradiol and folliculostatin secretion cease, progesterone secretion starts and the secondary FSH surge recruits the next follicles. Atresia occurs from all of these classes of follicles.

The model in Fig. 8 can be used to account for a number of experimental findings discussed in this chapter. The data on uni-

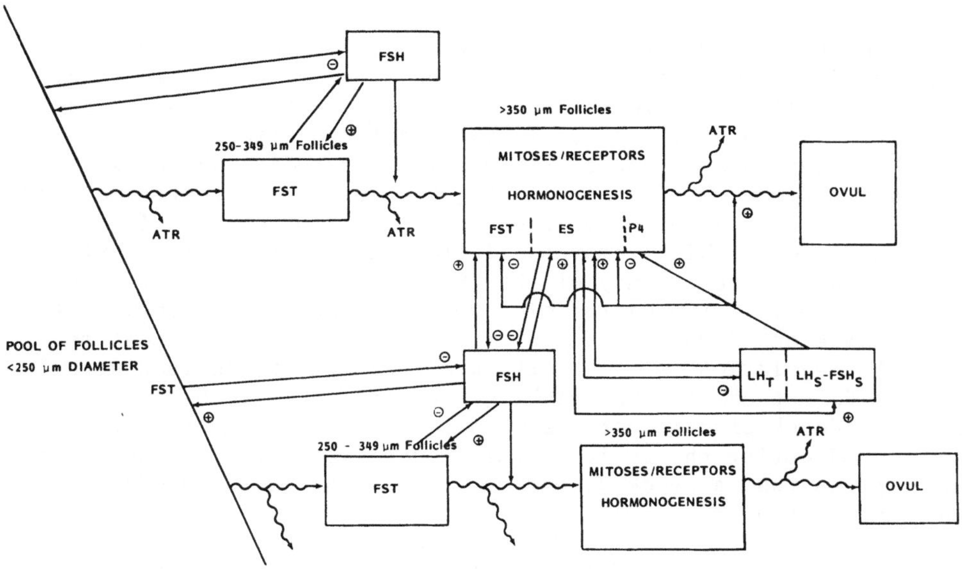

Fig. 8. Model of regulation of FSH secretion during sequential cycles in the rat. FST = folliculostatin; ES = estradiol; P4 = progesterone; LH_t = tonic LH secretion; LH_s and FSH_s = primary surges of the gonadotrophins; ATR = atresia.

lateral ovariectomy and follicle counts (Table 2; Welschen et al., 1978) indicate that the largest cohort of follicles (>350 μm; Fig. 8) is brought up to total volume in one ovary equal to that found in two ovaries, at that point when serum FSH returns to normal (28 h post-unilateral ovariectomy, Welschen et al., 1977). The responsiveness of serum FSH secretion to shifts in ovarian negative feedback is postulated to be due to tight negative feedback control by follicular secretion of folliculostatin, in proportion to numbers of granulosa cells. FSH rebounds in the serum after the primary surge (producing the secondary surge) or after exogenous folliculostatin, because growth of granulosa cells is initially suppressed resulting in lack of endogenous feedback; this leads to increased FSH secretion and follicular stimulation. FSH can be suppressed at any time of the cycle in the intact animal (Hoffmann et al., 1979; Schwartz and Channing, 1977; Marder et al., 1977; Grady et al., 1981; Fig. 5); in the bilaterally ovariectomized rat (Campbell and Schwartz, 1979) and the unilaterally ovariectomized rat (Welschen et al., 1978). Thus, the effect of folliculostatin in suppressing FSH can be expressed regardless of the point in the cycle at which follicle growth exists (Fig. 6).

The regulation of tonic levels of serum LH are not as well understood at this time. Following acute bilateral ovariectomy, the immediate LH response is dependent on the stage of the cycle (Fig. 1b), but even after ovariectomy at diestrus, serum LH does not really take off for three or four days (Brown-Grant and Grieg, 1975). Tonic LH secretion may result from combined estradiol and progesterone negative feedback (Goodman, 1978), but adrenal progesterone does not explain the delayed LH response to ovariectomy, since simultaneous adrenalectomy, which removes all progesterone from circulation, does not shorten the latency of the LH response (Lederhandler and Schwartz, 1978). Serum estradiol levels may exert a prolonged negative feedback on LH which fades only after four days. The net result of LH regulation, whatever the proximal causes, is that serum LH levels do not change appreciably even with total removal of gonadal feedback signals until several days elapse, and the major change in LH during the cycle is the short-lived burst on the afternoon of proestrus, a clock-dependent, as well as estradiol-dependent phenomenon (Schwartz, 1969).

In contrast, serum FSH levels are remarkably and immediately sensitive to alterations in gonadal feedback. The central importance of FSH in the regulation of granulosa cell LH receptor levels (Richards, 1980) parallels the elegant feedback system between the growing follicle and pituitary FSH secretion (Fig. 8), yielding a system of high adaptive significance.

Acknowledgements

I would like to acknowledge the excellent technical support of William L. Talley, Brigette R. Mann, Sheryl Veren, and Jack Modica in gathering the data reported here. I would also like to thank Dr. David Hoak who performed the hitherto unpublished experiments summarized in Table 1 and Figs. 4, 5, 6, and 7.

I would like to thank NIH, Endocrinology Study Section for the gonadotrophic hormones used as standard for the assays and for the FSH kit; Dr. L. E. Reichert, Jr., for the LH used for radioiodination; Dr. G. D. Niswender for antibody No. 244 for estradiol and antibody No. 869 for progesterone as well as antibody No. 15 for LH assay. This investigation was supported by PHS Research Grant HD-07504.

References

Ashiru, O. A., and Blake, C. A., 1979, Stimulation of endogenous follicle-stimulating hormone release during estrus by exogenous follicle-stimulating hormone or luteinizing hormone at proestrus in the phenobarbital-blocked rat, Endocrinology, 105:1162.

Ashiru, O. A., and Blake, C. A., 1980a, Effects of ovariectomy, estrogen, and LHRH on periovulatory increases in plasma gonadotropins, Biol. Reprod., 22:533.

Ashiru, O. A., and Blake, C. A., 1980b, Variations in the effectiveness with which rat follicle stimulating hormone can stimulate its own secretion during the rat estrous cycle, Endocrinology, 106:476.

Ayalon, D., Tsafriri, A., Lindner, H. R., Cordova, T., and Harell, H. R., 1972, Serum gonadotrophin levels in proestrous rats in relation to the resumption of meiosis by the oocytes, J. Reprod. Fert., 31:51.

Bast, J. D., and Greenwald, G. S., 1977, Acute and chronic elevations in serum levels of FSH after unilateral ovariectomy in the cyclic hamster, Endocrinology, 100:955.

Bronson, F. H., and Channing, C. P., 1978, Suppression of serum follicle stimulating hormone by follicular fluid in the maximally estrogenized ovariectomized mouse, Endocrinology, 103: 1894.

Brown-Grant, K., and Grieg, F., 1975, A comparison of changes in the peripheral plasma concentrations of luteinizing hormone and follicle stimulating hormone in the rat, J. Endocrinol., 65:389.

Butcher, R. L., 1977, Changes in gonadotropins and steroids associated with unilateral ovariectomy of the rat, Endocrinology, 101:830.

Campbell, C. S., and Schwartz, N. B., 1979, Time course of serum FSH suppression in ovariectomized rats injected with porcine follicular fluid (folliculostatin): Effects of estradiol treatment, Biol. Reprod., 20:1093.

Campbell, C. S., Schwartz, N. B., and Firlit, M. G., 1977, The role of adrenal and ovarian steroids in the control of serum LH and FSH, Endocrinology, 101:162.

Chappel, S., 1979, Cyclic fluctuations in ovarian FSH-inhibiting material in golden hamster, Biol. Reprod., 21:447.

Chappel, S. C., Holt, J. A., and Spies, H. G., 1980, Inhibin: differences in bioactivity within human follicular fluid in the follicular and luteal stages of the menstrual cycle, Proc. Soc. Exp. Biol. Med., 163:310.

De Jong, F. H., Welschen, R., Hermans, W. P., Smith, S. D., and Vander Molen, H. J., 1978, Effects of testicular and ovarian inhibin-like activity using in vitro and in vivo systems, Int. J. Androl. Suppl., 2:125.

De Paolo, L. V., Hirshfield, A. N., Anderson, L. D., Barraclough, C. A., and Channing, C. P., 1979a, Suppression of pituitary secretion of follicle-stimulating hormone by porcine follicular fluid during pro-oestrus and oestrus in the rat: effects on gonadotrophin and steroid secretion, follicular development and ovulation during the following cycle, J. Endocrinol., 83: 355.

De Paolo, L. V., Shander, D., Wise, P. M., Barraclough, C. A., and Channing, C. P., 1979b, Identification of inhibin-like activity in ovarian venous plasma of rats during the estrous cycle, Endocrinology, 105:647.

Dufy-Barbe, L., Dufy, B., and Vincent, J. D., 1978, Serum gonadotropin levels in the ovariectomized rabbit: effect of acute and chronic administration of estradiol, Biol. Rep., 18:118.

Erickson, G. F., and Hsueh, A. J. W., 1978, Secretion of 'inhibin' by rat granulosa cells in vitro, Endocrinology, 103:1960.

Goodman, A. L., Nixon, W. E., and Hodgen, G. D., 1979, Between ovary interaction in the regulation of follicle growth, corpus luteum function, and gonadotropin release after lutectomy in rhesus monkeys, Endocrinology, 105:69.

Goodman, R. L., 1978, A quantitative analysis of the physiological role of estradiol and progesterone in the control of tonic and surge secretion of luteinizing hormone in the rat, Endocrinology, 102:142.

Goodman, R. L., Pickover, S. M., and Karsch, F. J., 1981, Ovarian feedback control of follicle-stimulating hormone in the ewe, evidence for selective suppression, Endocrinology, 108:772.

Grady, R. R., Savoy-Moore, R. T., and Schwartz, N. B., 1981, Selective suppression of follicle-stimulating hormone by folliculostatin, a putative non-steroidal ovarian hormone, in: "Bioregulators of Reproduction," G. Jagiello and H. J. Vogel, eds., Academic Press, New York, pp. 359-369.

Hage, A. J., Groen-Klevant, A. C., and Welschen, R., 1978, Follicle growth in the immature rat ovary, Acta Endocrinology, 88:375.

Henderson, S. R., Baker, C., and Fink, G., 1977, Effect of oestradiol-17B exposure on the spontaneous secretion of gonadotrophin in chronically gonadectomized rats, J. Endocrinol., 73:455.

Hoak, D. C., and Schwartz, N. B., 1980, Blockade of recruitment of ovarian follicles by suppression of the secondary surge of follicle-stimulating hormone with porcine follicular fluid, Proc. Nat. Acad. Sci., (U.S.A.), 77:4953.

Hoffmann, J. C., Lorenzen, J. R., Weil, D., and Schwartz, N. B., 1979, Selective suppression of the primary surge of follicle-stimulating hormone in the rat: further evidence for folliculostatin in porcine follicular fluid, Endocrinology, 105:200.

Kawakami, M., and Higuchi, T., 1979, Effects of active and passive immunization with LH-RH on gonadotrophin secretion and reproductive function in female rats, Acta Endocrinologica, 91:616.

Larochelle, F. T., and Freeman, M. E., 1974, Superimposition of thyroid hormone regulation on gonadotropin secretion, Endocrinology, 96:379.

Lederhandler, M., and Schwartz, N. B.,1978, Simultaneous adrenalectomy and MER-25 treatment on acute LH and FSH response, Fed. Proc., 37:439 (abs).

Lorenzen, J. R., Channing, C. P., and Schwartz, N. B., 1978, Partial characterization of FSH suppression activity (folliculostatin) in porcine follicular fluid using the metestrous rat as an in vivo bioassay model, Biol. Reprod., 19:635.

Lorenzen, J. R., Dworkin, G. H., and Schwartz, N. B., 1981, Specific FSH suppression in the male rat by porcine follicular fluid. Am. J. Physiol., 240:E209.

Mandl, A. M., and Zuckerman, S., 1952, Cyclical changes in the number of medium and large follicles in the adult rat ovary, J. Endocrinol., 8:341.

Marder, M. L., Channing, C. P., and Schwartz, N. B., 1977, Suppression of serum follicle stimulating hormone in intact and acutely ovariectomized rats by porcine follicular fluid, Endocrinology, 101:1639.

McClintock, J. A., and Schwartz, N. B., 1968, Changes in pituitary and plasma follicle stimulating hormone concentrations during the rat estrous cycle, Endocrinology, 83:433.

Miller, K. F., Wesson, J. A., and Ginther, O. J., 1979, Changes in concentrations of circulating gonadotropins following administration of equine follicular fluid to ovariectomized mares, Biol. Reprod., 21:867.

Pedersen, T., 1970, Follicle kinetics in the ovary of the cyclic mouse, Acta Endocrinol., 64:304.

Peppler, R. D., and Greenwald, G. S., 1970, Influence of unilateral ovariectomy on follicular development in cycling rats, Am. J. Anat., 127:9.

Ramirez, V. D., and Sawyer, C. H., 1974, A sex difference in the rat pituitary FSH response to unilateral gonadectomy as revealed in plasma radioimmunoassays, Endocrinology, 94:475.

Richards, J. S., 1980, Hormonal control of ovarian follicular development: a 1978 perspective, Rec. Prog. Horm. Res., 35:343.

Roche, J. F., and Ireland, J. J., 1981, The differential effect of progesterone on concentrations of luteinizing hormone and follicle-stimulating hormone in heifers, Endocrinology, 108: 568.

Sasamoto, S., and Taya, 1980, Mechanism of the selective surge of follicle-stimulating hormone in dioestrous rats during the induction of ovulation by human chorionic gonadotrophin. J. Endocrinol., 86:489.

Savoy-Moore, R. T., Grady, R. R., and Schwartz, N. B., 1980a, Latency of action of folliculostatin in suppressing FSH: significance for bioassay, Soc. Study Reprod. Annual Meeting, p. 49A.

Savoy-Moore, R. T., and Schwartz, N. B., 1980, Differential control of FSH and LH secretion, in: "Reproductive Physiology III," R. O. Greep, ed., University Park Press, Baltimore, pp. 203-248.

Savoy-Moore, R. T., Schwartz, N. B., Duncan, J. A., and Marshall, J. C., 1980b, Pituitary gonadotropin-releasing hormone receptors during the rat estrous cycle, Science, 209:942.

Schwartz, N. B., 1969, A model for the regulation of ovulation in the rat, Rec. Prog. Horm. Res., 25:1.

Schwartz, N. B., and Channing, C. P., 1977, Evidence for ovarian "inhibin:" suppression of the secondary rise in serum follicle stimulating hormone levels in proestrous rats by injection of porcine follicular fluid, Proc. Nat. Acad. Sci., U.S.A., 74: 5721.

Schwartz, N. B., and Ely, C. A., 1970, Comparison of effects of hypophysectomy, antiserum to ovine LH, and ovariectomy on estrogen secretion during the rat estrous cycle, Endocrinology, 86:1420.

Schwartz, N. B., and Talley, W. L., 1978, Effects of exogenous LH or FSH on endogenous FSH, progesterone and estradiol secretion, Biol. Reprod., 17:820.

Schwartz, N. B., Krone, K., Talley, W. L., and Ely, C. A., 1973, Administration of antiserum to ovine FSH in the female rat: failure to influence immediate events of the cycle, Endocrinology, 92:1165.

Sherman, B. M., and Korenman, S. G., 1975, Hormonal characteristics of the human menstrual cycle throughout reproductive life, J. Clin. Invest., 55:699.

Stouffer, R. L., Coensgen, J. L., di Zerega, G. S., and Hodgen, G. D., 1981, Induction of defective corpus luteum function of administration of follicular fluid to monkeys during the follicular phase of the menstrual cycle, in: "Dynamics of Ovarian Function," N. B. Schwartz and M. Hunzicker-Dunn, eds., Raven Press, New York, pp. 185-190.

Uilenbroek, J. Th. J., Tiller, R., DeJong, F. H., and Vels, F., 1978, Specific suppression of follicle-stimulating hormone secretion in gonadectomized male and female rats with intrasplenic ovarian transplants, J. Endocrinol., 78:399.

Welschen, R., and Dulaart, J., 1976, Administration of antiserum against ovine follicle-stimulating hormone or ovine luteinizing hormone at proestrus in the rat: effects on follicular development during the oncoming cycle, J. Endocrinol., 70:301.

Welschen, R., Dullaart, J., and DeJong, F. H., 1978, Interrelationship between circulating levels of estradiol-17B, progesterone, FSH and LH immediately after unilateral ovariectomy in the cyclic rat, Biol. Reprod., 18:421.

Welschen, R., Hermans, W. P., Dullaart, J., and DeJong, F. H., 1977, Effects of an inhibin-like factor present in bovine and porcine follicular fluid on gonadotrophin levels in ovariectomized rats, J. Reprod. Fert., 50:129.

Welschen, R., Hermans, W., and DeJong, F., 1980, Possible involvement of inhibin in the interrelationship between numbers of antral follicles and peripheral FSH concentration in female rats, J. Reprod. Fert., 60:485.

PURIFICATION, CHARACTERIZATION, AND IN VITRO PRODUCTION OF INHIBIN

F. H. de Jong, E. H. J. M. Jansen,
W. P. Hermans,* and H. J. van der Molen

Departments of Biochemistry
(Division of Chemical Endocrinology)
and *Anatomy, Faculty of Medicine
Erasmus University Rotterdam
Rotterdam, The Netherlands

INTRODUCTION

The secretion of LH and FSH from the pituitary gland may vary independently in a number of physiological or experimental situations both in male and in female animals. In adult male animals, interference with spermatogenesis usually results in an increase in peripheral FSH concentrations, without concomitant increase in the concentration of LH (see review by Setchell et al., 1977). After gonadectomy of immature male rats FSH concentrations rise much faster than LH levels; this increase diminishes with increasing age of the animal. In contrast, the specific increase of peripheral FSH concentrations 8 h after ovariectomy in female rats becomes more pronounced with increasing age of the animal (Hermans et al., 1980). In adult females FSH concentrations seem to be inversely correlated with the number of developing large follicles (see Welschen et al., 1980). Finally, a specific suppression of FSH was observed after intrasplenic transplantation of ovaries into castrated male or female rats (Uilenbroek et al., 1978).

These observations might be explained by postulating the existence of a gonadal hormone, which can specifically suppress FSH secretion, without affecting the release of LH from the pituitary gland. Negative feedback of LH secretion would then be ex-

Correspondence to F. H. de Jong, Department of Biochemistry II, Faculty of Medicine, Erasmus University Rotterdam, P. O. Box 1738, 3000 DR Rotterdam, The Netherlands.

erted by the action of steroids both at the level of the pituitary gland and of the hypothalamus; this latter action would also influence FSH secretion through suppression of LHRH release. A substance with this specific effect on FSH secretion is present in fluids from the testis (Setchell and Sirinathsinghji, 1972) and the ovary (De Jong and Sharpe, 1976). The activity could not be extracted with organic solvents, and could be destroyed by treatment with proteases (see Setchell et al., 1977; De Jong et al., 1981). The name "inhibin" has been used most frequently to indicate this activity, because McCullagh (1932) coined this name for a water-soluble gonadal substance with an action on the gonadotrophic cells of the pituitary gland. Inhibin suppresses the pituitary release and production of FSH both in vivo (see Hudson et al., 1979) and in vitro, although in the presence of LHRH the in vitro release of LH is also affected (De Jong et al., 1979; Eddie et al., 1979; Scott et al., 1980). Finally, hypothalamic actions of inhibin-containing preparations have been observed (Le Lannou and Chambon, 1977; De Greef et al., 1982; Lumpkin et al., 1981).

The present paper describes the results of experiments, aimed at the purification and characterization of inhibin from bovine follicular fluid and extends the earlier work of Jansen et al. (1981) on this subject. Furthermore, experiments with an antiserum against a purified inhibin fraction and results of studies on the in vitro production of inhibin are described.

PURIFICATION OF INHIBIN

Several reports on the purification of inhibin from a variety of sources have been published. These results were recently reviewed by De Jong et al. (1981). The discrepancies between the results of various authors were remarkable; both the differing sources of inhibin and the variety of test systems used may have contributed to these differences. In the experiments described in the present paper, bovine follicular fluid (bFFl) was used as a source of inhibin (De Jong and Sharpe, 1976). The material was not treated with charcoal before the purification procedure, because loss of activity was observed after this treatment and steroids were not co-purified with the inhibin activity. For the detection of inhibin activity, an in vitro method with dispersed pituitary cells (De Jong et al., 1979) was used. In this method, the specific suppression of the spontaneous release of FSH from the pituitary cells is used as an endpoint in the estimation of inhibin. Specific activities of purified preparations were expressed relative to the activity in a standard charcoal-treated bFFl preparation, which contained 65 mg protein per ml.

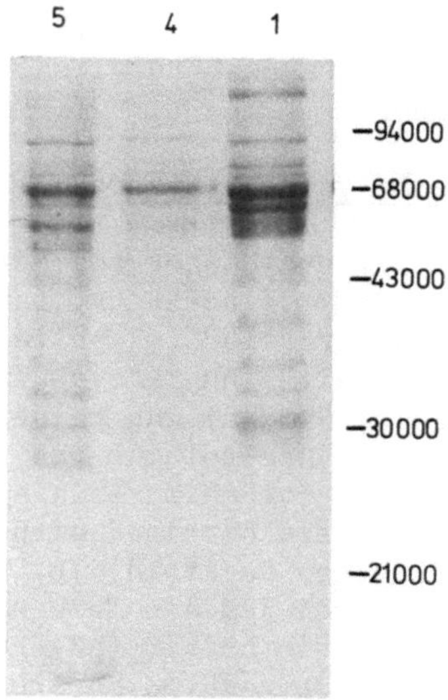

Fig. 1. Protein pattern after SDS-polyacryl amide electrophoresis of the fractions eluted from a column of deoxycholic acid-poly-(L-lysine)-Sepharose, which was loaded with the active fraction from a Matrex gel Red A column. Positions of molecular weight markers have been indicated.

Fraction 1: 5.4 mg protein eluted with 1.2 M NaCl in 25 mM Tris, pH 8.0 (no activity).

Fraction 4: 0.2 mg protein eluted with 50 mM borate buffer, pH 9.5 (no activity).

Fraction 5: 1.1 mg protein eluted with 8 M urea in 25 mM Tris, pH 8.0 (75% of the activity originally present in the sample, with a relative specific activity of 80-100).

1. Matrex gel Red A

In the original method published by Jansen et al. (1981), Matrex gel Red A (Amicon, Lexington, MA, U.S.A.) was used as the first column material for purification of the proteins from bFF1. Preelution of the column with 50 mM Tris buffer (pH 7.0), containing 350 mM KCl and 0.5 M urea, resulted in a 20-25-fold purification of the inhibin activity in the fraction which was subsequently eluted after increasing the KCl concentration to 1.2 M. In later experiments, it appeared that the purification factor on this column could be improved by increasing the pH of the Tris buffer from 7.0 to 7.5. The relative specific activity (RSA) of the latter fraction, in which more than 90% of the activity was eluted, increased to 40-50, i.e., twofold when compared with the procedure published previously. The protein pattern after SDS-polyacrylamide electrophoresis (Laemmli, 1970) showed an almost complete absence of BSA from this preparation.

2. Deoxycholic Acid-poly-(L-lysine)-sepharose

The dialysis or desalting step after the Matrex gel Red A column, which has a low recovery of activity, can be omitted from the procedure when the active fraction is applied subsequently to a column of hydrophobic material. The recovery from phenyl-sepharose CL-4B (Pharmacia, Uppsala, Sweden) was relatively low. Therefore, a number of other hydrophobic columns was tested (see De Jong et al., 1981); the use of these column materials did not improve the recovery. Better results were obtained with sepharose-immobilized deoxycholic acid (Pierce, Rockford, IL, U.S.A.). The active material from the Matrex gel Red A column was brought on this column in 1.2 M NaCl in 25 mM Tris buffer (pH 8.0). After elution with 50 mM Tris buffer (pH 8.0) and 50 mM borate buffer (pH 9.5), the column was eluted with 8 M urea in 25 mM Tris (pH 8.0). In this way, 75% of the original activity was recovered in a 80-100 fold purified fraction. The protein pattern of this fraction is shown in Fig. 1. The active fraction was desalted on a column of Sephadex G-25 coarse.

3. Lentil Lectin Sepharose

The final column material in the published purification procedure was Concanavalin-A-sepharose (Con A-sepharose, Pharmacia, Uppsala, Sweden). The recovery of activity from this column was rather low. Furthermore, after elution of the blank column material with 0.3 M α-methyl-D-mannoside in 50 mM Tris buffer, pH 7.0, a protein band with an apparent molecular weight of 32,000 daltons was present in the SDS-polyacrylamide electrophoresis pattern; this probably represents eluted Con A monomers (Zanetta and Gombos, 1976). For these reasons, another immobilized plant lectin, lentil lectin sepharose (Pharmacia) was tested. Application of an active fraction from Matrex gel Red A to a column of this material, and elution with 150 mM NaCl in 25 mM Tris buffer, pH 7.1, resulted in complete retention of the activity on the column. The activity was subsequently eluted with 0.5 M α-methyl-D-mannoside and 0.5 M α-methyl-D-glucoside. The protein recovery from the lentil lectin-column (81%) was appreciably higher than that from a Con A-sepharose column (44%), which was loaded with the same material.

4. Immobilized Antibodies, Raised against Bovine Plasma

Peripheral plasma from a two-weeks ovariectomized cow was run on a Matrex gel Red A column. The fraction, eluted with 1.2 M KCl and 0.5 M urea, which would contain the inhibin activity if bFFl

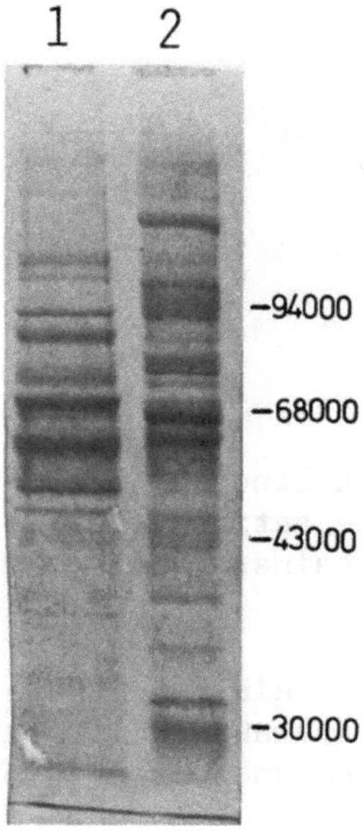

Fig. 2. Protein pattern after SDS-polyacryl amide electrophoresis of the fractions eluted from a column of an immobilized globulin fraction from an antiserum against a Matrex gel Red A fraction of peripheral plasma from an ovariectomized cow. The column was loaded with the active fraction from a Matrex gel Red A column. Positions of molecular weight markers have been indicated.

Fraction 1: 1.0 mg protein eluted with 0.1 M Hepes buffer, pH 7.5 (R.S.A. 38).

Fraction 2: 1.1 mg protein eluted with 0.1 M glycine buffer, pH 2.5 (no activity).

were used, was collected and used for the immunization of two rabbits. An amount of 150 µg of this protein in a phosphate-buffered saline (0.01 M, pH 7.0)/complete Freund's adjuvant emulsion was injected intramuscularly, followed by a second injection with the same amount of protein in incomplete Freund's adjuvant after 4 weeks. One week after the last injection, blood was collected. The presence of antibodies against bovine plasma proteins was ascertained by immunodiffusion (Ouchterlony, 1949). The globulin fraction of the plasma proteins was purified on a column of Protein A Sepharose (Pharmacia) and subsequently coupled to CNBr-activated Sepharose 4B (Pharmacia). A column of these immobilized antibodies was loaded with the active Matrex gel Red A fraction from bFF1, and eluted with buffer, followed by 0.1 M glycine (pH 2.5). SDS-polyacrylamide electrophoresis patterns of the two fractions, which do

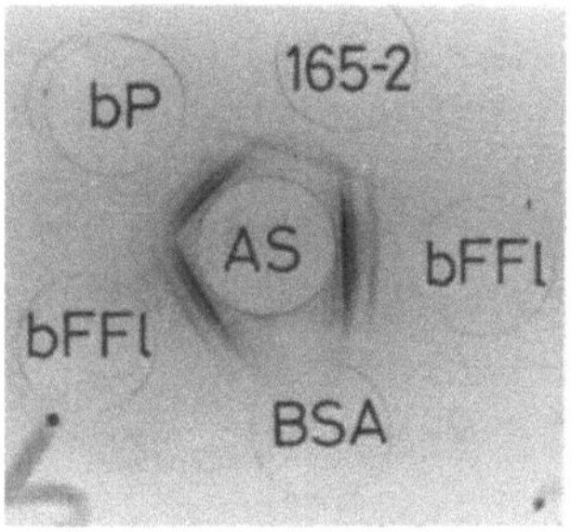

Fig. 3. Immunoprecipitation lines of bFFl, peripheral bovine plasma, BSA and the antigen (165-2), diffused against an antiserum raised against 165-2.

not contain corresponding protein bands, are shown in Fig. 2. After this procedure, the RSA of the unbound fraction was increased threefold; the recovery on this column was over 90%.

These results show the potential of the immunological approach in the purification of inhibin, as was previously indicated by Williams et al. (1979), who used immobilized antibodies against barrow plasma in the purification of inhibin from porcine folliculare fluid.

EXPERIMENTS WITH ANTIBODIES AGAINST A PARTIALLY PURIFIED INHIBIN PREPARATION

The partially purified inhibin preparation described earlier (fraction 165-2) (Jansen et al., 1981) was used for immunization of two rabbits. Amounts of 100 µg of this preparation were injected intramuscularly, using the immunization scheme described above. The animals were bled one week after each monthly booster injection; the experiments described here were performed with antisera obtained after the 5th and 6th bleeding. The following techniques were used to evaluate the possible presence of anti-inhibin antibodies in these antisera:

1. Double Immunodiffusion
(Ouchterlony, 1949)

Various dilutions of the antisera were diffused against bFFl, bovine plasma, BSA and fraction 165-2 against which the antisera were raised. Results are shown in Fig. 3. It can be concluded that the antiserum contains antibodies against plasma proteins, but not against BSA. No clear differences between the precipitation lines of plasma and bFFl can be observed. Finally, the fraction 165-2 caused a precipitation line which is not identical with lines in the bFFl and bovine plasma precipitates.

Table 1. Binding of Iodinated "Inhibin" to "Inhibin-Antiserum" (1:100), Which was or Was Not Pretreated with an Equal Volume of Peripheral Bovine Plasma (means of duplicate percentages, corrected for non-specific binding)

addition	non-pretreated antiserum	pretreated antiserum
none	37.0	18.2
10 μl antigen	15.9	1.3
25 μl bFF1	13.1	18.6
none	-	4.2
100 ng Concanavalin A	-	-1.0

2. Binding of Radioactively Labeled Material

Labeling of fraction 165-2, or fraction 242-2, which was obtained after a similar purification sequence, with ^{125}I (iodogen procedure) yielded good incorporation of iodine. After SDS-polyacrylamide electrophoresis of the labeled preparation and autoradiography of the gel, the pattern of labeled proteins was identical to that after Coomassie-blue staining of the proteins (Jansen et al., 1981): bands in the 60,000-70,000 daltons region and one with an apparent molecular weight of 32,000 daltons. This preparation was used as a label in a double antibody radioimmunoassay system.

Radioactivity could partly be displaced from the antibody after addition of bFF1 to the assay system (Table 1). Pretreatment of the antiserum with an equal volume of bovine plasma (bP) also decreased the binding of radioactivity; further addition of bFF1 did not influence the binding of the label. Finally, addition of the immunogen, fraction 165-2, displaced the remaining label from the bP-pretreated antibody; a similar effect could be obtained by addition of Con A (Table 1). These observations suggest that the antiserum contains antibodies against Con A, which might have leaked from the last column used in the preparation of the immunogen, fraction 165-2. As a final experiment to show this, a double immunodiffusion study was carried out in the presence of 25 mM α-methyl-D-mannoside to prevent glycoprotein-Con A precipitation. A strong precipitation line was obtained between Con A and the antiserum.

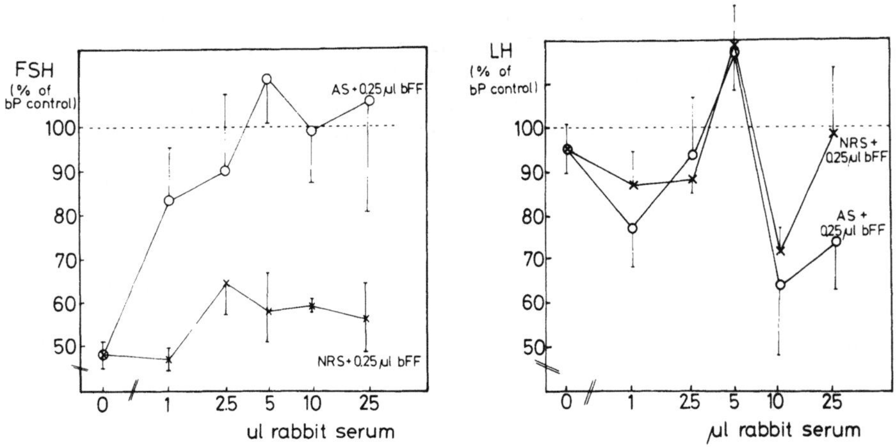

(a) (b)

Fig. 4. Concentrations of FSH (a) and LH (b) in the medium of pituitary cells, to which bovine plasma or bFF1 (0.25 µl) and various amounts of an antiserum against inhibin or normal rabbit serum were added. Results are expressed relative to those obtained with the bovine plasma-antiserum combination.

Nonimmune rabbit serum did not give rise to a precipitation line in this experiment.

3. Neutralization of Inhibin Activity in Vitro

Addition of increasing doses of bP-pretreated antiserum to the pituitary cells in vitro can abolish the FSH-suppressing activity of a maximally inhibiting dose of standard bFF1 (0.25 µl). As a control, the same amount of antiserum was added together with the same volume of bP, while also the combination of bFF1 and nonimmune serum was added in order to study effects of globulins in the rabbit sera on the second-antibody precipitation of FSH and LH in the ratioimmunoassay of these hormones. Results of these experiments for one of the antisera have been summarized in Fig. 4. With 1 µl of this antiserum, the effect of 0.25 µl of bFF1 can be abolished. The other antiserum showed the same effect with 10 µl of antiserum added. These results indicate that the antisera contain a low titer of anti-inhibin antibodies. The possibility of competition between antibody and the possible inhibin receptor in the pituitary cells has also to be taken into account.

Summarizing, it is concluded from these data that fraction 165-2, in addition to inhibin, contains plasma proteins and Con A. The antiserum raised against this fraction contains antibodies

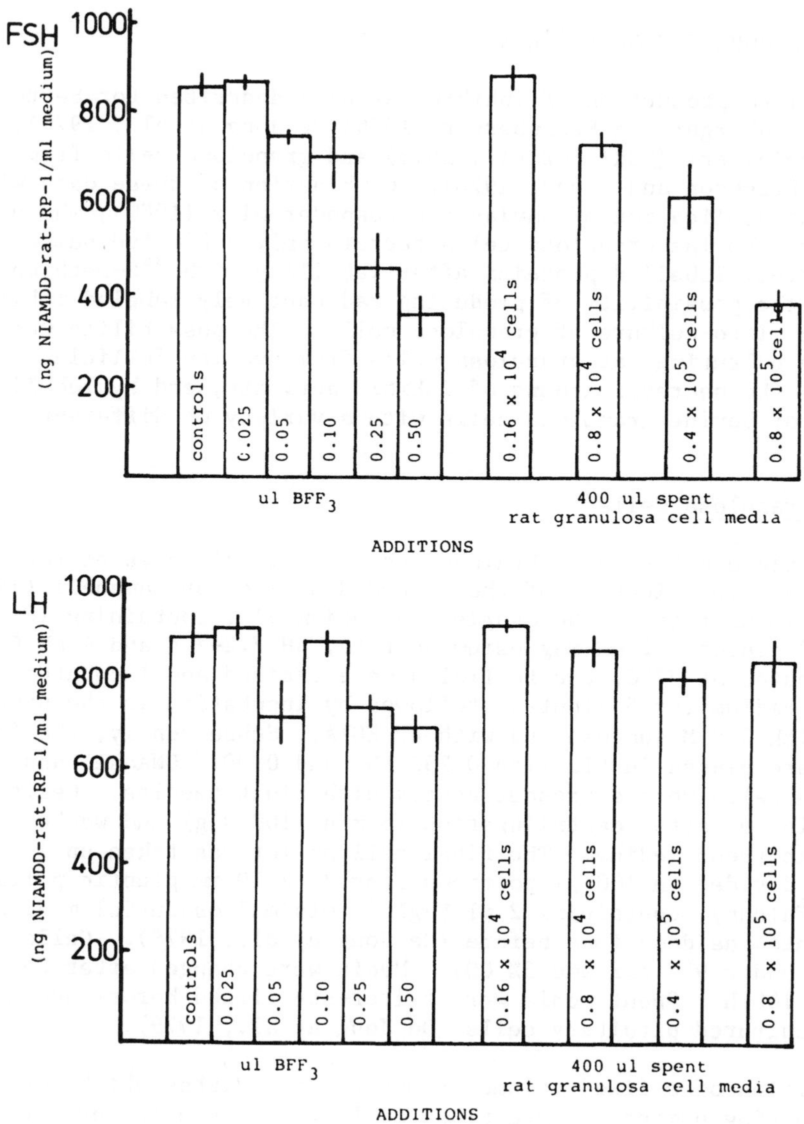

Fig. 5. Concentrations of FSH and LH in the medium of pituitary cells, cultured with a fixed volume (400 µl) of spent medium from cultured rat granulosa cells (day 1 of culture) (means ± s.e.m., n = 4).

against inhibin, against some plasma proteins with molecular weights between 60,000 and 70,000 which are not identical with BSA, and against Con A.

IN VITRO PRODUCTION OF INHIBIN

In vitro production of inhibin has been described for Sertoli cells (Steinberger and Steinberger, 1976; De Jong et al., 1979), tubuli seminiferi (Eddie et al., 1978) and granulosa cells from rat ovaries (Erickson and Hsueh, 1978). Combination of these data with the recent publication of Savian and Gospodarowicz (1980), who observed that bovine granulosa cells secrete only a limited number of radioactively labelled proteins after labelling with ^{35}S-methionine, suggests the possibility of producing radioactively labeled inhibin during in vitro culture of granulosa cells. The possibility was explored by culturing rat granulosa cells from ovarian follicles and measuring the secreted amount of inhibin activity, and by labeling cultures of bovine granulosa cells with a variety of different labels.

1. Rat Granulosa Cells

Granulosa cells were obtained from antral follicles of rat ovaries by a modification of the method described by Campbell (1979). After preincubation of the ovaries in medium F12, containing 10 mM Hepes, 1% glucose, 2 mM magnesium acetate, pH 7.2-7.4 and 4 mM EGTA for 5 minutes at 37°C, the follicles were excised and incubated in the same medium for 5 minutes, followed by incubation in the same medium with 0.2 M sucrose and without EGTA. Subsequently, the follicles were placed in F12 with 0.25% BSA and 0.001% DNAase, and granulosa cells were expressed gently with blunt needles. Cells were collected after centrifugation (5 min, 100 × g) and washed 3 times with fresh medium. The final cell pellet was taken up in F12, and divided in 100 µl portions over 35 × 10 mm plastic petri dishes (Falcon), containing 2 ml Eagle's minimal essential medium, supplemented as described before (De Jong et al., 1979). Cells were cultured under 95% air and 5% CO_2. Media were changed after 24 h, 96 h and 168 h. Spent media were filter-sterilized before adding them to cultured pituitary cells (De Jong et al., 1979).

Addition of a fixed volume of medium from dishes which contained varying numbers of granulosa cells which were cultured for 24 h, caused selective suppression of the release of FSH from pituitary cells (Fig. 5). The cell number-response curve was parallel to the dose-response curve for standard bovine follicular fluid. The time course of the release of the FSH-suppressing material from the granulosa cells was studied by using different volumes of medium from one batch of granulosa cells ($1.1 \cdot 10^5$ cells/dish). The media collected after 24 h and 96 h of culture caused

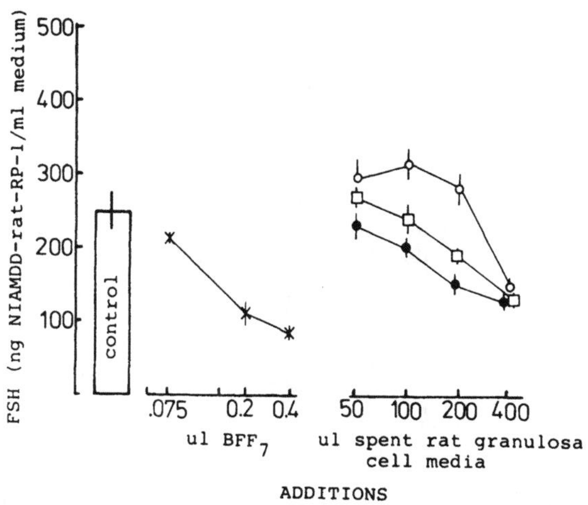

Fig. 6. Concentrations of FSH in medium from cultured pituitary cells, to which various volumes of standard bFF1 (x), or spent medium from a rat granulosa cell culture were added. The granulosa cell medium was collected after day 1 (●), or after culture periods from day 2-4 (□) or day 5-7 (O). (means ± s.e.m., n = 4).

Table 2. Concentrations of FSH and LH (μg/ml) in the Medium of Cultured Pituitary Cells, after Addition of Spent Media from a Culture of Bovine Granulosa Cells (bGCCM) (means ± s.d., n = 4)

Addition	FSH	LH
None	0.43 ± 0.10	0.48 ± 0.11
0.25 μl bFF1	0.21 ± 0.02*	0.43 ± 0.07
100 μl bGCCM	0.19 ± 0.03*	0.44 ± 0.08
idem, after Sephadex G 25	0.25 ± 0.05*	0.44 ± 0.08
idem, after Matrex gel Red A	0.24 ± 0.02*	0.60 ± 0.16

*Significantly different from control ($P < 0.05$, Student's t-test).

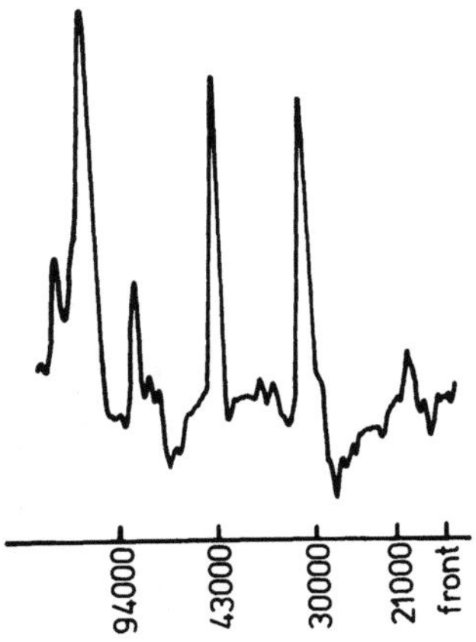

Fig. 7. Densitogram of an autoradiogram of tritiated proteins, secreted by bovine granulosa cells which were cultured in the presence of ^{3}H-leucine. The medium was desalted (Sephadex G-25) and run on a Matrex gel Red A column. The fraction with inhibin activity was collected, followed by electrophoresis in an SDS-polyacryl amide system. Molecular weights of marker proteins have been indicated.

Table 3. Incorporation of ^{3}H-Fucose and ^{35}S-Methionine into Proteins after Different Periods of Culture of Bovine Granulosa Cells. Radioactive Precursors were Added after 2 Days of Preculture

Day of culture	% incorporation		^{3}H/ ^{35}S
	^{3}H	^{35}S	(dpm/dpm)
3	1.16	2.16	0.63
4	2.50	2.00	1.52
5	4.55	2.57	2.02

dose-response curves parallel to that of standard bovine follicular fluid (Fig. 6). LH secretion was not affected (data not shown). Media collected during the period 96-168 h also suppressed the release of FSH, but only with the highest dose added; the dose-response curve was no longer parallel to the lines for the previous days or for standard bFFl.

From these data it can be concluded that release of inhibin takes place during the culture period; it is not clear if the decrease in activity from day 1 to the 3 day period 2-4 is due to a sharp decrease of the amount of activity produced or to contamination of the cells with minute amounts of follicular fluid, which would make the concentration after 24 h of culture artificially high. Finally, the medium of the 3rd culture period did not suppress FSH in parallel. This might be due to the increased amounts of steroid, produced by the luteinized granulosa cells.

2. Bovine Granulosa Cells

Granulosa cells were collected from bovine ovaries, which were collected in a local slaughterhouse and transferred to the laboratory on ice. The follicular fluid was removed from the follicles (0.5-2.0 cm diameter) by aspiration. Subsequently, the follicles were slit open with a scalpel, and the granulosa cells were scraped from the inside of the follicle with a small spatula. The cells were collected in 5 ml portions of medium 199, buffered with hepes. After collection, the cells were spun down and the medium was replaced by 10 ml fortified MEM (De Jong et al., 1979), containing 5% foetal calf serum. The cells were transferred to a tissue culture flask in this medium, and cultured for 2 days. After this period the medium was renewed. When ^{3}H-leucine incorporation was studied, the labeled amino acid (10 μCi) was added in fortified MEM without leucine with 5% foetal calf serum. Labeling with ^{35}S-methionine (10 μCi) and ^{3}H-fucose (10 μCi) was performed in the medium in which the cells were precultured. After a three-day culture period, media were collected and assayed for radioactivity and inhibin activity. The medium in which the cells were cultured between days 2 and 5 could selectively suppress the release of FSH from cultured pituitary cells (Table 2). Incorporation data for ^{35}S-methionine and ^{3}H-fucose are summarized in Table 3. It appears that the tritiated proteins are secreted later during the culture period than the ^{35}S-labeled macromolecular material.

The radioactively labeled proteins were subsequently desalted on a column of G-25-coarse, and chromatographed on a column of Matrex gel Red A, as previously described for bFFl. A densitogram of the SDS-polyacrylamide electrophoresis pattern of this fraction is shown in Fig. 7. The fraction, which contained bioactive inhibin, was added to the material from a Red A column, which had been loaded with 30 ml bFFl. This combined material was further subjected to a sequence of the different purification steps mentioned earlier in this report. During this purification procedure, the relative specific activity of the preparation increased in parallel with the increase in the ^{3}H/^{35}S ratio (Table 4).

Table 4. Relative Specific Activity (R.S.A) and $^{3}H/^{35}S$ Ratio (dpm/dpm) of ^{3}H-Fucose and ^{35}S-Methionine Labeled Bovine Granulosa Cell Proteins after Addition of the Labeled Proteins to Bovine Follicular Fluid and Purification of Inhibin

Purification step	RSA	$^{3}H/^{35}S$
Starting material	1	2.02
active fraction after		
- Matrex gel Red A	13.9	4.27
- Deoxycholic acid-Sepharose	35.5	5.08
- Sephadex G 25	28.0	5.22
- Lentil lectin-Sepharose	50.7	6.84
inactive fraction after		
- Lentil lectin-Sepharose	18.3	1.86

It is concluded from these experiments, that inhibin is co-purified with the ^{3}H-fucose-labeled bovine granulosa cell proteins. Furthermore, the radioactively labeled proteins can be used in monitoring procedural losses during the purification of inhibin. So far, it has not been possible yet to assign any of the radioactive bands to the protein with inhibin activity.

ACKNOWLEDGEMENTS

This work was supported in part by the Dutch Foundation for Medical Research (FUNGO). We thank Dr. J. Dullaart for the supply of gonadotropin antibodies, the NIAMDD for the standard gonadotropins, Miss J. Steenbergen, Mrs. E. C. M. van Leeuwen, and Miss M. H. M. Debets for their expert technical assistance and Miss W. Bakhuizen for typing the manuscript.

REFERENCES

Campbell, K. L., 1979, Ovarian granulosa cells isolated with EGTA and hypertonic sucrose: cellular integrity and function, Biol. Reprod., 21:773-786.

De Greef, W. J., de Jong, F. H., and de Koning, J., 1982, Levels of luteinizing hormone-releasing hormone in hypothalamus and hypophysial stalk plasma of dioestrous and ovariectomized rats: effects of treatment with bovine follicular fluids, in preparation.

De Jong, F. H., and Sharpe, R. M., 1976, Evidence for inhibin-like activity in bovine follicular fluid, Nature, 263:71-72.

De Jong, F. H., Smith, S. D., and van der Molen, H. J., 1979, Bioassay of inhibin-like activity using pituitary cells in vitro, J. Endocr., 80:91-102.

De Jong, F. H., Jansen, E. H. J. M., and van der Molen, H. J., 1981, Purification and characterization of inhibin, in: "Intragonadal Regulation of Reproduction," P. Franchimont and C. P. Channing, eds., Academic Press, London, pp. 229-250.

Eddie, L. W., Baker, H. W. G., Dulmanis, A., Higginson, R. E., and Hudson, B., 1978, Inhibin from cultures of rat seminiferous tubules, J. Endocr., 78:217-224.

Eddie, L. W., Baker, H. W. G., Higginson, R. E., and Hudson, B., 1979, A bioassay for inhibin using pituitary cell cultures, J. Endocr., 81:49-60.

Erickson, G. F., and Hsueh, A. J. W., 1978, Secretion of "inhibin" by rat granulosa cells in vitro, Endocrinology, 103:1960-1963.

Hermans, W. P., van Leeuwen, E. C. M., Debets, M. H. M., and de Jong, F. H., 1980, Involvement of inhibin in the regulation of follicle-stimulating hormone concentrations in prepubertal and adult, male and female rats, J. Endocr., 86:79-92.

Hudson, B., Baker, H. W. G., Eddie, L. W., Higginson, R. E., Burger, H. G., de Kretser, D. M., Dobos, M., and Lee, V. W. K., 1979, Bioassays for inhibin: a critical review, J. Reprod. Fert. Suppl., 26:17-29.

Jansen, E. H. J. M., Steenbergen, J., de Jong, F. H., and van der Molen, H. J., 1981, The use of affinity matrices in the purification of inhibin from bovine follicular fluid, Molec. Cell. Endocr., 21:109-117.

Laemmli, U. K., 1970, Cleavage of structural proteins during the assembly of the head of bacteriophage T4, Nature, 227:680-685.

Le Lannou, D., and Chambon, Y., 1977, Présence dans l'épididyme d'un facteur inhibant la synthèse hypothalamique de FSH-RH chez le rat, C.R. Soc. Biol., 171:1064-1067.

Lumpkin, M., Negro-Vilar, A., Franchimont, P., and McCann, S., 1981, Evidence for a hypothalamic site of action of inhibin to suppress FSH release, Endocrinology, 108:1101-1104.

McCullagh, D. R., 1932, Dual endocrine activity of the testes, Science, 76:19-20.

Ouchterlony, P., 1949, Antigen-antibody reactions in gels, Acta Pathol. Microbiol. Scand., 26:507-515.

Savion, N., and Gospodarowicz, D., 1980, Patterns of cellular peptide synthesis by cultured bovine granulosa cells, Endocrinology, 107:1798-1807.

Scott, R. S., Burger, H. G., and Quigg, H., 1980, A simple and rapid in vitro bioassay for inhibin, Endocrinology, 107:1536-1542.

Setchell, B. P., Davies, R. V., and Main, S. J., 1977, Inhibin, in: "The Testis, Vol. IV," A. D. Johnson and W. R. Gomes, eds., Academic Press, New York, pp. 189-238.

Setchell, B. P., and Sirinathsinghji, D. J. S., 1972, Antigonadotropic substance in rete testis fluid, a possible "inhibin", J. Endocr., 53:1x.

Steinberger, A., and Steinberger, E., 1976, Secretion of an FSH-inhibiting factor by cultured Sertoli cells, Endocrinology, 99: 918-921.

Uilenbroek, J. Th. J., Tiller, R., de Jong, F. H., and Vels, F., 1978, Specific suppression of follicle stimulating hormone secretion in gonadectomized male and female rats with intrasplenic ovarian transplants, J. Endocr., 78:399-406.

Welschen, R., Hermans, W. P., and de Jong, F. H., 1980, Possible involvement of inhibin in the interrelationship between numbers of antral follicles and peripheral FSH concentrations in female rats, J. Reprod. Fert., 60:485-493.

Williams, A. I., Rush, M. E., and Lipner, H., 1979, Isolation and preliminary characterization of inhibin-f, in: "Ovarian Follicular and Corpus Luteum Function," C. P. Channing, J. M. Marsh, and W. A. Sadler, eds., Plenum Press, New York, pp. 429-435.

Zanetta, J.-P., and Gombos, G., 1976, Affinity chromatography of brain membrane glycoproteins on Concanavalin A — "Sepharose" in the presence of SDS, in: "Concanavalin A as a Tool," H. Bittiger, and H. P. Schnebli, eds., J. Wiley and Sons, London, pp. 389-398.

HORMONAL CONTROL OF INHIBIN SECRETION

Larry D. Anderson and Dennis J. Hoover

Departments of Anatomy and Physiology
University of Maryland School of Medicine
Baltimore, Maryland 21201

INTRODUCTION

The studies in this manuscript are mainly concerned with the question: What determines the output of inhibin from the ovary? In an attempt to answer this question we have chosen to examine inhibin output using two approaches, (1) Quantification of the amount of inhibin secreted into the ovarian vein of the rat in vivo during various reproductive states and (2) secretion of inhibin from the granulosa cell compartment of the pig ovary in vitro.

Our routine method for detecting and quantifying inhibin activity is the rat pituitary culture bioassay as described in the following section.

Bioassay of Inhibin Activity

Inhibin activity is bioassayed by determining the degree of inhibition of basal (i.e., non-LHRH stimulated) 24 h secretion of FSH by dispersed rat anterior pituitary cells in primary, monolayer culture. For this culture, anterior pituitary glands are taken from randomly cyclic female Sprague-Dawley rats (250-300 g body weight). After mincing with scissors, the glands are incubated with stirring for: (1) 50 min with collagenase (4500 units per 20 glands) and hyaluronidase (4000 units per 20 glands) in HEPES buffer with Hanks salts at 37°C and then (2) 15 min with viokase (10 ml of 10% per 20 glands) in Hanks-HEPES buffer at 37°C. After washing, the dispersed cells are incubated in Falcon multiwell culture plates for 72 h in Dulbecco's modified Eagles's medium with 10% pooled female rat serum, and 5.5% horse serum (growth medium) in a humidified incubator at 37°C with 5% CO_2 in air. Each pituitary yields six culture wells. After 72 h, the medium is replaced by fresh growth medium supplemented with 10^{-6} M progesterone and 10^{-8} M 17β-estradiol containing the unknowns to be assayed for inhibin activity. In addition, a dose response curve with the laboratory inhibin standard (charcoal-treated, pooled porcine follicular fluid) is run in each assay (14 doses in triplicate) along with triplicate growth medium controls in each culture plate. After 24 h of incubation, the conditioned medium is recovered for RIA (rFSH).

In general, a culture will yield a working linear range of 0.01% to 0.05% for the follicular fluid standard. The coefficient of determination ($r^2 \times 100$) for acceptable assays usually exceeds 90% and the index of precision (λ = residual variance/slope) for the standard curve ranges between 0.3 and 0.06. The average coefficient of variation for the triplicate determinations at each dose of FFl standard is in the range of 7 to 12%.

OVARIAN SECRETION OF INHIBIN IN RATS

Inhibin Secretion during the Estrous Cycle

Most reports on inhibin have been concerned with identifying the presence of FSH suppressing activity in follicular fluid or ovarian extracts, purification of the non-steroidal substance, or investigating the effects of injecting inhibin preparations into animals during various reproductive states. However, in order for inhibin to regulate the pituitary gland and/or hypothalamus and affect FSH secretion, it must leave the ovary via the venous blood. Two recent reports have addressed this question by transplanting the ovary either into the kidney or spleen. In the first study, transplantation of a mouse ovary beneath the kidney capsule caused a marked suppression of serum FSH even in the presence of 100 μg estradiol. Also, the injection of charcoal-treated pFFl into ovariectomized mice implanted with 100 μg estradiol capsules significantly decreased FSH serum levels as compared to saline or pig serum injected controls (Bronson and Channing, 1978). Using a similar approach, Uilenbroek et al. (1978) reasoned that if an ovary was transplanted to the spleen (in castrated male and female rats) splenic venous blood would pass through the liver, thereby metabolizing the majority of steroids; however, proteins would remain in the venous blood. In ovariectomized rats with intrasplenic-ovarian transplants, LH levels increased whereas FSH levels decreased. This is in contrast to intrarenal-ovarian transplants, which suppressed both LH and FSH levels to that present in intact male and female rats. These transplantation studies suggest that some ovarian product other than estradiol or other steroids, possibly inhibin, is responsible for the suppression of FSH levels.

However, the concept of inhibin regulation of FSH secretion requires that changes in inhibin secretion during the reproductive cycle of intact animals should be accompanied by changes in FSH secretion. More direct evidence for the ovarian secretion of inhibin-like material in the rat has been provided by De Paolo et al. (1979a). Inhibin activity was assessed in vitro by the addition of ovarian vein plasma (OVP) to our dispersed pituitary cell culture system. The studies revealed that the amount of inhibin activity present in the OVP changes depending upon the stage of the estrous cycle and varies inversely with plasma FSH concentrations. Thus, when OVP inhibin activity is high (diestrous 1, diestrous 2, and proestrous morning), plasma FSH levels are low. During the preovulatory, proestrous LH and FSH surges, inhibin activity declines rapidly in OVP and is further reduced on estrous morning when plasma FSH is elevated (Fig. 1). In addition, these results suggest that inhibin secretion may be regulated directly by gonadotropins.

Effect of Hormones on Inhibin Secretion in vivo

Previous studies demonstrated that the administration of LH or FSH to phenobarbital-blocked proestrous rats induces an endogenous secondary rise in FSH similar to that observed on estrus in cycling rats with no changes in LH (Ashiru and Blake, 1979). Using a similar protocol, we investigated whether this secondary rise in FSH on 0900 estrus (after administration of exogenous FSH or LH to phenobarbital-treated rats on 1300 proestrus) is correlated with the amount of inhibin activity present in the ovarian vein at this time period (Shander et al., 1980). The results show that while the secondary FSH surge was suppressed by the administration of phenobarbital, the OVP from phenobarbital-treated rats had the highest amount of inhibin activity of all the treatment groups. With injection of 1 μg FSH or 8 μg LH, inhibin activity was decreased and restored to that observed in control, untreated-rats (Fig. 2). These results suggest that either gonadotropin can suppress the amount of inhibin activity leaving the ovary at 0900 on the expected estrus. Furthermore, blocking the endogenous gonadotropin surges on proestrus causes increased inhibin secretion on the following day. It remains to be determined whether these actions are direct effects of the gonadotropins or whether other hormones or factors mediate the gonadotropic action on inhibin activity.

Indirect evidence by Sasomoto and Taya (1980) suggests that LH may be the gonadotropin releasing the pituitary gland from the negative feedback action of inhibin upon FSH secretion during various reproductive states. A selective surge in serum FSH was observed during the period of ovulation after injection of hCG in rats on day 15 of lactation. Similar surges of FSH were noted in 29 day-old, immature rats and adult cycling rats on 0900h diesterus following hCG injection accompanied by ovulation. In contrast, the same amount

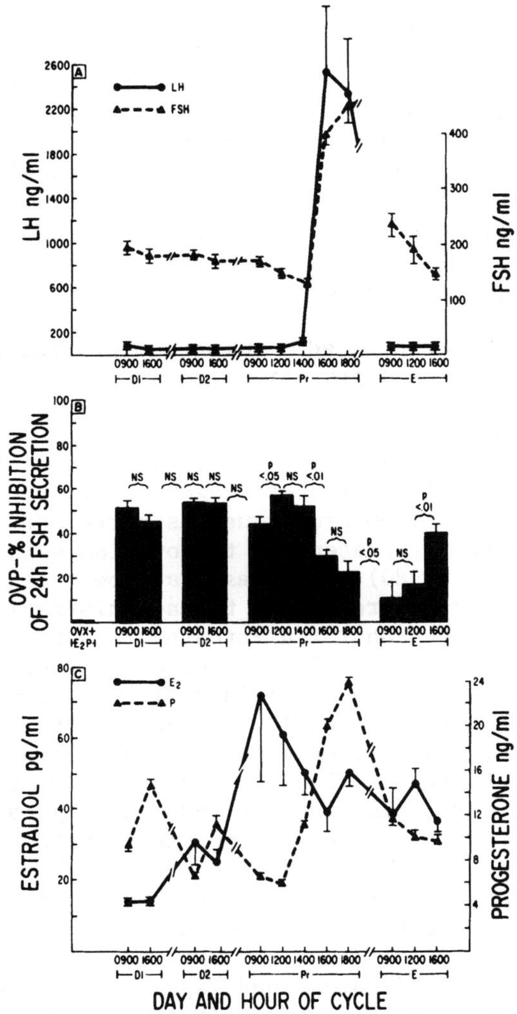
A
LH
FSH
LH ng/ml
FSH ng/ml
B
OVP-% INHIBITION OF 24h FSH SECRETION
NS
p <.05
p <.01
OVX+ E2P4
C
E2
P
ESTRADIOL pg/ml
PROGESTERONE ng/ml
0900 1600 0900 1600 0900 1200 1400 1600 1800 0900 1200 1600
D1 D2 Pr E
DAY AND HOUR OF CYCLE

Fig. 1. Changes in peripheral plasma concentrations of FSH, LH, estradiol (E_2) and progesterone (P) during the 4-day rat estrous cycle are correlated with changes in inhibin activity in ovarian venous plasms (OVP). All OVP and peripheral plasma samples were obtained from the same animals. A (top panel): plasma LH and FSH concentrations. B (middle panel): changes in percent of inhibin activity during the rat estrous cycle. As the proestrous preovulatory surge in LH and FSH occurs (1400-1800 h) inhibin activity declines significantly ($p < 0.01$) in OVP. At 0900h estrous, peripheral plasma FSH is elevated and inhibin activity has further declined. When plasma FSH approaches basal D1 levels, inhibin activity again is increased in OVP. C (bottom panel): peripheral plasma E_2 and P concentrations during the rat estrous cycle. Plasma E_2 levels increase between D2 and Pr to reach peak concentrations at 0900h Pr. Following the preovulatory LH and FSH surge, plasma E_2 declines and P increases in the peripheral circulation. On estrous morning P plasma levels are basal but E_2 values are somewhat elevated above D1 concentrations. Thus, as inhibin activity declines in OVP on Pr, E_2 also declines and P rises in peripheral plasma.

Abbreviations: D1, D2, diestrous days 1, 2; Pr, proestrous; E, estrous; OVP, ovarian vein plasma. (Data reprinted with permission from DePaolo et al., 1979a.)

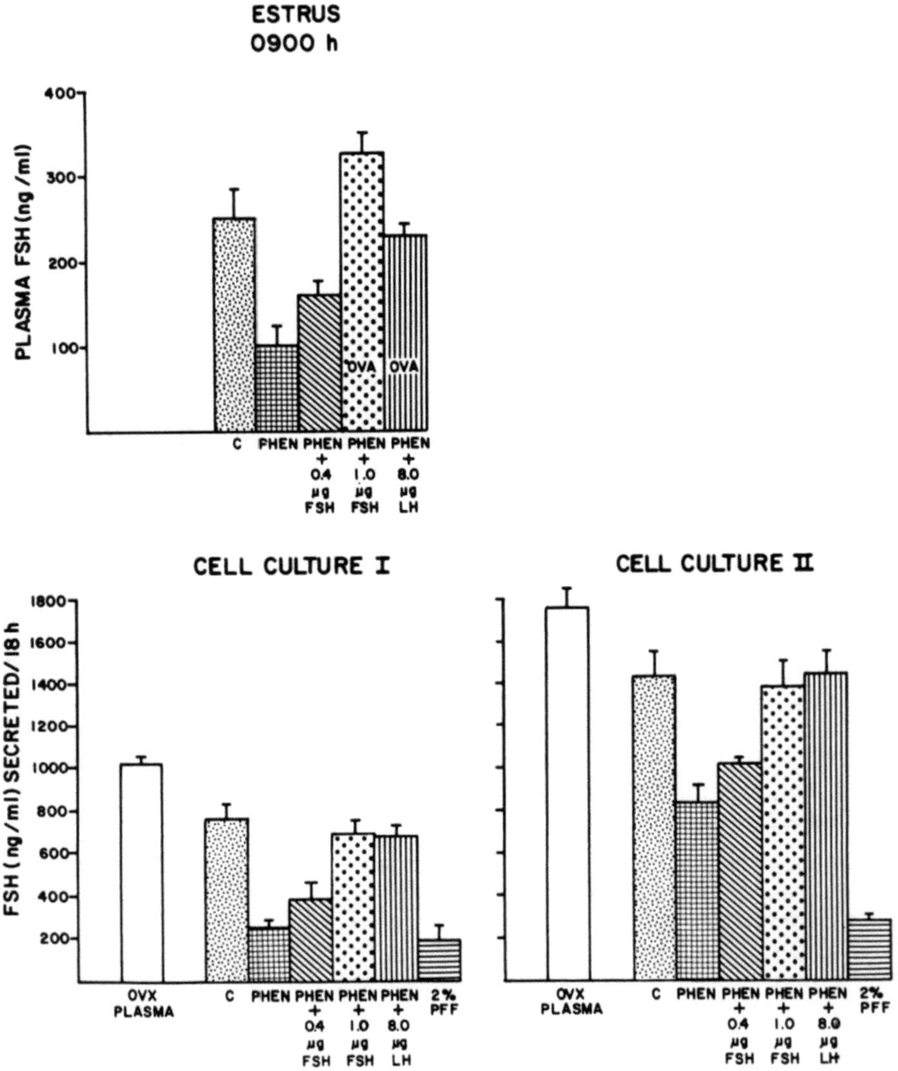

% INHIBITION 18h FSH SECRETION	C	PHEN	PHEN + 0.4 µg FSH	PHEN + 1.0 µg FSH	PHEN + 8.0 µg LH	2% PFF
Cell Culture I	25.7	75.1	62.5	31.6	33.3	81.2
Cell Culture II	17.6	52.3	41.4	20.0	16.6	82.9

Fig. 2. The effect of exogenous FSH and LH upon the endogenous release of pituitary FSH and the ovarian secretion of inhibin in rats. Between 1215-1300h proestrus all rats were anesthetized with ether and polyethylene cannulae were inserted into the right external jugular veins to the level of the right atria. Immediately after cannulae insertion, rats were divided into 5 treatment groups: Group 1 (n = 8) received only 0.9% phosphate-buffered saline (PBS) i.v. via jugular cannulae, while Groups 2-5 were reanesthetized with phenobarbital at 1245h to block the spontaneous FSH and LH surges. At 1300h the following substances were injected i.v. in 0.4 ml of 0.9% PBS:

Group 2, PBS only (n = 8)
Group 3, 0.4 μg highly purified rat $FSH\text{-}1_3$ (n = 12)
Group 4, 1.0 μg highly purified rat $FSH\text{-}1_3$ (n = 17)
Group 5, 8.0 μg highly purified rat $LH\text{-}I_5$ (n = 14)

Blood samples (0.5 cc) were taken via jugular cannulae from each rat at 1400 and at 1700h proestrous and assayed for FSH, LH, and progesterone (data not shown). At 0900h estrous all rats were anesthetized with ether, the original cannulae were removed and new polyethylene cannulae were coursed, via the external jugular veins and vena cava into the left ovarian vein. Ovarian vein blood was collected for 20 min. (2.0 cc) after which peripheral blood samples were taken from the inferior vena cava. In all estrous rats the oviducts were removed and examined for the presence of ova. Upper panel represents the concentration of FSH in the peripheral plasma at 0900h estrous. Lower panels represent the relative amounts of inhibin activity present in the ovarian blood as bioassayed in two separate rat pituitary cell cultures. (Data reprinted with permission from Shander et al., 1980a.)

of hCG did not induce a selective FSH surge, or ovulation, in 26 day-old immature rats or on day 3 of lactation. This may suggest that the population of follicles in these nonresponsive rats was not capable of responding to hCG with ovulation or changes in inhibin secretion. Perhaps injecting FSH into rats on day 3 of lactation or at 26 days of age might have induced a subsequent selective FSH surge.

It is quite possible that FSH and LH may have preferential effects on inhibin secretion depending upon the maturational state of the ovarian follicles. Both FSH and LH may decrease inhibin activity present in the ovarian vein by causing luteinization of the granulosa compartment and/or ovulation of the preovulatory follicles. On the other hand, inhibin secretion from non-preovulatory follicles may be gradually increased as the follicles mature under the influence of FSH and LH.

Inhibin Secretion after Unilateral Ovariectomy (ULO)

Since gonadotropin injections in our above experiment were done only at proestrus (when the ovary contains numerous preovulatory follicles), the LH and FSH effect on inhibin secretion should be examined at other times of the cycle to determine if similar changes in inhibin secretion occur in response to endogenous gonadotropins from ovaries without preovulatory follicles. We have made an initial attempt to examine the gonadotropin action on non-proestrous ovaries by determining the output of inhibin activity from the remaining ovary following unilateral ovariectomy at diestrous day 1 (De Paolo et al., 1981). Unilateral ovariectomy results in compensatory hypertrophy of the remaining ovary. This has been shown to occur in rats (Welschen et al., 1977, 1978), and in hamsters (Chiras and Greenwald, 1978); apparently the cause of this phenomenon appears to be an acute rise in serum FSH (Bast and Greenwald, 1977; Welschen et al., 1978). Furthermore, compensatory hypertrophy can be prevented in mice by the administration of follicular fluid (Sato et al., 1978; Sato and Ishibashi, 1978; Sato et al., 1980), while the acute rise in FSH following ULO can be suppressed by follicular fluid (Welschen et al., 1978).

In the present study, one ovary was removed from rats at 0800h of diestrous day 1. The remaining ovary was cannulated, and ovarian vein plasma was sampled at various time periods post-surgery. In addition, peripheral blood samples were obtained from the abdominal aorta at each time period. The amount of inhibin activity present in OVP was assessed by its ability to suppress basal FSH secretion in our pituitary cell bioassay system. The results show that plasma FSH concentrations were significantly higher ($p < 0.05$) by 4 h in rats that underwent ULO at 0800h on D1 when compared to FSH levels in sham-operated rats (Fig. 3A). However, plasma FSH at 4 h in ULO rats was similar to levels prior to ULO surgery. Peak concentra-

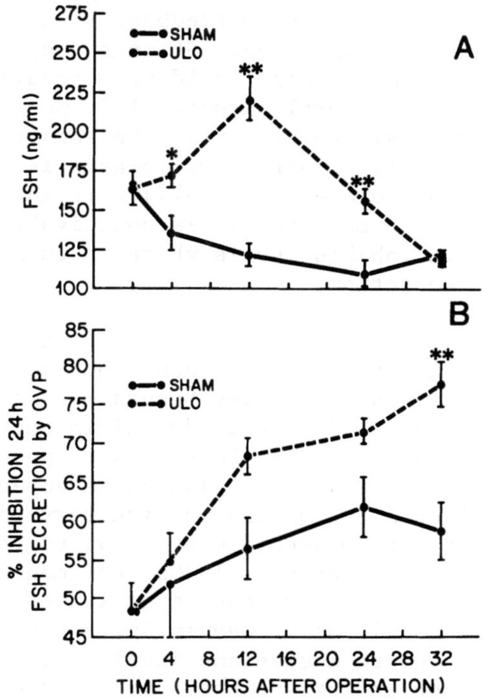

Fig. 3. Changes in plasma FSH concentrations (A) and the percent inhibition of basal FSH secretion by OVP — inhibin activity (B) as assessed in a dispersed pituitary cell bioassay system following ULO or sham operation. Whereas both FSH and inhibin activity increase between 4 and 12 h after ULO, FSH declines and inhibin activity in OVP remains elevated between 12 and 32 h after ULO. * = $p < 0.05$, ** = $p < 0.01$ between sham- and ULO-animals. (Data reprinted from De Paolo, et al., in press.)

tions of FSH were reached 12 h after ULO. While plasma levels of FSH declined by 24 h, they were still statistically elevated ($p < 0.01$) above levels observed in sham-operated rats but not significantly different from values measured before ULO. Interestingly, FSH levels in plasma gradually declined between 0 and 32 h after sham-ULO such that levels by 12 h were lower ($p < 0.01$) than levels measured before sham surgery. Like FSH, inhibin in ovarian vein plasma significantly increased ($p < 0.05$) between 4 and 12 h following ULO (Fig. 3B). However, in contrast to the decline in FSH levels between 12 and 32 h after ULO, inhibin remained elevated.

These data demonstrate a temporal relationship between FSH and inhibin activity in ovarian vein plasma after ULO. As well, the operation of an FSH-inhibin negative feedback system following this experimental alteration is suggested. Stronger support for the existence of such a feedback system would have required the determination of inhibin in peripheral plasma. This necessitates the development of a highly sensitive RIA or a more sensitive bioassay for inhibin, since its activity in peripheral plasma is undetectable in our bioassay system (De Paolo et al., 1979). If inhibin could be measured in peripheral plasma, then it is conceivable that ULO may result in a decrease in inhibin levels which in turn may act to increase pituitary FSH secretion.

Interestingly, the increase in FSH levels between 4 and 12 h after ULO was accompanied by a simultaneous increment of inhibin in OVP. Thereafter, inhibin activity was maintained at these levels while FSH levels declined. If increased inhibin secretion resulting in rising peripheral plasma inhibin levels is responsible for suppressing plasma levels of FSH, than a lag phase appears necessary for the inhibitory effects of inhibin on pituitary gonadotrophs to commence. This finding is in accord with previous reports (Campbell and Schwartz, 1979; De Paolo et al., 1979b; Hoffman et al., 1979) which show a delay of 2 to 5 h between injection of rats with FFL and inhibin of plasma FSH levels. Seemingly, a time interval may be required after ULO before mechanisms involved in the suppression of pituitary FSH release and/or synthesis become operable. Alternatively, the initial increase in inhibin may have occurred later than 4 h after ULO following the initial increment in FSH levels since the exact temporal relationship between FSH and inhibin during the interval from 4 to 12 h after ULO could not be determined from the data. Finally, since ovarian rather than peripheral plasma inhibin activity was assessed, it is highly possible that the lag phase may simply reflect the time required for peripheral inhibin levels in plasma to achieve a threshold concentration sufficient to suppress FSH release.

The decline in plasma FSH concentrations and the increase in inhibin in OVP observed between 0800h on D1 0800h on D2 in sham-operated rats also is an interesting finding. Elevations in FSH levels on the morning of D1 have been previously documented (Butcher et al., 1974; De Palo et al., 1979a), and it is possible that these high levels may act to increase inhibin production by the ovary. Therefore, an FSH-inhibin feedback system not only may function after ULO and during estrus, but may operate during diestrous days of the rat reproductive cycle as well.

Summary of Inhibin Secretion in vivo

The following points condense the findings discussed above:

(1) Throughout diestrus and during early proestrus and late estrus, inhibin output from the ovary remains at maximal levels regardless of the composition of the follicular population.

(2) The proestrous-estrous decline in inhibin output from the ovary coincides with the gonadotropin surges at that time. This effect can be reproduced by administration of either LH or FSH to phenobarbital-blocked, proestrous rats.

(3) The sequence of changes in FSH and inhibin secretion following unilateral ovariectomy during diestrus suggests the existence of a long-loop feedback system involving FSH and inhibin.

(4) The interaction of gonadotropins and inhibin appears to change with the stage of the estrous cycle. At the ovarian level, the characteristics of the follicular population may govern the inhibin secretory response to gonadotropin input.

Based on the present evidence, the following sequence of events may occur during the periovulatory period of the rat estrous cycle. On proestrous afternoon, the release of LH and FSH caused by the hypothalamic discharge of LHRH (Sarkar et al., 1976) precipitates a decline in inhibin activity present in the ovarian vein. Thereafter a second, more sustained rise in plasma FSH commences on late proestrus which can conceivably be due to a further decrease in peripheral plasma inhibin levels between proestrus and estrus accompanied by slight (possibly insignificant) increases in LHRH release (Wise et al., 1979). As a consequence of elevated FSH levels on estrus, selected follicles are stimulated to mature and release inhibin which subsequently inhibits further FSH release until the next proestrus.

Secretion of Inhibin from Granulosa Cells, in vitro

One of the shortcomings with in vivo studies of ovarian physiology is that we are dealing with a complex organ with many structures which may be involved in the production of inhibin and/or the regulation of its synthesis and secretion. Naturally, the next step in this investigation is to examine inhibin production by isolated compartments of the ovary, in vitro. In the female, granulosa cells of the ovarian follicle appear to be a source of inhibin. Rat granulosa cells isolated from preovulatory follicles secrete an inhibin-like substance into culture media that suppresses FSH secretion when the

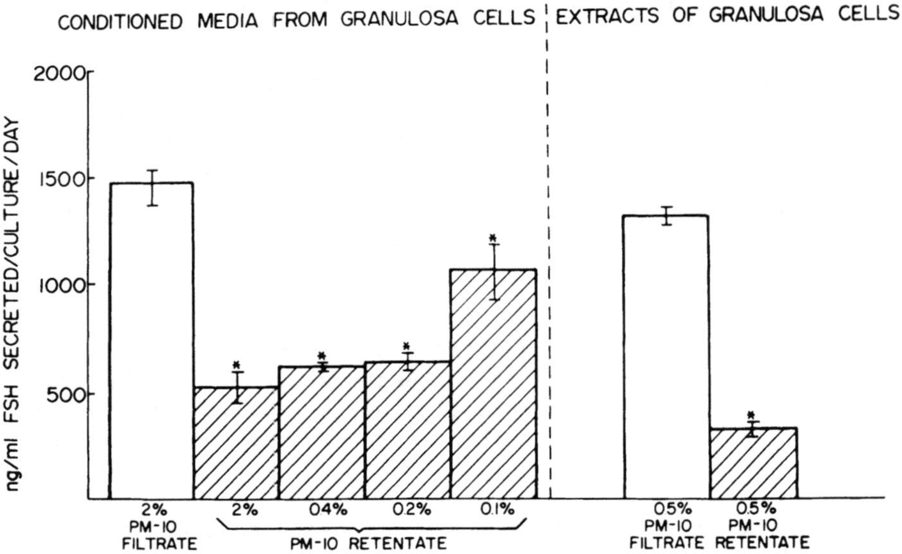

Fig. 4. The effects of PM-10 retentate and filtrate fractions of granulosa cell extracts and conditioned media on basal FSH secretion from rat, anterior pituitary cells, in vitro. Granulosa cells were isolated from medium-sized, antral follicles (3-5 mm diameter) of porcine ovaries obtained at a local slaughterhouse, as previously described (Channing and Ledwitz-Rigby, 1975). These granulosa cells were homogenized or incubated in siliconized glass vials at 37°C in TC 199 + 10% pig serum for 2 days at a concentration of $1 \cdot 10^6$ viable cells/ml of media. Homogenized extracts or conditioned media were treated with 10% (v/v) activated charcoal and 1% dextran, to remove free steroids, followed by ultrafiltration through an Amicon PM-10 membrane. The high molecular weight substances retained by the membrane (<10,000 daltons) and filtrates (<10,000 daltons) were added at various concentrations to media. These media were then added to anterior pituitary cells (obtained from female, Sprague-Dawley rats) using a primary monolayer system. The data represents the amount of FSH (ng/ml) secreted from pituitary cells during a 24 h period of culture, (mean ± S.E); *$p < 0.01$ as compared to the PM-10 filtrate additions. The secretion of LH during this time interval was not affected by any additions to the pituitary cell media. Gonadotropin concentrations were determined by radioimmunoassay as previously described by Shander et al. (1980a).

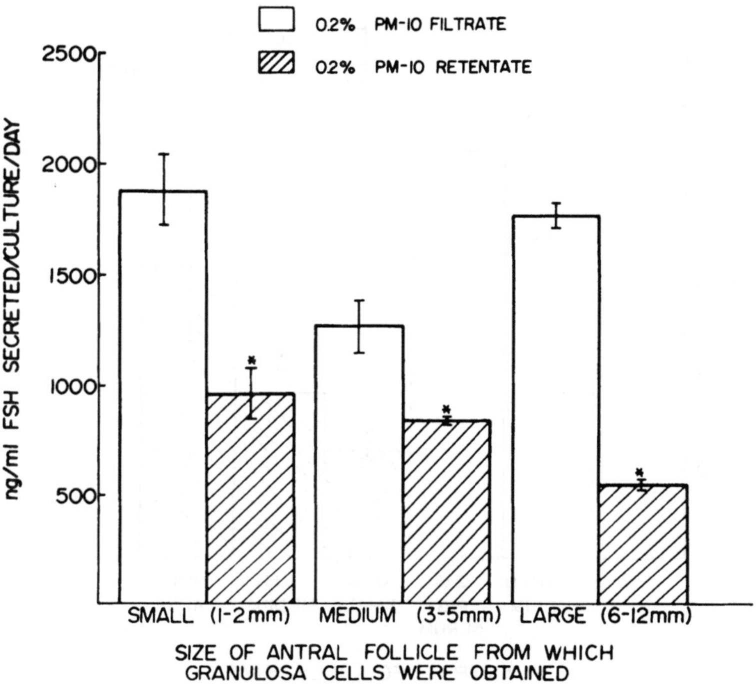

Fig. 5. The effect of PM-10 retentate and filtrate fractions of conditioned media from granulosa cells of various-sized, antral follicles on basal FSH secretion from rat anterior pituitary cells, in vitro. Granulosa cells were isolated from small (1-2 mm), medium (3-5 mm) and large (6-12 mm) antral follicles of porcine ovaries obtained at a local slaughterhouse. The granulosa cells were incuabated for 2 days and the conditioned media was treated as outlined in Fig. 4. The data represents secretion of FSH (ng/ml), mean ± S.E., $*p < 0.01$ as compared to PM-10 filtrates. The secretion of LH during this time interval was not affected by any of the additions to the pituitary cell culture.

media was added to rat pituitary cells in vitro (Erickson and Hsueh, 1978). In this regard, we investigated whether this inhibin activity, present in rat granulosa cell-conditioned media, was present in the high molecular weight fraction of conditioned media from granulosa cells of the pig. In these studies, we utilized the pig ovary, since porcine granulosa cells can be easily obtained in large numbers from antral follicles in various stages of maturation. The granulosa cell was chosen for initial studies since it, of all other cell types of the ovary, appears to be the most active in its protein synthetic capability. In addition, the presence of inhibin activity in follicular fluid further implicates the granulosa cell as a source of inhibin.

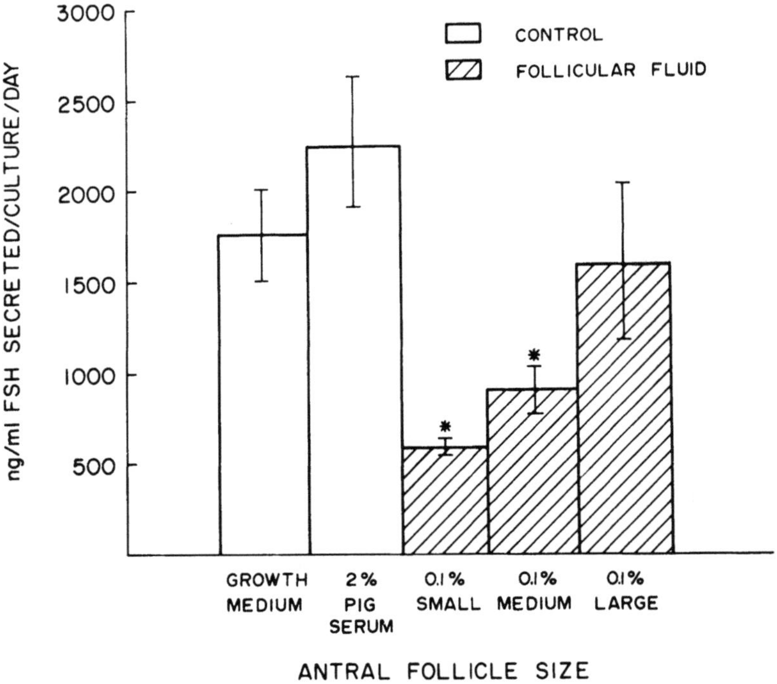

Fig. 6. The effect of follicular fluid from various-sized, antral follicles on basal FSH secretion from rat anterior pituitary cells, in vitro. The follicular fluid (obtained from the various follicles aspirated for granulosa cells isolation in Fig. 5) was treated with 10% (v/v) activated charcoal and 1.0% dextran. After centrifugation (90,000 × g) of the charcoal-follicular fluid mixture, the supernatant was sterilized by passing it through a 0.22 μ millipore filter. The fluid was added to the media incubating rat anterior pituitary cells, in vitro and the incubation continued for 24 h. The secretion of LH was not affected during this time interval by any of the additions. The data represent the secretion of FSH (ng/ml), mean ± S.E., *$p < 0.01$ as compared to growth media alone or 2% pig serum addition.

As shown in Fig. 4, only the retentates from both conditioned media and granulosa cell extracts contain inhibin activity as assessed by their ability to suppress basal secretion of FSH in a dispersed pituitary cell culture. Previously, we have shown that retentates of granulosa cell conditioned media and porcine follicular fluid do not suppress basal LH secretion (Channing et al., 1980). The Amicon PM-10 membranes used in this study retain substances >10,000 daltons, while the filtrates contain substances <10,000 daltons. The presence of inhibin activity only in the retentates is in agreement with the many reports that inhibin is >10,000 daltons in molecular weight. The reader is referred to other papers in this book for more information on the chemical nature and purification of inhibin. The filtrates of either conditioned media or extracts did not alter basal secretion of FSH as compared to media incubated in the absence of granulosa cells and treated similarly (data not shown). However, the filtrates do contain oocyte maturation inhibitor (OMI) activity; they inhibit spontaneous nuclear maturation of cumulus-enclosed porcine oocytes in vitro (Anderson et al., 1981; refer to the paper by Channing et al., in this book on the chemical properties of OMI). The amount of inhibin activity present in porcine granulosa cell-conditioned media and follicular fluid from antral follicles of various sizes is shown in Figs. 5 and 6, respectively. These data indicate that as the follicle increases in size and follicular maturation progresses, the granulosa cells acquire a greater ability to secrete inhibin. However, the amount of inhibin activity present in the various follicular fluids, obtained from the same follicles as the granulosa cells, decreases during follicular development. Lorenzen et al. (1978) also observed lower inhibin activity in follicular fluid from large follicles of the pig by assessing its activity in the ovariectomized, metestrous rat. This has been further corroborated by Henderson and Franchimont (1981) using bovine follicular fluid assayed for inhibin using the dispersed pituitary cell culture bioassay. This suggests that as the follicle grows larger in size, the granulosa cells secrete more inhibin which may rapidly exit the follicular compartment to contribute to the activity seen in the ovarian venous blood. Alternatively, it could indicate that the granulosa cells may secrete large amounts of inhibin, but the proteases in the follicular fluid rapidly degrade the protein. Granulosa cells of small follicles may secrete inhibin into the follicular antrum, but it may remain within the follicular compartment contributing little to the inhibin present in the ovarian vein until the follicle gradually matures in response to hormones. The possibility also exists that inhibin may be secreted by the granulosa cells as a larger molecule (prohormone) or it may be bound to another moiety within follicular fluid thereby preventing its release from the follicle. In this instance, its exit from the follicle may be regulated by hormones and/or other factors such as an enzyme. A likely candidate, for enzymatic regulation would be plasmin since plasminogen activator has been shown to be stimulated from granulosa cells by FSH and not LH. This factor or enzyme could conceivably modify either the inhibin molecule, the binding protein or the follicular wall itself to become more permeable to various substances. Also, it may be simply that larger follicles have a richer blood vasculature than smaller follicles, allowing inhibin to be transported out via the venous blood as soon as it is secreted by the granulosa cells.

Therefore, the previous hypothesis proposed can be restated to explain the changes in inhibin activity occurring in ovarian venous

Table 1. Inhibin Activity in Molecular Filtration Fractions of Conditioned Media and Follicular Fluid

	>10,000 to <30,000	>30,000
Conditioned Media, Small GC	884.5 ± 31.4*	755.0 ± 58.1
Conditioned Media, Medium GC	936.7 ± 6.2	1002.3 ± 39.3
Conditioned Media, Large GC	—	527.7 ± 4.0
Blank Medium 199	—	1171.7 ± 28.9
10% Pooled FFL Standard in Medium 199	—	487.1 ± 18.9

Control FSH secretion: 1150.7 ± 21.3
2% Follicular fluid standard FSH secretion: 509.6 ± 29.2
*Values represent the mean ± S.E. of FSH (ng/ml) secreted basally during 24 h of incubation (n = 3 culture wells).

Porcine granulosa cells from small, medium and large antral follicles were incubated in suspension culture at 37°C in 95% air and 5% CO_2. After 48 h, the cells and conditioned media were recovered and centrifuged at 300 × g for 5 minutes. The resulting cell-free, conditioned media were removed and passed through a Millipore membrane which retains substances >30,000 daltons. Additional conditioned media samples were passed through 10,000 dalton membranes, also the <30,000 dalton filtrates were passed through 10,000 dalton membranes. Proteins in the various fractions (>30,000, between 30,000 and 10,000 and >10,000 daltons) were precipitated with cold ethanol (—20°C), at a final concentration of 86% v/v and stored overnight at —20°C. The precipitates were washed with cold acetone and lyophilized. The lyophilized powders were reconstituted in pituitary cell media-DMEM (without serum) and added to the pituitary cell culture bioassay at a dose of 50 μl equivalent of the starting material. Porcine follicular fluid and granulosa cell culture media without cells were treated in a similar fashion and also added to the pituitary cell culture bioassay. For this culture, control FSH secretion is shown at the bottom of Table 1 along with the FSH secretion from cells incubated with 2% pFFl standard.

blood throughout the estrous cycle. During the periovulatory surge of gonadotropins, LH and FSH causes a decline in inhibin secreted, as well as luteinization of granulosa cells in the large preovulatory follicles, and possibly atresia of other antral follicles, decreasing the output of inhibin present in the ovarian vein. However, with the sustained high levels of FSH throughout proestrus and early estrus, which result from low inhibin levels and changing steroid concentrations, the secretion of inhibin is stimulated by the remaining, non-ovulated follicles; increasing the ovarian output of inhibin until the next proestrus. Similarly, during unilateral ovariectomy, the loss of one ovary decreases the peripheral serum levels of inhibin. The resulting elevations in FSH secretion, in turn stimulates follicles of the remaining ovary to increase their inhibin synthesis and secretion into the ovarian vein. Of course, the validity of this proposed hypothesis requires further investigation.

Heterogeneity of the Inhibin Molecule in Conditioned Media

Since there have been conflicting reports on the size of the inhibin molecule, ranging from less than 30,000 to greater than 100,000 daltons, we reasoned that if such size heterogeneity exists, then it might explain the differential ability of the molecule to exit the follicular antrum, i.e., passage of a larger molecule through the basement membrane into the vascular compartment might be impeded. We wished to determine if the inhibin activity present in conditioned media had heterogeneity in size characteristics. To test this, we subjected media conditioned by granulosa cells from small, medium, and large antral follicles to membrane filtration using millipore immersible filters with molecular size cutoffs of 10,000 and 30,000 daltons. These very preliminary data (Table 1) show that inhibin activity present in media conditioned by granulosa cells from small and medium follicles is heterogeneous in size, in that inhibin activity was found in the >30,000 dalton retenate and in the fraction containing substances between 10,000 and 30,000 daltons. So far, we have tested only the 30,000 dalton retentates of conditioned medium from granulosa cells of large follicles and follicular fluid. Consistent with the larger estimates of the molecular weight of inhibin, activity was found in both of these higher molecular weight samples. The 30,000 dalton retentate of blank medium 199 had no effect on FSH secretion. We are presently pursuring this area of investigation further.

Hormonal Effects on Inhibin Secretion by Granulosa Cells, in vitro

All of these studies have been somewhat peripheral to the question of whether hormones directly affect the secretion of inhibin by granulosa cells in culture. We were spurred on to study this question more directly both by the in vivo work which was discussed earlier and by work which showed that the secretion of an oocyte maturation inhibitor from granulosa cells in culture was affected by hormones (Anderson et al., 1981). The data we will discuss here, on hormonal control of inhibin secretion by cultured granulosa cells, is from our initial studies (Table 2). For control incubations, this experiment confirms our earlier observation that granulosa cells taken from increasing sized antral follicles secrete greater amounts of inhibin. For granulosa cells from small and medium follicles, none of the hormonal additions affected the inhibin activity

Table 2. Amount of Inhibin Units[1] Secreted by Granulosa Cells from Small, Medium, and Large Antral Follicles in Response to Various Hormones

Size of Follicle	Hormone Treatment						
	Control	FSH	LH	P_4	E_2	A	DHT
Small	2338[2] ±461	2498 ±348	2192 ±726	2181 ±605	1918 ±654	2013 ±30	1875 ±514
Medium	3280 ±183	3475 ±307	3593 ±877	3877 ±319	2583 ±360	3611 ±474	4386 ±497
Large	4587 ±406	>5206[3]	>5206[3]	4253 ±673	4664 ±188	4186 ±641	>5206[3]

[1]One inhibin unit in conditioned media is that amount of activity equivalent to one nanoliter of the follicular fluid standard (AL 36).
[2]Mean ± S.E. of inhibin units secreted by 10^6 viable cells. The volumes of conditioned media added to the pituitary cell bioassay were 150, 100, and 50 µl. Only the 50 µl dosage was within the dose response range of the inhibin standard curve.
[3]The amount of inhibin units present in 50 µl of conditioned media was greater than the dose response range.
Granulosa cells were harvested from small, medium, and large follicles, and initially preincubated in suspension culture for 2 h with control media containing no exogenous hormones, or with the hormones indicated in the table. Each treatment group was done in duplicate. Following preincubation the cells were washed three times with control media to remove unbound hormone. The preincubation time was performed to allow binding of the gonadotropins to their receptors and initiate the hormonal responses. Then, the free gonadotropins were washed away to prevent their interference in the pituitary cell bioassay. The control, FSH, and LH treated cells were returned to control media after washing, and the steroid treated cells were returned to media containing the same steroid inclusions as during the preincubation period. Then, following 48 h of incubation in suspension culture, the conditioned media were removed, charcoal treated and added to the pituitary cell bioassay at a dose of 10% final concentration in the culture growth medium. After dextran-charcoal treatment, the amount of progesterone and estrogen remaining in the media was:

Table 2. (Continued)

	Before Treatment	After Treatment	% Steroid Removed
Progesterone	1.4 μg/ml	13 ng/ml	99%
Estradiol	1.4 μg/ml	22 ng/ml	98.4%

Also, there was no detecable FSH present in the final incubation of the cells, therefore washing the granulosa cells 3 times is sufficient to remove the unbound FSH from the conditioned media. Furthermore, the blanks (media incubated for 48 h without cells) of growth media with steroid additions showed no detectable inhibin activity present as demonstrated in the following table:

Growth Media Only	690.0 ± 20.0 ng/ml of FSH, n = 12
P_4	684.6 ± 42.9 ng/ml of FSH, n = 3
E_2	690.1 ± 42.0 ng/ml of FSH, n = 3
A	697.2 ± 54.9 ng/ml of FSH, n = 3
DHT	631.2 ± 42.9 ng/ml of FSH, n = 3

Statistical analysis of the various additions of 50 μl of these blank media to the pituitary cell bioassay show no effect upon FSH secreted into the media, therefore, no detectable inhibin activity is present in the blank media. Statistical analysis was a one way ANOVA on $\log_e$ transformed values.

Table 3. Amount[1] of Estrogen Secreted by Granulosa Cells from Small, Medium, and Large Antral Follicles in the Presence of Androstenedione (1.4 μg/ml of media)

Small	$16.56^{b} \pm 3.30$ ng/10^{6} viable cells
Medium	$20.17^{b} \pm 5.59$ ng/10^{6} viable cells
Large	*$43.40^{a} \pm 5.75$ ng/10^{6} viable cells

Statistical analysis was by one-way ANOVA and Duncan's new multiple range test, $p < 0.05$ with values of different lettered superscripts.

[1]The amount of estrogen secreted is expressed as ng of estrogen per 10^{6} viable cells per 500 μl of conditioned media. The values represent the mean ± S.E., n = 2 replicate granulosa cell cultures.

Granulosa cells were incubated in suspension culture for 48 h. In the absence of aromatizable androgen, granulosa cells secreted negligible amounts of estrogen into the conditioned media. However, in the presence of androstenedione, granulosa cells did retain their ability to aromatize the androgen to estrogens.

present in conditioned media. In the case of the granulosa cells from large follicles, it is tempting to speculate that the two gonadotropins and DHT significantly increase the amount of inhibin activity present in conditioned media. However, since these values exceeded the range of the standard curve for inhibin, these samples must be reassayed before this speculation can be confirmed.

We are considering the possibility that incubation of granulosa cells in suspension culture may not be appropriate for examining the effect of hormones on inhibin secretion. In fact, these calls are not as steriodogenically active as cells incubated in monolayer culture. Granulosa cells in suspension culture do convert androgens to estrogens (Table 3). However, the total amount of estrogen secreted is substantially less than that routinely observed from cells in monolayer culture. This is also true for the amount of progesterone produced by these cells (Table 4). LH did stimulate secretion of progesterone from granulosa cells of medium and large follicles. However, the absolute amounts of progestertone are far lower than that observed routinely from cells in monolayer culture. In addition, the inhibitory effect of estradiol on progesterone secretion that is seen in monolayer culture, was not observed in suspension culture. Androstenedione did not inhibit progesterone secretion by cells from large follicles as is observed in monolayer culture. DHT did not augment progestertone secretion as observed in monolayer culture, and finally FSH did not stimulate progesterone secretion from cells of any size follicle; whereas in monolayer culture FSH stimulates progestertone secretion by granulosa cells from all sized follicles.

Table 4. Amount of Progesterone[1] Secreted by Granulosa Cells from Small, Medium, and Large Antral Follicles in Response to Various Hormones

Size of Follicle	Hormone Treatment: Control	FSH	LH	E_2	A	DHT
Small	24.05^{cde} [2] ±1.40	21.30^{cdef} ±0.33	23.00^{cde} ±0.85	27.07^{c} ±1.21	* 14.25^{g} ±1.46	24.68^{cde} ±3.81
Medium	17.19^{fg} ±2.13	18.93^{ef} ±0.42	* 25.84^{cd} ±0.28	20.82^{def} ±1.03	* 7.51^{h} ±1.12	* 22.85^{cde} ±12.21
Large	109.21^{b} ±8.90	105.82^{b} ±8.16	* 146.03^{a} ±5.63	101.24^{b} ±4.52	102.63^{ab} ±13.79	119.14^{b} ±2.95

*Represents the hormone treatment which progesterone secretion was significantly different ($p < 0.05$) from the amount secreted by granulosa cells incubated in control media.

[1]The amount of progesterone secreted during 48 h of incubation in suspension culture is expressed as ng of progesterone per 10^6 viable cells per 500 µl of conditioned media. The values represent the mean ± S.E., n = 2 replicate granulosa cell cultures.
[2]Statistical analysis was performed by two way ANOVA $\log_e$ transformed values. Means with one or more identical letter superscripts are not significantly different ($p < 0.05$) by Duncan's new multiple range test.

The apparent lack of hormonal effects may stem from the fact that granulosa cells in suspension culture do not undergo mitosis as is observed in monolayer culture. Granulosa cells from small, medium, and large follicles do show differences in hormone secretion when in suspension culture. This may reflect the capability that they acquired in vivo. We are presently comparing the hormonal effects on inhibin secretion by granulosa cells incubated in suspension versus monolayer culture to ascertain whether inhibin and/or steroid secretion are linked in some way to cell division. We are also exploring the possibility that hormonal effects on inhibin secretion may occur at times earlier than the 48 h period examined in this experiment.

Other investigators are currently examining hormonal control of inhibin secretion, in vitro. In the male, Steinberger and her colleagues have shown that either FSH, testosterone, cAMP, or MIX increased the amount of inhibin secreted into media conditioned by rat Sertoli cells. The addition of testosterone with either one of the above substances augmented the amount of inhibin secreted into the media (Steinberger, 1981). These results are similar to those observed with the administration of hormones to hypophysectomized, immature rats with subsequent isolation of the Sertoli cells. However, in vivo, LH increased Sertoli-cell inhibin secretion, while LH addition in vitro had no effect on the amount of inhibin secreted into the media. It is likely that the LH effect seen in vivo was mediated via testosterone production.

In the female, a recent report by Henderson and Franchimont (1981) has shown that androgens increase the amount of inhibin present in media conditioned by bovine granulosa cells isolated from preovulatory follicles; while the addition of progesterone to these cells suppressed inhibin secretion. Neither FSH nor LH had any effect upon inhibin secretion from granulosa cells incubated in media devoid of serum. This is not surprising, since granulosa cells not exposed to hormones and incubated in serum-free media failed to secrete detectable amounts of inhibin into the media (with concomitantly low progesterone secretion). However, gonadotropin addition to cells in serum-free media did restore progesterone secretion to that observed for untreated cells incubated in serum containing media. Interestingly, these authors report that bovine luteal cells did not secrete detectable amounts of inhibin in culture. Thus, it appears that granulosa cells lose their capacity to secrete inhibin once they luteinize, in vivo.

Conclusions

The following conclusions can be drawn from the findings discussed in this paper:

(1) In vivo studies with the cyclic rat suggest that inhibin secretion by the ovary can be affected by the gonadotropins, and that the inhibin secretory response to gonadotropin stimuli may be a function of the characteristics of the follicular population on the ovary.

(2) In vitro studies on the secretion of inhibin by porcine granulosa cells suggest that:

(a) Inhibin may be present in conditioned media in more than one molecular form.

(b) The capacity of granulosa cells to secrete inhibin in suspension culture increases with the stage of maturity of the follicle from which the granulosa cells are taken. This pattern is the reverse of that observed for inhibin activity in porcine follicular fluid.

(c) Gonadotropins and steroids (P_4, E_2, A, DHT) do not affect inhibin secretion by suspension cultured granulosa cells from small and medium follicles. FSH, LH, and DHT may increase inhibin secretion by suspension cultured granulosa cells from large follicles, but other hormones tested do not affect inhibin secretion by these cells.

Based on the current status of research in the area of hormonal control of inhibin secretion, we propose the following hypothesis as stimuli for future investigation.

Ovarian Output of Inhibin

Inhibin output from the ovary in vivo may be a function of:

(1) The numbers and types of follicles on the ovary,

(2) the hormonal status of the animal, or

(3) both of these factors.

Follicular Output of Inhibin

A number of factors may affect the output of inhibin by individual follicles:

Inhibin Kinetics:

(1) Exit of inhibin from the follicular antrum may be controlled by the permeability of the follicular wall.

(2) Since inhibin activity may reside in more than one molecular form, inhibin kinetics may be additionally controlled by changing the proportion of the various forms. Size heterogeneity of inhibin may be explained by one or more of the following possibilities:

(a) An inhibin binding protein,

(b) secretion of proinhibin(s) subsequently altered to smaller form(s) by enzymatic action in the follicular antrum (either the enzyme, an activator of the enzyme, or both may be secretion products of follicle cells),

(c) secretion of more than one form of inhibin.

(3) All of the above effects may be controlled by hormones.

Maturation Effects on Basal Secretion of Inhibin by Granulosa Cells:

(1) Changes in the capacity of granulosa cells to secrete inhibin during follicular maturation may be due to quantitative differentiation of the cells which results from cellular division.

(2) Hormonal effects on inhibin secretion by granulosa cells may be due in part to their actions on granulosa cell mitosis.

Direct Action of Hormones on Inhibition Secretion:

Gonadotropins, ovarian steroids or other factors may directly control the rate of synthesis and/or secretion of inhibin by granulosa cells and other ovarian sources.

ACKNOWLEDGEMENTS

The research reported in this paper was supported, in part, by the Population Division of the Rockefeller Foundation (GA PA 8012; to L. D. Anderson). D. J. Hoover was supported by NICHHD Training Grant HD 07170-02.

REFERENCES

Anderson, L. D., and Stone, S. L., and Channing, C. P., 1981, Hormonal control of an inhibitor of oocyte meiosis secreted by porcine granulosa cells: Possible involvement in oocyte maturation during atresia and ovulation, Gamete Research, (submitted for publication).

Ashiru, O. A., and Blake, C. A., 1979, Stimulation of endogenous follicle-stimulating hormone release during estrus by exogenous follicle-stimulating hormone or luteinizing hormone at proestrus in phenobarbital-blocked rats, Endocrinology, 105: 1162.

Bast, J. D., and Greenwald, G. S., 1977, Acute and chronic elevation in serum levels of FSH after unilateral ovariectomy in the cycle hamster, Endocrinology, 100:955.

Bronson, F. H., and Channing, C. P., 1978, Suppression of serum follicle-stimulating hormone by follicular fluid in the maximally estrogenized, ovariectomized mouse, Endocrinology, 103:1894.

Butcher, R. L., Collins, W. E., and Fugo, N. W., 1974, Plasma concentration of LH, FSH, prolactin, progesterone, and estradiol-17β throughout the 4-day estrous cycle of the rat, Endocrinology, 94:1704.

Campbell, C. S., and Schwartz, N. B., 1974, Time course of serum FSH suppression in ovariectomized rats injected with porcine follicular fluid (folliculostatin): effect of estradiol treatment, Biol. Reprod., 20:1093.

Channing, C. P., Schaerf, F. W., Anderson, L. D., and Tsafriri, A., 1980, Ovarian follicular and luteal physiology, in: "International Review of Physiology, Vol. 22," R. O. Greep, ed., pp 117-201, University Park Press, Baltimore.

Chiras, D. D., and Greenwald, G. S., 1978, Acute effects of unilateral ovariectomy on follicular development in the cyclic hamster, J. Reprod. Fert., 52:221.

De Paolo, L. V., Shander, D., Wise, P. M., Channing, C. P., and Barraclough, C. A., 1979a, Identification of inhibin-like activity in ovarian venous plasma of rats during the estrous cycle, Endocrinology, 105:647.

De Paolo, L. V., Wise, P. M., Anderson, L. D., Barraclough, C. A., and Channing, C. P., 1979b, Suppression of the pituitary follicle-stimulating hormone secretion during proestrus and estrus in rats by porcine follicular fluid: Possible site of action, Endocrinology, 104:402.

De Paolo, L. V., Anderson, L. D., and Hirshfield, A. N., 1981, Possible existence of a long-loop feedback system between FSH and inhibin in female rats, Am. J. Physiol., in press.

Erickson, G. F., and Hsueh, A. J. W., 1978, Secretion of "inhibin" by rat granulosa cells in vitro, Endocrinology, 103:1960.

Henderson, K. M., and Franchimont, P., 1981, A relationship between bovine follicular fluid concentrations of inhibin and steroids, and the regulation of inhibin production by bovine ovarian cells in vitro, J. Reprod. Fert., in press.

Hoffman, J. C., Lorenzen, J. R., Weil, T., and Schwartz, N. B., 1979, Selective suppression of the primary surge of follicle-stimulating hormone in the rat: further evidence for folliculostatin in porcine follicular fluid, Endocrinology, 105:200.

Lorenzen, J. R., Channing, C. P., and Schwartz, N. B., 1978, Partial characterization of FSH suppressing activity (folliculostatin) in porcine follicular fluid using the metestrous, ovariectomized rat, Biol. Reprod., 19:635.

Sarkar, D. K., Chiappa, S. K., Fink, G., and Sherwood, N. M., 1976, Gonadotropin releasing hormone surge in pro-oestrus rats, Nature, 264:461.

Sasamoto, S., and Taya, K., 1980, Mechanism of the selective surge of follicle-stimulating hormone in dioestrous rats during the induction of ovulation of human chorionic gonadotropin, J. Endocrinol., 86:489.

Sato, E., and Ishibashi, T., 1978, Partial purification of the gonadotropin inhibiting substance found in bovine follicular fluid, Jap. J. Zootech. Sci., 49:313.

Sato, E., Ishibashi, T., and Iritani, A., 1980, Effect of inhibin-like substance isolated from porcine follicular fluid on the follicle-stimulating hormone (FSH) level in mouse serum and on FSH binding to porcine granulosa cells, Fert. Steril., 34:55.

Sato, E., Miyamoto, H., Ishibashi, T., and Iritani, A., 1978, Identification, purification, and immunohistochemical detection of the inhibitor from porcine ovarian follicular fluid to compensatory ovarian hypertrophy in mice, J. Reprod. Fert., 54:263.

Shander D., Anderson, L. D., and Barraclough, C. A., 1980, Follicle stimulating hormone (FSH) and luteinizing hormone (LH) affect the endogenous release of pituitary FSH and the ovarian secretion of inhibin in rats, Endocrinology, 106:1047.

Steinberger, A., 1981, Regulation of inhibin secretion in the testis, in: "Intragonadal Regulation of Reproduction, P. Franchimont and C. P. Channing, eds., Academic Press, London, in press.

Uilenbroek, J. T. J., Tiller, R., de Jong, F. H., and Vels, F., 1978, Specific suppression of follicle-stimulating hormone secretion in gonadectomized male and female rats with intrasplenic ovarian transplants, J. Endocrinology, 78:399.

Welschen, R., Dullaart, J., and de Jong, F. H., 1978, Interrelationships between circulating levels of estraiol-17β, P, FSH, and LH immediately after unilateral ovariectomy in the cyclic rat, Biol. Reprod., 18:421.

Welschen, R., Hermans, W. P., Dullaart, J., and de Jong, F. H., 1977, Effects of an inhibin-like factor present in bovine and porcine follicular fluid on gonadotropin levels in ovariectomized rats, J. Reprod. Fert., 50:129.

Wise, P. M., Rance, N., Barr, G. D., and Barraclough, C. A., 1979, Further evidence that luteinizing hormone-releasing hormone also is follicle-stimulating hormone-releasing hormone, Endocrinology, 104:940.

CHARACTERISTICS OF INHIBIN: A COMPARATIVE STUDY USING BULL AND HUMAN SEMINAL FLUIDS AND PORCINE FOLLICULAR FLUID

M. R. Sairam, K. Kato, P. Manjunath,
K. Ramasharma, W. M. Miller*,
E. S. R. Haung*, and H. G. MadhwaRaj†

Reproduction Research Laboratory
Clinical Research Institute of Montreal
110 Pine Avenue West
Montreal, Quebec H2W 1R7, Canada

Introduction

The gametogenic and steroidogenic functions of the ovary and testis are both well known and extensively studied in a number of species including primates. Besides sustaining the growth and secretory activity of the accessory sex organs, the gonadal steroids also exert important positive and negative feedback effects on pituitary gonadotropin secretion in both sexes. The suggestion that the testis (gonads) may also secrete additional hormonal factors that are soluble in aqueous solutions and different from steroids, was made a long time ago. It was further suggested that this factor(s) may be involved in the regulation of pituitary gonadotropin secretion. The name inhibin was thus coined for this hypothetical substance (McCullagh, 1932). The hypothesis fell into disrepute and remained dormant until the 1970's. The pertinent literature covering this has been authoritatively reviewed, to which the interested reader is referred (Setchell and Main, 1974; Main et al., 1979; Franchimont et al., 1979b). Although earlier studies connected inhibin with the regulation of FSH secretion primarily in the male, there is now convincing evidence to show its existence in the female. Elsewhere in this volume, Schwartz (1982)

*Department of Biochemistry, North Caroline State University, Raleigh, North Carolina, U.S.A.
†Department of Obstetrics and Gynecology, University of North Carolina, Chapel Hill, U.S.A.

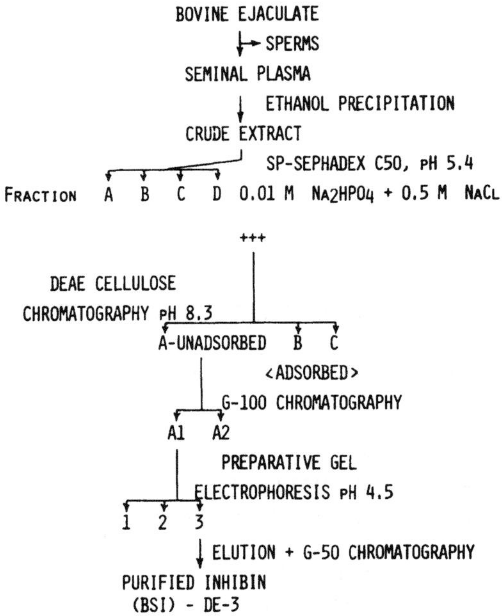

Fig. 1. Scheme for the isolation of inhibin protein from bull ejaculates (Sairam et al., 1981a).

has discussed the role and implications of such a substance in the female, which they refer to as folliculostatin in the ovary. For the purposes of discussion of the present chapter, it suffices to mention that inhibin like factors have been found in testicular extracts, rete testicular fluid, bull and human ejaculates, ovarian follicular fluid or ovarian extracts and additionally in culture media of ovarian granulosa cells and testicular Sertoli cells.

The inhibin hypothesis cannot rest on firm foundation until the substance(s) has been isolated in pure form to homogeneity and its structure eventually identified. This is a crucial step in further exploitation of our knowledge of gonadotropin control mechanisms and regulation of fertility. There is little doubt now that inhibin activity is associated with a peptide or a protein (see Franchimont et al., 1979b). What is most contentious, bitterly debated and still poorly understood is the molecular nature and its size. Widely differing molecular weights ranging from 1500-150,000 even from within the same species has been reported by different laboratories. These and other problems related to inhibin can only be resolved by its isolation in pure form. With this in view, our laboratory has been for the past several years focusing attention on developing isolation methods for inhibin from bull and human seminal plasma and recently extended it to porcine foblicular fluid.

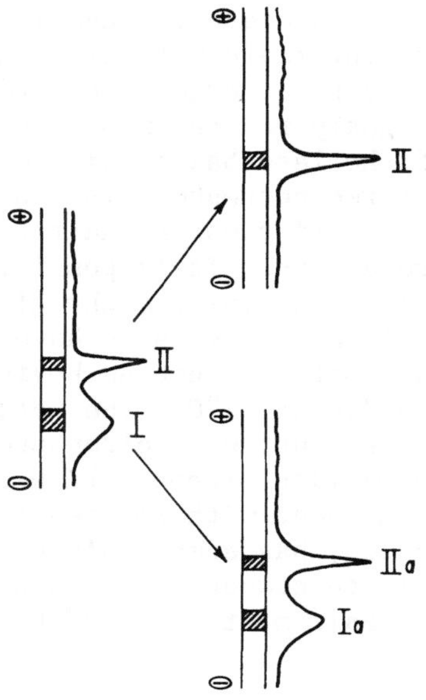

Fig. 2. Disc gel electrophoretic pattern (pH 4.5) of purified bovine seminal plasma inhibin. The pattern of the DE-3 fraction (see Fig. 1) is shown on the left. The solid line to the right of the gel pattern shows the scan of the gel at 280 nm. Note that the peaks of absorbance coincide with the stained bands. When the two bands from an unstained gel on the left are cut out, eluted, and rerun at pH 4.5, the patterns shown on the right are obtained. Peak 1 gives rise to bands Ia and IIa again, while rerun of band II does not show transformation into I.

Thus, within the same laboratory, this will provide an opportunity to compare the behavior and properties of inhibin substance(s) from the three sources.

Properties of Bull Seminal Plasma Inhibin

In a recent report from this laboratory we have described the purification and properties of the inhibin protein from bull ejaculates (Sairam et al., 1981a). The purification scheme at its present stage of development is depicted in Fig. 1. As with all evolving procedures, this method will undoubtedly undergo modifications with the acquisition of additional information concerning the properties of the active material. Despite the introductin of the preparative disc gel electrophoresis step at pH 4.5, the eluted fraction still exhibited multiplicity of bands, which was puzzling to us. This phenomenon was seen repeatedly in several batches of the purified preparation. Data presently available suggest that the more basic component migrating rapidly towards the cathode at pH 4.5 at the bottom has a tendency to undergo transformation into a less basic component with a slightly reduced mobility (Fig. 2). However, there is no conversion of the less basic component at the top to the faster migrating band upon reelectrophoresis. Since, the purified protein lacks carbohydrate, variation in their content can be ruled out as a reason for heterogeneity. It is possible that the presence of a contaminating proteolytic activity might give rise to altered molecules. This however, is unlikely as we have been unable to detect any proteolysis of a substrate such as bovine serum albumin, even after prolonged incuation at 37°C. The heterogeneity (conversion?) could have resulted during electrophoresis or the subsequent handling during the elution process which included lyophilization also. That the component with reduced mobility which is a minor but persistent fraction, has some inhibin activity is shown by the fact that it is able to compete in binding assays involving sheep pituitary membranes (Sairam et al., 1981b).

Gel Filtration Behavior

Molecular weight determination of the bull seminal plasma inhibin fraction by SDS gel electrophoresis and gel filtration on Sephadex G-100 has given us (Sairam et al., 1981a) differing estimates. By the former technique the molecular weight was assessed to be in the range 15,000-20,000 daltons, while gel filtration revealed a slightly higher value of 27,000 daltson. As well documented, the presence of carbohydrate in the molecule could prodce a slightly higher molecular weight by gel filtration but this explanation is untenable because the fraction lacks carbohydrate.

The gel filtration behavior of the seminal plasma inhibin is not altered by denaturing agents such as 4 M urea, or guanidine HCl or when the experiment is carried out at acidic pH's (Viz., 0.05 M sodium acetate, pH 4.6). It is also not retained by columns of immobilized concanavalin A (conA-Sepharose).

Biological Activity

The characteristic of inhibin activity is the suppression of pituitary gonadotropin (FSH and/or LH) secretion. Using several in vivo and in vitro tests involving both direct and indirect measurement of gonadotropins, we have demonstrated activity in the purified protein fraction (Sairam et al., 1981a).

A persistent and perplexing behavior of the inhibin like activity which was indicated briefly in our previous reports (Sairam et al., 1980; Sairam et al., 1981a) deserves mention. We have consistently seen throughout our inhibin purification work that at

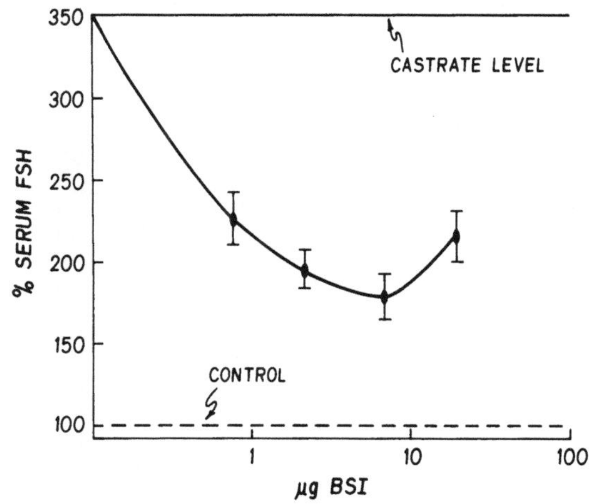

Fig. 3. Effect of increasing doses of purified BSI on castration induced rise in serum FSH levels in 34 day old male rats (Sairam et al., 1981a). At higher concentrations a decrease in inhibition is seen.

very high concentrations of the test material, there is actually a reduced inhibition of FSH release, in vivo as well as in vitro experiments, the evidence in the latter being more pronounced. Two examples of this phenomenon are shown in Figs. 3 and 4.

In the 34 day old castrate male rats, the purified inhibin fraction brought about a suppression of the castration induced rise in FSH at doses ranging between 0.75-6.6 μg. However, at a higher dosage, such as 20 μg, a reduction in inhibition is observed. There was no inhibition of LH release by the inhibin fraction at any of the concentrations (Sairam et al., 1981a).

The pattern of inhibition of FSH release in vitro is also dependent on the concentration of inhibin added. Effective inhibition occurs in presence of low doses but not at very high doses (Fig. 4). In these experiments, the active fraction shows preferential inhibition of LHRH induced FSH release when mouse pituitaries are pre-exposed in the inhibin. It may be noted that the in vitro effect occurs in presence of ng quantities of the purified fraction as compared to μg amounts needed in the in vivo situations (Fig. 3).

The precise mechanism of this paradoxical effect of our inhibin fraction is unclear. But several possibilities may be suggested. The target (pituitary) may become refractory in presence of high concentrations of the active fraction or there could be a rebound effect in these situations resulting in an increase of FSH

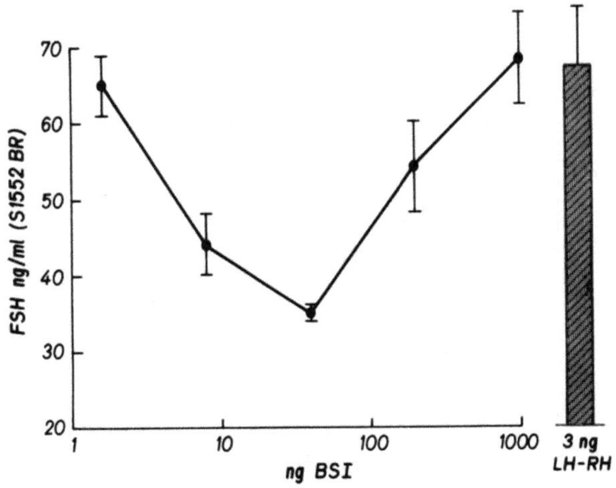

Fig. 4. Effect of increasing doses of purified BSI on LHRH response of the 34 day old mouse pituitary incubated in vitro (Sairam et al., 1981a and Ramasharma et al., 1981a). The whole pituitary tissue in these experiments (n = 5) has been pre-exposed to the indicated concentration of BSI for 60 min before the addition of LHRH (3 ng synthetic). The total volume of incubation was 1 ml. After 3 additional hours of incubation, the FSH released into the medium was estimated directly by specific radioreceptor assays. S1552BR is a highly purified ovine FSH preparation from our laboratory.

release which contributes to a reduced inhibin response. Alternatively, the presence of FSH like material with the increase in dosage of the active fraction injected in vivo or added to the in vitro incubations may be reflected as a false increase in FSH levels in the circulation (assayed by radioimmunoassay) or medium (as assayed by radioreceptor assays). But this reasoning is improbable because the active fraction has no influence on the binding of ^{125}I-labeled ovine FSH or rat FSH to their specific rabbit antisera or specific bovine testicular receptors (Sairam et al., 1981a). This also rules out the possibility that destruction of the label added in our RIA or RRA systems, could have produced these artificial results. Further work is clearly necessary to carefully analyze these paradoxical results. As we shall document below this is not peculiar to bull seminal plasma inhibin fraction as it has also been observed with human seminal plasma and porcine follicular fluid material.

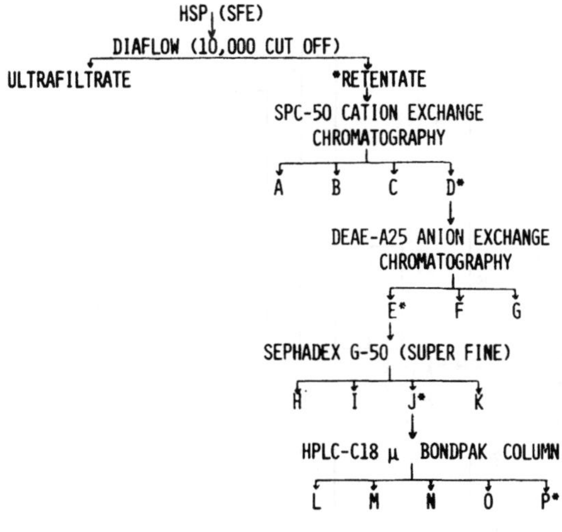

Fig. 5. Scheme for the purification of human seminal plasma inhibin. * denotes the location of inhibin activity in the fractions (Ramasharma et al., 1981). SFE = Sperm free extract.

Properties of Human Seminal Plasma Inhibin

The finding of inhibin like activity in human seminal plasma (HSP) (Franchimont et al., 1979a) presents an alternative source to the human ovarian follicular fluid (Chari et al., 1979) which is difficult to procure in adequate quantities for the isolation work. Furthermore, the demonstration that the levels of inhibin substance(s) in the human seminal plasma appears to be directly related to activity of the germinal epithelium and inversely related to circulating FSH levels, has made its study interesting (Scott and Burger, 1980; Asch et al., 1980).

At a conference on inhibin in 1978, data from two different laboratories showed association of activity with regions of different molecular weight. Based on their gel filtration data, Franchimont et al. (1979a) concluded that HSP-inhibin was a rather smaller molecular weight compound in the vicinity of 5000 daltons, while Vaze et al. (1979) described the isolation of a larger size protein of 19,000 daltons.

Data arising from our own laboratory and presented at a recent conference (Ramasharma and Sairam, 1981) have revealed that the nature of human seminal plasma inhibin is such that differing estimates of molecular weight are obtained depending upon the condition

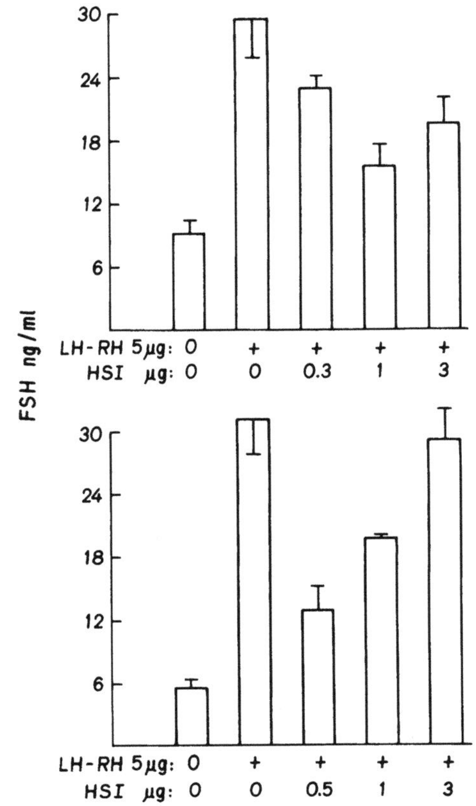

Fig. 6. The effect of human seminal plasma inhibin (fraction J of 2 different batches are shown) on the acute effect of LHRH in 34 day old male rats. The test fraction was injected subcutaneously at 0 h and +3 h. At +6 h LHRH (5 µg) was administered. Blood samples collected 1 h later were assayed for FSH by specific RIA.

in which the material is investigated. We noted that when crude sperm free extracts were investigated, inhibin activity was associated with a larger protein fraction either on Sephadex G-100 or G-75. However, after purification by a scheme (Fig. 5) which also included steps such as chromatography on strongly cationic ion-exchangers such as sulfopropyl-Sephadex-C50, the activity migrated as a small molecular weight substance upon subsequent gel filtration on Sephadex G-50. Based on SDS-gel electrophoresis and gel filtration analysis of the purified fraction we concluded that human seminal plasma inhibin fraction purified by our method (Fig. 5) exhibited a molecular weight of about 5000 daltons. This is in agreement with the report of Franchimont et al. (1979a).

In addition to this, attention should be called to yet another report (Scott and Burger, 1980) which has found activity (by analysis of crude sperm free extract) to be associated with fractions of molecular weights 100,000, 45,000, and 15,000 daltons. The relationship of the smaller molecular weight fraction (Fig. 5) to other larger species, some of which are not yet purified, can only be clarified in the future by much closer chemical scrutiny.

Biological Activity of Human Seminal Plasma Inhibin

The 5000 dalton fraction of HSP-inhibin acts in much the same way as the purified bull seminal plasma inhibin in vivo and in vitro experiments (Ramasharma and Sairam, 1981) exhibiting preference for the inhibition of FSH secretion.

The dose-response relationship of the HSP-inhibin is similar to the BSP-inhibin in its peculiar characteristic of showing decreased inhibition of FSH secretion when higher doses are employed (Fig. 6). In this instance also, the decrease in inhibition cannot be accounted for on the basis of an apparent FSH-like activity or the destruction of the added FSH label in the analytical procedures.

Studies on Inhibin From Porcine Follicular Fluid

Porcine ovarian follicular fluid is a rich source of potent inhibin activity which is associated with a non-steroidal proteinaceous fraction. Although convincing arguments for its existence and role in regulation of FSH secretion in the female have been presented (see Schwartz in this volume) the active substance has not been obtained in a homogeneous form. Methods have been described in the literature for the partial purification and purification and characterization of inhibin from follicular fluid of bovine (de Jong et al., 1980), human (Chari et al., 1979) and pig (Williams et al., 1978) ovaries. Additional data from two laboratories which are active in this area have also been presented in this volume (de Jong et al., 1982; Williams and Lipner, 1982).

We have examined inhibin activity present in follicular fluid of porcine ovaries collected in two ways. In one method of collection which was used for our experiments prior to 1981, the ovaries were collected on ice from the local slaughter house within 20 min of death. The ovaries were on ice for about 4-6 h before actual processing commenced. After being brought to the laboratory, the ovaries were passed through an electric press to squeeze out the fluid. Thus fluids from all sizes and stages of follicles were obtained in a pool, including some entrapped blood components. However, contribution by blood was regarded to be minimal as the ovaries had been first rinsed out with saline after being trimmed.

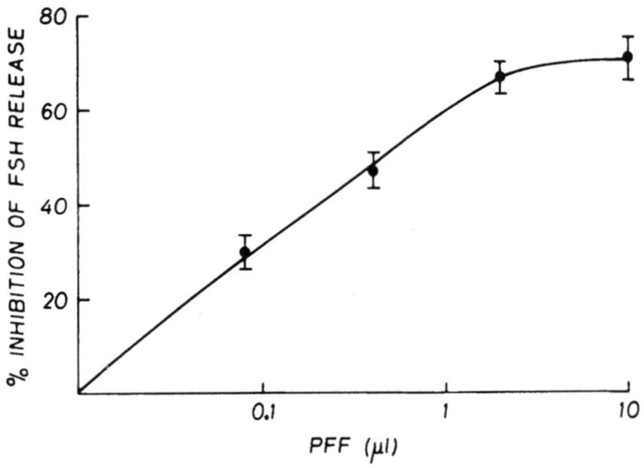

Fig. 7. Inhibition of LHRH induced FSH release in vitro by porcine follicular fluid. The pituitaries were exposed to the indicated concentration of fluid (NIH batch SPO-44, unprocessed) in a total volume of 1 ml (see Fig. 4 for details). % inhibition was calculated taking the net FSH release (25-35 ng/ml) by 3 ng LHRH as reference. n = 5 pituitaries per group.

The fluid was filtered thru several layers of cheese cloth and centrifuged to eliminate debris. The supernatant thus obtained was either used immediately or frozen at −70°C.

The second type of follicular fluid was obtained thru the courtesty of the Contraceptive Development Branch of the NIH, Bethesda, Md., U.S.A. This is also a pooled fluid from pig ovaries, with the difference that it has been obtained by aspiration regardless of size of the follicle. The data reported below has been derived from one batch labeled SPO-44.

Before proceeding to outline aspects of purification it may be pertinent to discuss the effect of inhibin activity in this fluid in the in vitro mouse pituitary incubation system that has been employed in our laboratory (Ramasharma et al., 1981). As shown in that report, the 34 day old mouse pituitary responds readily to added synthetic LH-RH in a dose-dependent manner by releasing both FSH and LH into the medium during a 3-4 h incubation at 37°C. These levels are directly measurable by specific radioreceptor assays. If the pituitary is pre-exposed for 1 h at 37°C to inhibin containing preparations, the subsequent stimulation induced by added LH-RH is curtailed. That this effect is not due to destruction of LH-RH or the gonadotropins has been ascertained (Ramasharma et al., 1981). We have been particularly attracted to this incubation system as it is relatively inexpensive (compared to in vivo or culture assays)

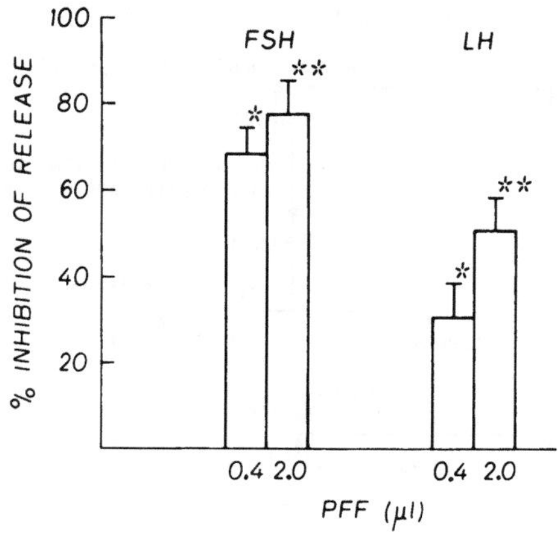

Fig. 8. Relative inhibitory effect of untreated porcine follicular fluid (NIH-SPO-44) on LHRH induced FSH and LH release by the mouse pituitary in vitro. * and ** denote statistically significant difference ($p < 0.05$ or < 0.01).

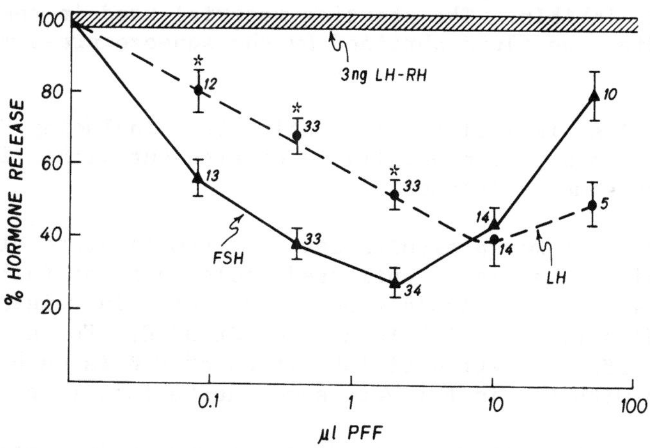

Fig. 9. Effect of increasing doses of untreated porcine follicular fluid on LHRH induced gonadotropin release by the mouse pituitary (see Fig. 4 legend for details). The values near each point indicate the number of pituitaries used. The inhibition obtained in different experiments have been normalized using the net hormone release by 3 ng LHRH as 100%. The * denotes that the inhibitory effects on FSH and LH at these doses show statistically significant differences. Note at the higher doses of PFF, FSH inhibition decreases while that of LH still persists. At the doses employed, the PFF did not have any influence on the measurement of FSH and LH in the medium by radioreceptor assays.

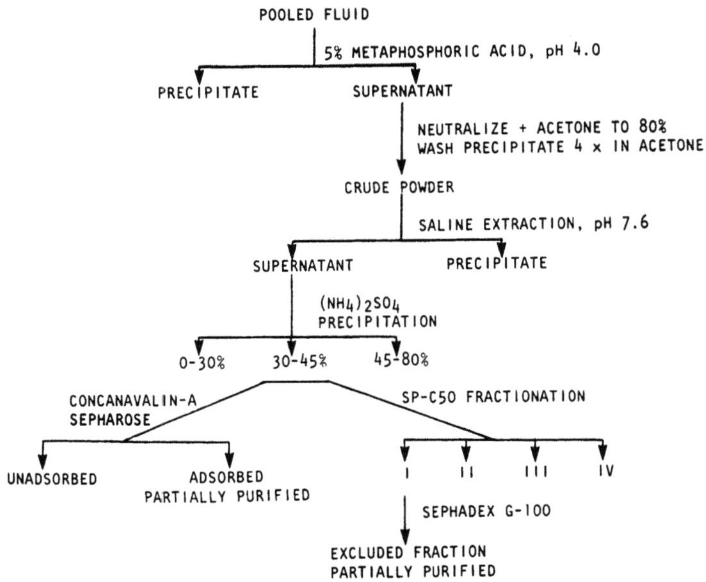

Fig. 10. Scheme for the partial purification of porcine follicular fluid inhibin. The starting material used in these studies was fluid obtained by the squeeze press method.

and because of its simplicity. The entire test including FSH and LH radioreceptor assays can usually be carried out within 24 h by one person with some assistance.

Pre-exposure of mouse pituitaries in vitro to extremely low (< 0.05% by volume) levels of untreated pooled porcine follicular fluid (PFF) results in an inhibition of gonadotropin release (Fig. 7, showing FSH) during the 4 h incubation at 37°C. For a given concentration of PFF, the degree of inhibition of FSH is much greater than that seen with LH, in the very same incubations (Fig. 8).

When higher concentrations (1-5% by volume of incubation) of PFF are tested in the in vitro short term incubations, the paradoxical decrease of FSH release with a similar tendency for LH release is observed (Fig. 9). We have ruled out that FSH and LH, present in the PFF added (up to 50 µl used) could not account for the observed decrease of inhibition.* There are indications that this paradoxical effect may not all be due to the presence of an intrinsic LH-RH like activity (Ying et al., 1981) reported to be

*We have indeed found FSH and LH in PFF which are active in radioreceptor and radioimmunoassays (Sairam et al., unpublished results). The characteristics of such activity are being studied and will be reported separately.

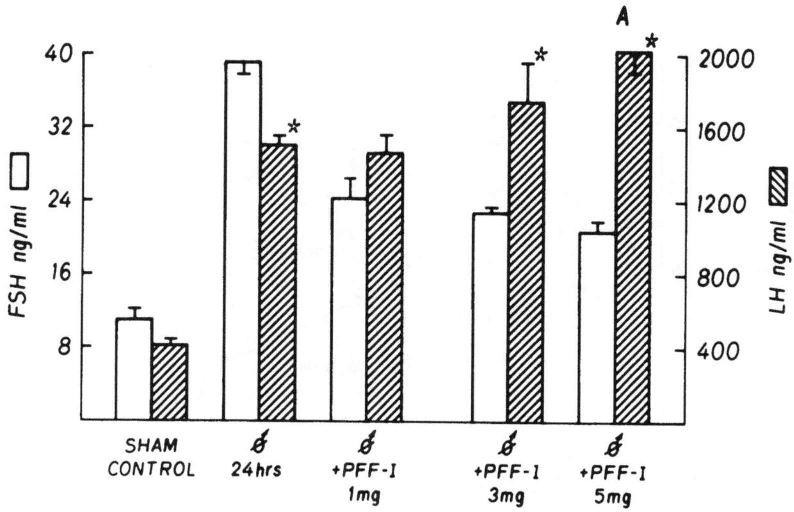

Fig. 11. Effect of partially purified porcine inhibin (30-45% $(NH_4)_2SO_4$ ppt of Fig. 10) on gonadotropin levels in the acutely castrated 34 day old male rat. * indicates that the difference between the levels in 24 h castrated controls and the treated group is statistically significant. The suppression of FSH is significant at all levels.

in the follicular fluid, whose effect might predominate at higher concentrations. This is reasoned from the data which shows that dialyzed PFF (using a membrane cut off of < 3500) still retained this strange behavior. Whether or not the LH-RH like activity present in the PFF used is less than 3500 and effectively removed by our dialysis condition (which was for 48 h against water at 4°C) is uncertain. If it indeed is due to this, the LH-RH like factor may be a molecule greater than 6000 daltons as our purified inhibin fractions (see below) which have undergone extensive dialysis using such membranes to allow their diffusion, still show the paradoxical effect in vivo and in vitro.

Our studies on the purification of PFF-inhibin must be regarded as being in a stage of exploration and infancy. With these reservations we have shown in Fig. 10 our working scheme of purification. As one of the main objectives has been to locate fractions with activity close attention has not yet been paid to recovery studies.

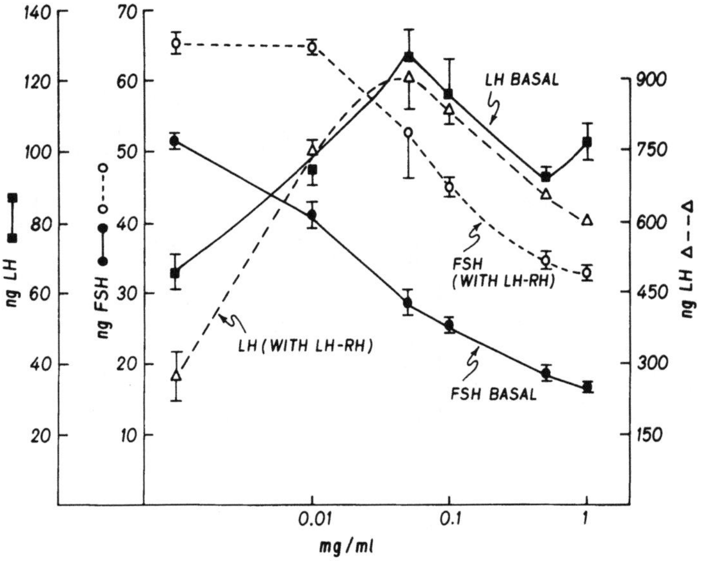

Fig. 12. Effect of partially purified porcine inhibin fraction (concanavalin-A adsorbed fraction, Fig. 10) on gonadotropin release in sheep pituitary cells. The sheep pituitary cells were obtained and placed in culture according to the method described recently (Huang and Miller, 1980). FSH and LH in the medium were analyzed by specific RIA and are expressed as ng per 10^6 cells.

Using PFF from the squeeze press method, a partially purified inhibin fraction which shows good inhibitory activity (Fig. 10) has been obtained. Many workers have invariably employed charcoal treatment of their follicular fluid with intention to remove steroids. In our hands, this procedure although achieving almost complete removal of steroids, has also resulted in very significant loss of inhibin activity, up to one third or even higher. A similar loss of activity by charcoal treatment has been noted in the purification of inhibin from rete testis fluid (Baker et al., 1981). Thus at this time, we have reservations of including such treatment in our method.

The scheme depicted in Fig. 10 has not yet been investigated using PFF obtained by the aspiration method. After the $(NH_4)_2SO_4$ precipitation step, two possibilities of further fractionation have been studied. Both steps, viz., concanavalin-A Sepharose chromatography and fractionation on SP-Sephadex C-50 appear to be useful. With additional experience, a combination of these steps may prove beneficial in purification. Both with the $(NH_4)_2SO_4$ fraction (Fig.

Table 1. Failure of Antiestrogen to Reverse the Effects of Partially Purified Porcine Follicular Fluid Inhibin Fraction in Sheep Pituitary Cells In Vitro

	HORMONE RELEASE, ng/10^6 cells/6 hr, Mean ± S.E.			
	FSH		LH	
	Basal	*LH-RH*	*Basal*	*LH-RH*
Control	122.5 ± 1.2	171.2 ± 2.2	167.0 ± 3.4	634 ± 24.6
PFF-I 50 µg/ml	60.1 ± 0.2	109.9 ± 1.2	329 ± 11.5	1116 ± 29.7
Tamoxifen 10^{-6} M	109.1 ± 4.7	148.1 ± 1.6	179.9 ± 12.4	705.9 ± 12.7
PFF-I + Tamoxifen	57.9 ± 1.1	104.6 ± 2.6	350.5 ± 15.8	1027.6 ± 25.7

Sheep pituitary cells were cultured as described by Huang and Miller (1980). After 2 days of incubation in presence of indicated substances or combinations they were challenged with 2 x 10^{-8} M D-Lys^6-LH-RH during a 6 hr period. FSH and LH in the medium were analyzed by specific RIA. PFF-I used is the 30-45% $(NH_4)_2SO_4$ fraction from Fig. 10.

10) and the PFF batch SPO-44 (directly analyzed), the inhibin activity was fully retained on concanavalin-A Sepharose, when care was taken to prevent overloading of the column.

In the 34 day old castrate male rat the elevation of FSH was partially but significantly suppressed by the purified fraction (Fig. 11). No inhibition of LH was evident at low concentration, but at higher concentrations a significant elevation of serum LH over and above the castrate levels was seen. The same material when tested in cultures of sheep pituitary cells obtained according to previously published protocols (Huang and Miller, 1980) showed divergent results. When the secretion patterns of FSH and LH were examined, the basal as well as the LH-RH stimulated release of FSH were both inhibited by the added PFF-inhibin. In marked contrast to this the basal and LH-RH stimulated release of LH in vitro was however stimulated (Fig. 12). Such a discordant effect on FSH and LH release patterns seemed to be typical of that evoked by estrogenic compounds as shown by Miller et al. (1981) in sheep pituitary cells. Radioimmunologic estimation of estradiol-17β revealed that the steroid level ($<10^{-12}$ M) in our partially purified inhibin fraction was too low to be responsible for the observed effects. To confirm this, the experiments were repeated with the inhibin fraction in presence of an antiestrogenic compound tamoxifen which when added completely reverses all the divergent effects of estrogens. This compound has been shown to be the most potent of all tested in

Table 2. Summary of Characteristics

	Nature	*Carbohydrate*[1]	*Molecular Weight*	*Membrane Binding*[2]	*Specificity of Inhibition*
BULL SEMINAL PLASMA	Protein	None	15-20,000	Yes	specific for FSH[5]
HUMAN SEMINAL PLASMA	Peptide	None	5,000[3] (> 100,000)	Yes	Specific for FSH[5]
PORCINE FOLLICULAR FLUID	Protein	present	>90,000[4]	Yes	Inhibits[6] FSH & LH -variable

1. *The purified fractions from bull or human seminal plasma did not have sugars as assessed by chemical methods and failure to be retained on concanavalin-A Sepharose. The presence of carbohydrate in PFF inhibin has been inferred solely by its binding ability to concanavalin-A Sepharose.*
2. *Membrane binding. This was assessed by using membrane fractions from sheep pituitary and labeled bovine seminal plasma inhibin as described by Sairam et al. (1981b).*
3. *Data of Franchimont et al. (1979) and Ramasharma et al. (1981) indicate 5000 daltons. A value of 19,000 is shown in the report of Vaze et al. (1979).*
4. *As shown by gel filtration of the fraction shown in Fig. 10.*
5. *We have rarely seen an inhibition of LH release including that stimulated by LH-RH at concentrations where FSH inhibition occurs.*
6. *In the mouse pituitary incubations PFF inhibited both FSH and LH; but in sheep pituitary cells LH was not affected.*

this regard (Miller et al., 1981). Tamoxifen at 10^{-6} M concentration showed no effeet on the activity of PFF inhibin fraction in sheep pituitary cells (Table 1). The conclusion that can be drawn from these experiments is that the divergent effects we have observed in the PFF-inhibin fraction above cannot be due to estrogen. Similar results have been obtained with an active inhibin fraction retained and eluted from concanavalin-A Sepharose columns.

The chromatographic characteristics of PFF inhibin (flow sheet in Fig. 10) appear to be different from similar activity found in bovine seminal plasma (Sairam et al., 1981a; Fig. 1) or human seminal plasma (Ramasharma and Sairam, 1982; Fig. 5) in several respects. First of all, the activity in PFF can be retained on concanavalin-A Sepharose whereas that in seminal plasma passes through unadsorbed on this affinity column. Unlike the seminal plasma inhibin(s) which is strongly retained by the cation exchanger SP-Sephadex C-50 at pH 5.4, the PFF-I activity remains in the unadsorbed fraction. This suggests a more acidic character to the PFF-I than seminal plasma inhibin. These results are in general agreement with that of other laboratories using follicular fluid for inhibin isolation (see de Jong et al., Williams and Lipner, this volume). Finally, the PFF-I appears to have a molecular weight much larger than that described for seminal plasma inhibin. On Sephadex G-100, the inhibin activity of porcine follicular fluid emerges following the exclusion volume of the column indicating that it is perhaps greater than 90,000 daltons. In view of our experience with human seminal plasma (Ramasharma and Sairam, 1981), we should perhaps reserve judgment as to the molecular weight figures for ovarian inhibin until more pure and better preparations are available. At this time, we must also add the caveat that the actual molecular weight of any inhibin preparation isolated can only be arrived at by a determination of its structure. Obviously consideration must also be given to the possibility that what is isolated after extensive manipulation could be somewhat different from what is in the secreting cell and what is actually secreted.

Concluding Comments

We have investigated inhibin activity in two different fluids from three species (bovine, human, and porcine) with the hope of understanding some of their characteristics. The main features of this comparative study along with inferences drawn from published work from other laboratories are summarized in Table 2. The molecular weight of inhibin, as determined by using extracts or fluids, although informative, should be viewed with considerable caution. The literature (Franchimont et al., 1979b) is replete with reports of very small to very large molecular weights for inhibin substance(s). It is our experience that the presence of other proteins may affect evaluation of this parameter significantly.

The relationship, if it is of a precursor type, among the small and large molecular weights cannot be clarified at present. Exploration of these possibilities would undoubtedly be facilitated by continued intensive investigations directed at developing isolation procedures. If the large form(s) can be isolated in pure form, meaningful limited proteolytic digestion experiments can be initiated to search for active core(s) or fragment(s), if any. Meanwhile, chemical analysis of the smaller molecular inhibin(s) substances may yield a useful base for further work.

A perplexing feature of the seminal plasma inhibin preparations as well as porcine follicular fluid material, both in crude or semi-purified forms, has been the nature of the dose-response curve with respect to inhibition. In many of the systems we have studied for activity, inhibitory activity manifests at very low concentrations which is either reduced or disappears at higher doses. We are not certain if opposing factors, in this case, stimulating agents akin to LH-RH like substances, being identified in the ovary (Ying et al., 1981) or testis (Sharpe et al., 1982), could account for the paradoxical observations. We feel that this cannot be the sole reason because the PFF-inhibin at almost all stages of the fractionation scheme (Fig. 10) exhibited this paradoxical effect in the sheep pituitary cells. Furthermore, since all of our fractions had been dialyzed using a membrane with a cut off of <6000 daltons, we would expect ovarian LH-RH like activity to be eliminated as judged from the report of Ying et al. (1981). We hope that continued purification work in progress in several laboratories including our own will resolve many of the outstanding problems.

Acknowledgements

This research work has been supported by grants from the MRC of Canada, WHO 79058, and NIH (HD 12254-01). The assistance of Messrs. M. R. Ranganathan, R. Carrière, and Mrs. Carole Chagnon-Labelle is greatly appreciated. The procurement of pooled porcine follicular fluid was made possible by the efforts of Drs. G. Hodgen and G. Bialy at NIH, to whom we are thankful.

References

Asch, R. H., Vaze, A. Y., Thakur, A. N., and Sheth, A. R., 1980, Concentrations of immunoreactive inhibin in seminal plasma and serum from normospermic, oligospermic, vasectomized, Klinefelter's syndrome and Sertoli-cell-only syndrome subjects, J. Androl., 1:252.

Baker, H. W. G., Eddie, L. W., Higginson, R. E., Hudson, B., and Niall, H. D., 1982, Studies on the purification of ovine inhibin, in: "Cell Biology of the Testis Conference," Ann. New York Acad. Sci., in press.

Chari, S., Hopkinson, C. R. N., Daume, E., and Sturm, G., 1979, Purification of inhibin from human ovarian follicular fluid, Acta Endocrinol. Copenh., 90:157.

de Jong, F. H., Hermans, W. P., Jansen, E. H. J. M., Steenbergen, J., and van der Molen, H. J., 1980, Biochemical and physiological aspects of inhibin-like activity, Proc. Int. Congr. Endocrinol., 1980, pp. 255.

de Jong, F. H., Jansen, E. H. J. M., Hermans, W. P., and van der Molen, H. J., 1982, Purification, characterization, and in vitro production of inhibin, chapter in this volume.

Franchimont, P., Demoulin, A., Proyard, J. V., Hagelstein, M. T. H., and Turnbridge, W. M. G., 1979a, Identification in human seminal fluid of an inhibin like factor which selectively regulates FSH secretion, J. Reprod. Fertil., Suppl. 26:123.

Franchimont, P., Proyard, J. V., Hagelstein, M. T. H., Renard, Ch., Demoulin, A., Bourguignon, J. P., and Hustin, J., 1979b, Inhibin: From concept to reality, Vitamins and Hormones, 37:243.

Huang, E. S. R., and Miller, W. L., 1980, Effects of 17-β-estradiol on basal and luteinizing hormone-releasing hormone induced secretion of luteinizing hormone and follicle stimulating hormone by ovine pituitary cell culture, Biol. Reprod., 23:124.

Main, S. J., Davies, R. V., and Setchell, B. P., 1979, The evidence that inhibin must exist, J. Reprod. Fertil., Suppl. 26:3.

McCullagh, D. R., 1932, Dual endocrine activity of testes, Science N.Y., 76:19.

Miller, W. M., and Huang, E. S. R., 1981, Antiestrogens and ovine gonadotropins: antagonism of estrogen induced changes in gonadotropin secretions, Endocrinology, 108:96.

Ramsharma, K., and Sairam, M. R., 1982, Isolation and characterization of inhibin from human seminal plasma, in: "Cell Biology of the Testis Conference," Ann. New York Acad. Sci., in press.

Ramasharma, K., Sairam, M. R., and Ranganathan, M. R., 1981, Effect of inhibin like factors on gonadotropin release by the mouse pituitary in vitro, Acta Endocrinol., 98:496.

Sairam, M. R., Ranganathan, M. R., Sharma, K. R., and Lamothe, P., 1980, Isolation and characterization of a basic protein suppressing the release of pituitary follitropin (FSH), VI Int. Congr. Endocrinol. 1980, Abst. 562.

Sairam, M. R., Ranganathan, M. R., Ramasharma, K., and Lamothe, P., 1981a, Isolation and characterization of a bovine seminal plasma protein inhibiting pituitary FSH secretion, Molec. Cell. Endocrinol., 22:231.

Sairam, M. R., Ranganathan, M. R., and Ramasharma, K., 1981b, Binding of an inhibin like protein from bull seminal plasma to ovine pituitary membranes, Molec. Cell. Endocrinol., 22:251.

Schwartz, N. B., 1982, Role of ovarian inhibin (folliculostain) in regulating FSH secretion in the female rat, chapter in this volume.

Scott, R. S., and Burger, H. G., 1980, Inhibin is absent from the azoospermic semen of infertile men, Nature, 285:246.

Scott, R. S., and Burger, H. B., 1980, The measurement of human seminal plasma inhibin in azoospermic patients, VI Int. Congr. Endocrinol., Melbourne, Abst. 561.

Setchell, B. P., and Main, S. J., 1974, Bibliography (with review) on inhibin, Bibliogr. Reprod., 24:245-252 and 361-367.

Sharpe, R. M., Fraser, H. M., Cooper, I., and Rommerts, F. F. G., 1982, The secretion, measurement and function of a testicular luteinizing hormone releasing hormone (LHRH)-like factor, in: "Cell Biology of the Testis Conference," Ann. New York Acd. Sci., in press.

Vaze, A. Y., Thakur, A. N., and Sheth, A. R., 1979, Development of a radioimmunoassay for human seminal plasma inhibin, J. Reprod. Fertil., Suppl. 26:135.

Williams, A. T. Rush, M. E., and Lipner, H., 1978, Isolation and preliminary characterization of inhibin-f, in: "Ovarian Follicular and Corpus Luteum Function," Advances in Experimental Medicine and Biology, Vol. 112, C. P. Channing, J. M. Marsh, and W. W. Sadler, eds., p. 429, Plenum Press, New York.

Williams, A. T., and Lipner, H., 1982, Partial purification of porcine follicular fluid gonadostatin, chapter in this volume.

Ying, S. Y., Ling, N., Böhlen, P., and Guillemin, R., 1981, Gonadocrinins: peptides in ovarian follicular fluid stimulating the secretion of pituitary gonadotropins, Endocrinology, 107:1206.

PARTIAL PURIFICATION OF PORCINE FOLLICULAR FLUID GONADOSTATIN

A. T. Williams and H. Lipner

Department of Biological Science
The Florida State University
Tallahassee, Florida 32306

INTRODUCTION

Inhibin, a hormone first suggested by McCullagh (1932), has only come under intensive study in the last decade. The generalization of this testicular factor to that of a gonadal factor by de Jong and Sharpe (1976) extended its significance and increased the need to learn more about it. Inhibin (McCullagh, 1932) has been named inhibin-m or inhibin-f (Williams et al., 1978), folliculostatin (Schwartz and Channing, 1977) and gonadostatin (Ying et al., 1981). We prefer and suggest the general adoption of the name gonadostatin (GnS) because it is independent of sex, it implies target (gonadotrophic cell) as well as origin (the gonads). The most pressing need in gonadostatin research is to isolate and characterize this molecule and develop assay procedures which will allow direct study of this putative hormone. At present much that we have learned of its physiology is indirect and derives from its one clearly defined effect, depression of plasma FSH. Initially bioassays used whole animals such as immature castrate males (Nandini et al., 1976) and metestrous ovariectomized female rats (Marder et al., 1977). The extension of bioassays to dispersed pituitary cell culture (DPCC) systems has made possible further refinements in the studies of the hormone (de Jong et al., 1979; Scott et al., 1980). Perhaps most importantly the DPCC systems make possible isolation studies not easily performed with whole animals.

Gonadostatin is present in both sexes but there is no evidence yet that the male and female hormones are similar molecular entities though they have similar effects. The various sources used for starting material are testicular extracts, testicular lymph, rete testis fluid, seminal plasma and follicular fluid. Perhaps as a consequence, no consistent molecular entity has yet been described. The confusion is obvious in a consideration of the data summarized by de Jong (1979), and expanded in Table 1. Although most workers use relatively similar protein isolation procedures, the assumed molecular weight ranges between 1400 and 180,000. That the hormone is a protein is based on the observations of its sensitivity to heat denaturation and proteolytic enzyme activity (Table 2).

Table 1. Gonadostatin: Reported Molecular Weights*

Author	Source of Protein	Purification Steps	Detection Methods	Molecular Weight (Daltons)
Cahoreau et al., 1979	ovine rete testis fluid	ethanol extraction ultrafiltration Sephadex G-200	infusion into cryptorchid ram	>100,000
Chari et al., 1978	bull seminal plasma	ethanol precipitation Sephadex G-100 (urea) CM cellulose	ovarian weight,[a] *in vivo* FSH suppression	19,000
Chari et al., 1979	human follicular fluid	ammonium sulphate fractionation Sephadex G-200 ion-exchange chromatography Sephadex G-100 ultrafiltration SDS page	ovarian weight, *in vivo* FSH suppression	∿ 23,000
Davies et al., 1978	ovine rete testis fluid	Sephadex G-100 DEAE Sephadex Sephadex G-200 Sephadex G-75 (urea)	*in vivo* FSH suppression, hermipituitaries *in vitro* (LHRH)[b]	90,000 20,000 < 5,000
Franchimont et al., 1978	human seminal plasma	Sephadex G-100 Sephadex G-25	pituitary cells *in vitro* (LHRH), *in vivo* FSH suppression	5,000
	ovine rete testis fluid	Sephadex G-100 Sephadex G-25 Sephadex G-200	pituitary cells *in vitro* (LHRH), *in vivo* FSH suppression, ovarian weight	< 5,000
Godbout, 1980	porcine follicular fluid	Sephadex G-200	pituitary cells *in vitro* (basal)	∿200,000
de Jong & Sharpe, 1976	bovine follicular fluid	ultrafiltration	*in vivo* FSH suppression	> 10,000
de Jong et al., 1979	bovine follicular fluid	equilibrium dialysis Sephadex G-75 ethanol extraction ultrafiltration	pituitary cells *in vitro* (basal)[c], pituitary cells *in vitro* (LHRH)	>10,000 <10,000

Table 1 (Continued)

Author	Source of Protein	Purification Steps	Detection Methods	Molecular Weight (Daltons)
Murthy et al., 1979	ovine testes	60 C-H_2O bath, ammonium sulphate precipitation, Sephadex G-75, DEAE-Cellulose, CM-cellulose	uterine weight[d]	20,000
Sairam et al., 1978	bull seminal plasma	Sephadex G-100 Sulfopropyl Sephadex G-50 DEAE cellulose	*in vivo* FSH suppression, uterine weight	∿ 15,000
Sheth et al., 1979	ovine testis	cytosol extraction or metaphosphoric acid precipitation, ultrafiltration Sephadex G-75 & G-25	ovarian weight	1,400-1,500
Vaze et al., 1979	human seminal plasma	ethanol precipita- Sephadex G-100 electrophoresis DEAE cellulose PAGE	*in vivo* FSH suppression, ovarian weight, pituitary cells *in vitro* (basal)	19,000
Vijayalakshmi et al., 1980	ovine ovaries ovine testes	Sephadex G-100 Amicon UM-2-Retetate Sephadex G-25 CL Sepharose 6-B		1,500 1,500
Ward et al., 1980	porcine follicular fluid	Gel filtration	X-ray inactivation	180,000
Williams et al., 1979	porcine follicular fluid	Affi-Gel-Blue Sephadex G-150 affinity chromatography, isoelectric focusing	*in vivo* FSH suppression	20,000
Williams and Lipner, 1981	porcine follicular	Sephacryl S-300 Sephacryl S-300 8 M urea	*in vitro* FSH suppression	120,000 26,000

*in part from F. H. de Jong, 1979

[a]Inhibition of ovarian weight gain induced by hCG in immature rats
[b]Inhibition of response to LHRH challenge
[c]Inhibition of basal gonadotropin release
[d]Inhibition of uterine weight gain induced by hCG in immature rats or mice

In our purification procedure we have avoided methods that might induce aggregation of molecules, e.g., ethanol precipitation (86%) (Franchimont et al., 1975), Amicon ultrafiltration (Davies et al., 1979), and ammonium sulfate precipitation (Sato et al., 1978). Our gonadostatin containing preparation (porcine follicular fluid SPO-40) was generously provided by Dr. Gabrial Bialy of the NIH. This material has gonadostatin activity indistinguishable

Table 2. Treatment of Inhibin Containing Preparations

Reference	Source of Material	Temp/ °C time	Enzyme	Activity
Nandini, et al. (1976)	ovine testes extract	60°/30-45 min		retained
Sheth, et al. (1979	ovine testes extract	100°/30 min		destroyed
			Trypsin	destroyed
Demoulin, et al. (1979)	mouse testes culture media	?/?		partial loss
			Trypsin	partial loss
Hermans, et al. (1980)	bovine follicular fluid		Pronase	destroyed
Franchiment, et al. (1979)	human seminal fluid	60°/60		destroyed
			Trypsin	destroyed
Hopkinson, et al. (1977)	ovarian extracts		Trypsin	destroyed
Lorenzen, et al. (1978)	porcine follicular fluid	60°/20		retained
		80°/30		destroyed
			Trypsin	destroyed
Setchell & Sirinathsinghji (1972)	ovine rete testicular fluid	100°/60		destroyed
Vijayalakshmi	ovine ovary ovine testis	100°/30		retained
			Trypsin	destroyed

from our own earlier collected porcine follicular fluid (pFF) in the metestrous ovariectomized rat assay (Williams et al., 1979). The 50% depression of serum FSH in the metestrous ovariectomized rat assay (Marder et al., 1977) was achieved by 150 µl of pFF (a volume equivalent to approximately 1.8 mg of crude dry solids).

MATERIALS AND METHODS

Animals. Sprague-Dawley female rats weighing 127-175 gms were obtained from Charles River Breeding Lab, Wilmington, Mass. The animals were maintained in air conditioned quarters, fed *ad libitum* with Purina Rat Chow and subjected to a 12:12 light:dark schedule. The rats were ovariectomized shortly after arriving and were used 1-2 weeks after surgery.

Dispersed Pituitary Cell Culture Bioassay. The assay is a modification of that described by Scott et al. (1980). Ovariectomized rats were decapitated and their anterior pituitaries removed and placed in HEPES buffered saline (HBS, 25 mM HEPES, 137 mM NaCl, 5 mM KCl, 0.7 mM Na_2HPO_4, 10 mM glucose, 0.36 mM $CaCl_2$; pH 7.4). The pituitaries were transferred to fresh HBS, diced and washed 3 times at unit gravity in 30 ml HBS. The tissue was incubated for 40 min at 37°C in 20 ml of 0.35% collagenase (CLS, millipore) and 0.1% hyaluronidase (Type III, Sigma) per 25 pituitaries. Clumps were dispersed by repeated pipetting.

The tissue suspension was washed once by centrifugation (125 × g, 10 min, 25°C) in 30 ml HBS. The pellet was resuspended in 20 ml (per 25 pituitaries) of 0.25% pancreatin (4 × NF, Sigma) and incubated for 5 min with continuous pipetting. The cells were washed 3 times with 30 ml HBS each time (125 × g, 10 min, 25°C). The pellet was resuspended in Dulbecco's Modified Eagle Medium pH 7.45 with 25 mM HEPES, 44 mM $NaHCO_3$, 100 IU/ml Penicillin G, 100 μg/ml streptomycin sulfate, 2.5 μg/ml Amphotericin B, 5% horse serum (Gibco), 10% ovariectomized rat serum, 10^{-7} M progesterone and 10^{-8} estradiol-17β. Yields averaged 1.5×10^6 cells/pituitary and viability was greater than 95% by trypan blue exclusion.

Cells were plated in multiwell plates (#3008, Falcon) at 5×10^5 cells/ml/well. Standards and test materials were added in 20 μl volumes 6-12 h after plating. Incubation was continued for a total of 48 h in 5% CO_2, 95% air at 37°C. At the end of the incubation the medium was removed and replaced with 0.5 ml DME (without sera) containing 10^{-8} M LHRH (Beckman). Standards and test materials were again added in 20 μl volumes. The cells were incubated as before for 6 h. The media were then collected, centrifuged at 1100 × g for 10 min (4°C) and stored at −20°C until assayed. Gel permeation chromatography:

Porcine Follicular Fluid (pFF) (SPO-40, NIH) was incubated with 30 mg Norit per ml at 4°C overnight. The carbon was removed by centrifugation and a final filtration through a fine sintered glass filter. The pFF (100 ml) was centrifuged at 20,000 × g for 30 min at 4°C, 250 mM NaCl was added and the pFF was chromatographed on Sephacryl S-300 (10 × 103 cm). The eluting buffer was 0.02 M phosphate pH 7.6, 0.50 M NaCl, 0.02% NaN_3. The fractions were collected in 200 ml volumes; 600 μl aliquots of each fraction were desalted on G-25 M, filter sterilized (0.22 μ, 13 mm Swinnex, millipore) and assayed in the DPCC assay.

The Sephacryl S-300 column was calibrated for determination of molecular weight with ferritin (mw 440,000, horse splenn, Sigma); catalase (mw 220,000, Sigma); bovine serum albumin (mw 67,000, Sigma); ovalbumin (mw 45,000, Sigma), and α-chymotrypsinogen-A (mw 25,000, Sigma). In addition the Kav of the porcine serum albumin and γ-globulin in the pFF were checked against the calibration curve and served as internal markers. The active fractions were concentrated and desalted by molecular filtration (PM-10, Amicon) and lyophilized.

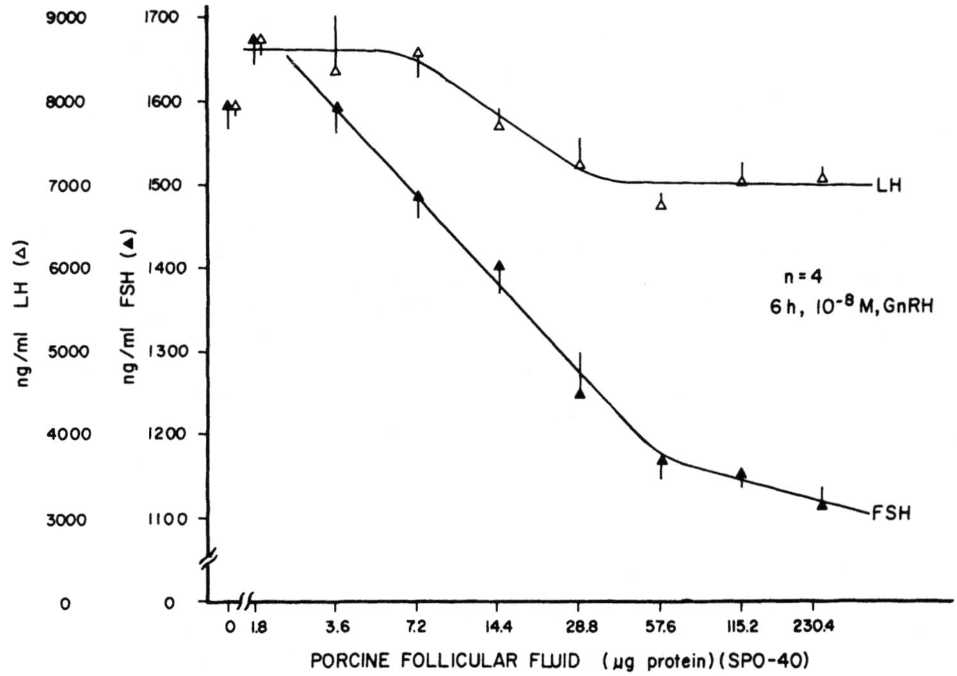

Fig. 1. The effect of crude porcine follicular fluid pFF (SPO-4, NIH) on the release of FSH and LH from dispersed pituitary cells. Each well contained 5×10^5 cells which were incubated in medium containing the test material (added after 6 to 12 h) for 48 h at 37°C. The initial growth medium Dulbecco's Modified Eagle Medium with horse serum and ovariectomized rat serum) was discarded at the end of 48 h. Incubation was continued for 6 h with DME containing 10^{-8} M LHRH (Beckman) without horse or rat serum and in the continued presence of each of the test materials. The data from a single such experiment is illustrated.

Isoelectric Focusing of pFF

Isoelectric focusing was carried out in a LKB 8101 column (110 ml) at 4°C with 2% carrier ampholytes pH 3-10 (Pharmacia). The column was stabilized in a linear sucrose gradient in the presence of 5 M urea to decrease precipitation. Phosphoric acid (0.04 M) and triethylamine (0.02 M) were used as the anolyte and catholyte solutions respectively.

The sample was SPO-40 (2 ml) mixed into the dense and light gradient solutions. The column was focused for 28 h and then drained at the rate of 25 ml h^{-1} into 40 fractions (2.5 ml/fraction). The

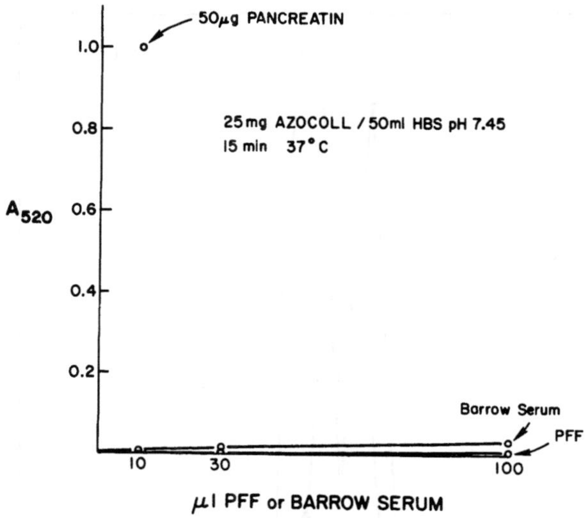

Fig. 2. Assay for protease activity in porcine follicular fluid (pFF) and barrow serum (BS). The activity was determined using Azocoll (a nonspecific horse hide proteolytic enzyme substrate) and pancreatin as a control. Protease activity was determined over a range from 10 through 100 μl of both pFF and BS.

pH of each fraction was measured at 22°C. Aliquots of each fraction (400 μl) were chromatographed on G-25 M (0.8 × 4.0 cm) to remove ampholines, sucrose and urea. The G-25 eluates from each fraction were filter sterilized and 2.5 μl was examined for biological activity in the DPCC assay.

Analyses

Protein content of the pFF, barrow serum and various fractions were determined by the method of Lowry et al. (1951). The gonadotrophins were assayed with R-R kits generously provided by NIAMDD based on requests submitted to the National Pituitary Agency.

Rat serum pools were run in duplicate at several points in each assay reported herein, the mean FSH and LH values being 622 and 124 ng/ml. The intra-assay coefficients of variation for FSH and LH assays were 3.0% and 4.9% respectively. The interassay coefficients of variation were 5.9% and 23% respectively.

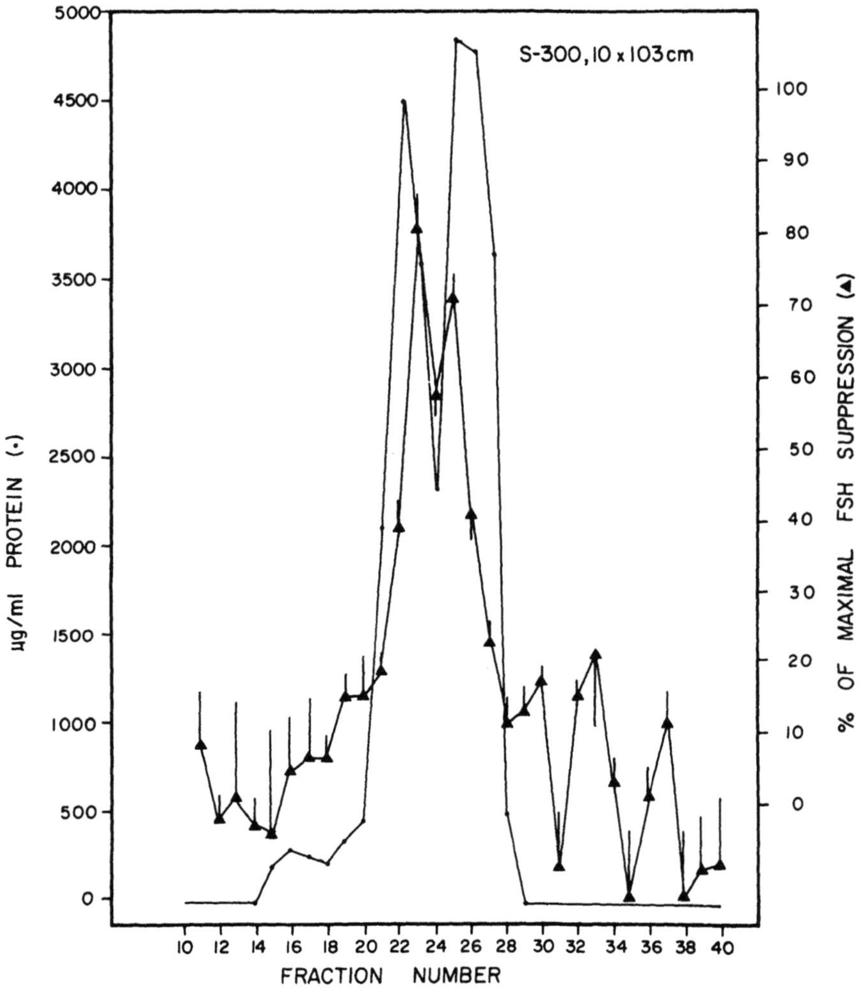

Fig. 3. The elution profile of 100 ml of SPO-40 from a Sephacryl S-300 column (10 × 103 cm). The fractions were collected in volumes of 200 ml. The ordinate represents protein determined by Lowry procedure with BSA as the standard. 99.7% of the added protein was recovered in the 40 fractions. The eluting buffer was 0.02 M phosphate buffer pH 6.7, 0.5 M NaCl, 0.02% NaN_3. The biological activity of each of the fractions is superimposed on the elution profile. The ordinate on the left expresses the depression as percent of maximal FSH. The major biological activity is localized in fractions 23, 24, and 25.

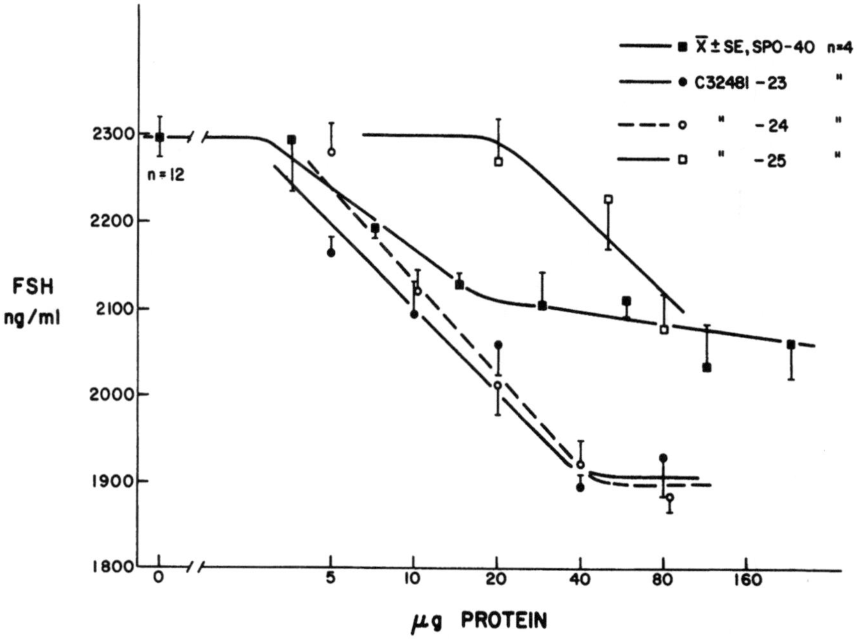

Fig. 4. Dispersed putiutary cell culture assay for several fractions obtained from gel permeation chromatography and Sephacryl S-300. ■) SPO-40; ●) fraction 23; O) fraction 24; □) fraction 25. The major activity is contained in fractions 23 and 24.

Statistics

The data were subjected to a one way analysis of variance. The level of significance of the difference between means was set up at $p < 0.05$.

RESULTS

Validation of the DPCC Bioassay

Porcine follicular fluid (pFF) was assayed in the dispersed pituitary cell culture (DPCC) system. The depression of FSH and LH secretion after incubation with pFF for 48 h and during a 6 h challenge with 10^{-8} M LHRH is illustrated in Fig. 1. Secretion of FSH was depressed over a range of 3.6 to 57.6 μg of pFF protein representing a depression of 29%. Secretion of LH was depressed over a range of 7.2 to 28.8 μg of pFF protein representing a depression of 16%. Between the values of 57.6 and 230.4 μg pFF protein the FSH curve changed slope but continued to show a depression. This latter

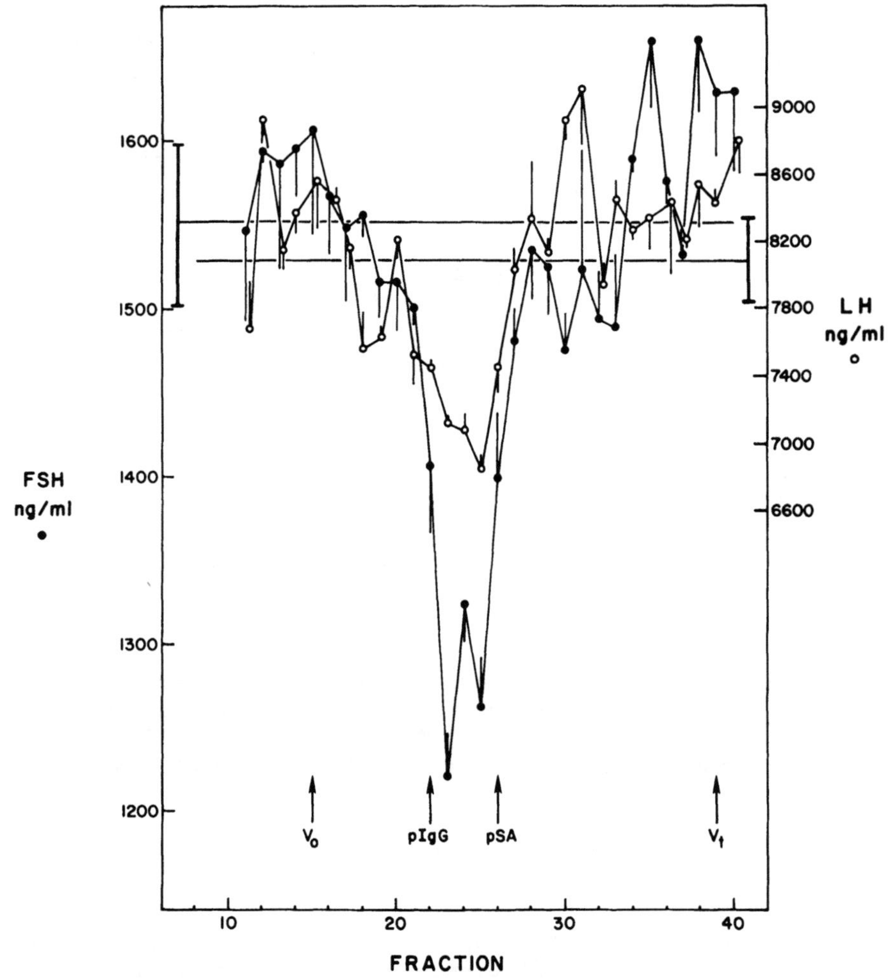

Fig. 5. Elution profile of gonadostatin activity obtained on Sephacyl S-300. The gonadostatin activity against secretion of FSH and LH is present in the same fractions.

observation has been noted repeatedly and may be due to contaminants in the pFF or due to still unexplained mechanisms involved in the control of secretion. Over the wider range of 28.8 to 230.4 μg pFF the LH curve showed a slope of 1 indicating a failure to inhibit the further secretion of LH.

Gel Permeation Chromatography

Because freezing and thawing of pFF causes the appearance of a precipitate, the pFF was centrifuged at 20,000 × g just prior to ap-

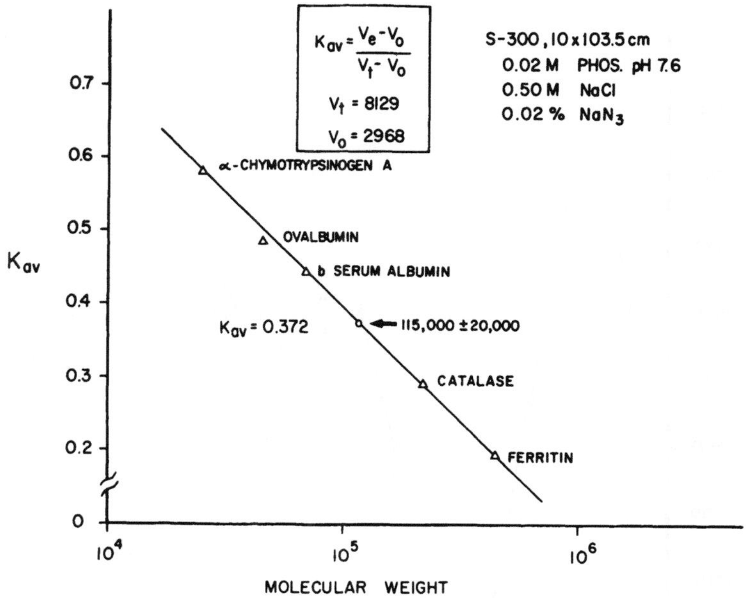

Fig. 6. Partition coefficients of marker proteins on the Sephacryl S-300 column. The proteins used were α-chymotrypsinogen (mw 25,000); ovalbumin (mw 45,000); bovine serum albumin (mw 67,000); catalase (mw 220,000); ferritin (mw 440,000). The molecular weight of the peak inhibin activity is 115,000 ± 20,000.

plication on the column. The elution was performed at room temperature. We could find in the pFF no evidence of protease activity with Azocoll (Sigma) as the non-specific protease substrate (Fig. 2). Bacterial activity was limited by the use of sodium azide in the eluting buffer. The pFF elution profile in Fig. 3 represents the separation of 100 ml of SPO-40. The elution profile divided into two major peaks separated by a single trough. Assay of each of the fractions indicated that the biological activity was contained in fractions 23, 24, and 25 (Fig. 3). Approximately 900 mg of dry weight material was recovered in fractions 23 and 24. No activity was found in any other fractions. A comparison of the 3 active fractions in a dose: response assay indicated that the major gonadostatin activity was contained in fractions 23 and 24 (Fig. 4). An effect of GnS on LH secretion is also present in the same fractions (Fig. 5).

The elution of marker proteins from the column was highly reproducible (Fig. 6). A plot of the molecular weight against the partition coefficient (Kav) indicates linearity. The Kav for the active fractions had a mean of 0.372 with a corresponding molecular

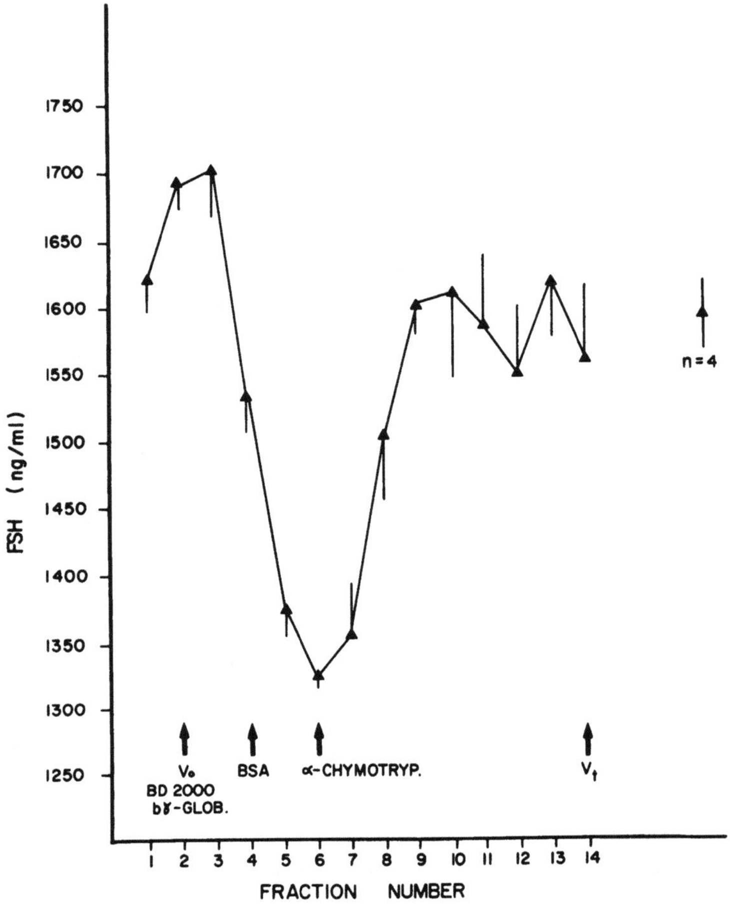

Fig. 7. The elution profile of gonadostatin activity obtained from gel permeation chromatography of pFF (SPO-40, NIH) on a Sephacryl S-300 (1.5 × 100 cm) column. The fractions were collected in 4.0 ml volumes. Each point represents a mixture of 2 fractions. The pFF was equilibrated against the eluting buffer (0.02 M PO_4, pH 7.4; 0.5 M NaCl; 8 M urea) by passage through a Sephadex G-25 column.

weight of 115,000 ± 20,000. This value was obtained in two sepations.

pFF was equilibrated with an 8 M urea containing buffer (0.02 M PO_4, pH 7.4; 0.5 M NaCl) and subjected to gel permeation chromatography on Sephacryl S-300 (1.5 × 100 cm). The elution profile of the pFF is indicated in Fig. 7. The column was calibrated between two protein markers (α-chymotrypsinogen-A, mw 25,000, and BSA,

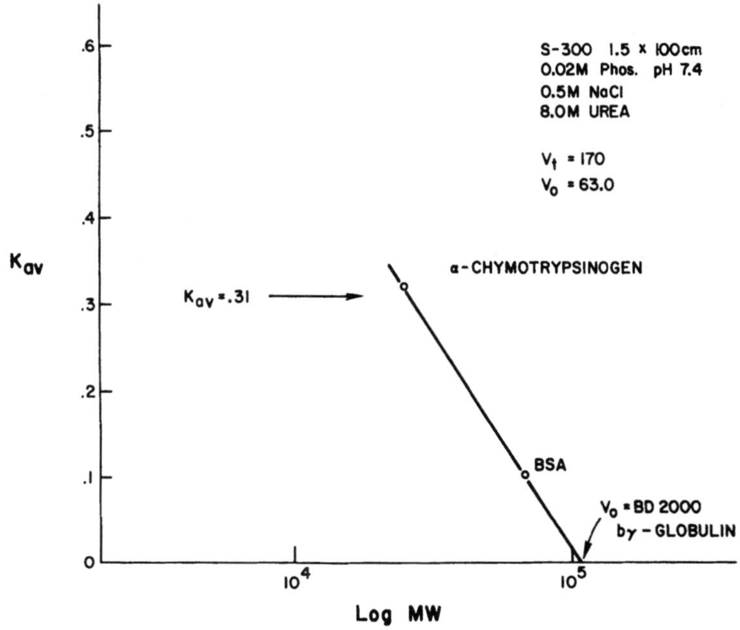

Fig. 8. The Sephacryl S-300 calibration curve using two proteins, α-chymotrypsinogen-A mw 25,000 and BSA mw 67,000. The Kav of the peak fraction was 0.310 corresponding to a molecular weight of 26,000 daltons.

mw 67,000). The fraction with maximum gonadostatin activity was represented by a partition coefficient (Kav) of 0.310 with a molecular weight equivalent to 26,000 daltons. The larger molecule appears therefore to be an aggregate (Fig. 8).

Isoelectric Focusing of pFF

Gonadostatin activity was identified in 4 fractions over a pH range of 4.45 to 6.05 (Fig. 9). The depression of the FSH concentration by these fractions in the DPCC assay was significant at $p < 0.05$ as compared to the FSH concentration after treatment with barrow serum. The failure to obtain gonadostatin activity at a single pH may be due to heterogeneity of the molecule.

DISCUSSION

The dispersed pituitary cell culture assay system (Scott et al., 1980) provides a bioassay for gonadostatin that allows assay of small quantities of material, is sensitive, accurate, and has small variation. The adoption and standardization of a dispersed pituitary

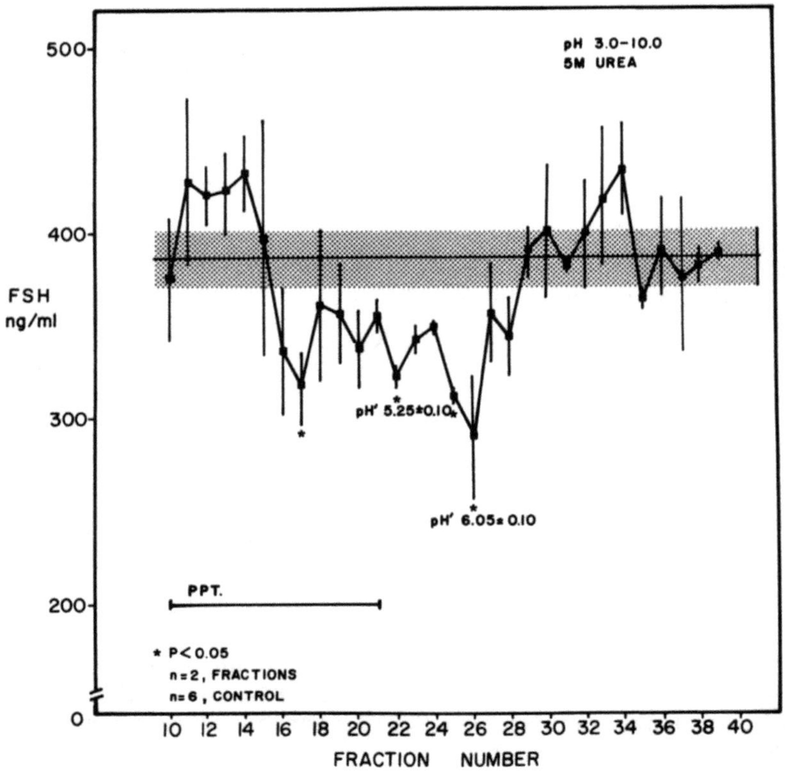

Fig. 9. Biological activity contained in fractions resulting from an isoelectric focusing profile of pFF (SPO-40) performed at 4°C with 2% carrier ampholines (pH 3-10). The column was stabilized in a linear sucrose gradient containing 5 M urea. The pFF was distributed in the dense (40% sucrose + 5 M urea) and in the light solution (5 M urea). Each fraction contained 2.5 ml.

assay for gonadostatin has made it possible to screen many fractions simultaneously. The advantages of the DPCC assay are readily apparent in terms of cost effectiveness and conservation of materials.

Assay of crude pFF results in an initial increased secretion of FSH followed by a dose related decreased secretion of both gonadotrophins. The initial stimulation of FSH secretion is suggestive of a stimulating factor, but is unexplained at present. At very high doses of pFF the inhibition of FSH secretion showed a change in slope (Figs. 1 and 4). Furthermore, crude pFF fails to depress FSH secretion as effectively as does the partially purified GnS (Fig. 4). The lack of uniform inhibition of secretion of both gonadotrophins may reflect a disproportionate sensitivity of the gonadotrophs to gonadostatin. Alternatively, pFF may contain a factor(s) that opposes the inhibin exerted by GnS. The factor may be gonadocrinin (Ying et al., 1981), the GnRH-like material in follicular fluid.

We have tried several procedures to isolate gonadostatin (Williams et al., 1979). Affi-Gel Blue chromatography (a Cibacron blue dye coupled to Sepharose, with a high affinity for albumin) was an early attempt that has been abandoned due to a relatively large loss of activity. We also tried absorbing the plasma derived follicular fluid proteins to polymerized barrow serum antiserum. This approach was based on the rationale that the barrow (a long term castrate boar) serum would be devoid of gonadostatin and that an antiserum to the barrow serum would leave behind the gonadostatin. This procedure proved impractical because of the small volumes of pFF that could be processed. We also developed a rabbit anti-pFF antiserum and absorbed the non-specific antibodies with barrow serum. The resulting antiserum was unable to cause elevation of FSH in test rats.

The plethora of reports on the molecular weight of isolated and supposedly purified gonadostatin necessitates a reexamination of the procedures used in attempted isolations. Protein isolation technology is reasonably standardized with most of the differences based on the sequence of procedures used. Most workers use gel permeation chromatography followed by ion exchange chromatography and isoelectric focusing with verification of homogeneity of the substance by electrophoresis. Several have employed a precipitation step with ethanol or ammonium sulfate. Surprisingly there is little agreement among the various reports as to molecular weight.

Our early attempts to isolate gonadostatin from pFF suggested a molecular weight of 20,000 daltons with an isoelectric point of 5.4. This value confirms the reports of Chari et al. (1978, 1979), Sairam et al. (1978), Murthy et al. (1979), and Vaze et al. (1979). Chromatographic studies, however, support the observation that the molecular weight is in a range of 100,000 to 140,000 daltons. Gel permeation chromatography in the presence of 8 M urea results in a molecular weight of approximately 26,000 daltons. This study requires replication. Since urea causes unfolding of large protein molecules by splitting hydrogen bonds and dissociation of large molecules that are noncovalently bound, one must assume that the large molecular weight entity is an aggregate, and that the true molecular weight is represented by the lower value.

"The apparent isoelectric point (pIapp) of proteins determined by isoelectric focusing in density gradients of sucrose and in the presence of ampholytes should be corrected for the primary medium effect and the pH measuring cell effect to give isoelectric points (pI) in water" (Gelsma et al., 1979). The factors which must be evaluated are concentration of non-electrolyte, temperature and chemical nature of the ampholyte. The correction term we have selected from Gelsma et al. (1979) is approximate since the final concentration of sucrose in the fraction was not determined. The ampholyte was 2% ampholine and the non-electrolyte was 5 M urea, the temperature of the pH measurement was 22°C. Gelsma et al. (1979) give values of pIapp — pI for these conditions which lie between 0.44 and 0.48. Applied to the two isoelectric points least affected by the precipitation of protein we obtain pI = 5.75-0.46 or 5.39 and pI = 6.05-0.44 = 5.61. Although these pI may be correct and suggest microheterogeneity of GnS they need to be replicated.

ACKNOWLEDGMENTS

The authors wish to thank: The National Pituitary Agency and the NIAMDD for generously providing assay kits for FSH and LH; Dr. Janice R. Lorenzen and Dr. M. E. Freeman for their reviews of the manuscript; Mrs. Pushpahrani Dhanarajan for her excellent technical assistance. This study was funded in part by US Public Health Service Grant HD 14032, NSF Grant PCM 7823382, and Ford Foundation Grant 801-0294.

REFERENCES

Cahoreau, C., Blanc, M. R., Dacheux, J. L., Pisselet, Cl., and Courot, M., 1979, Inhibin activity in ram rete testis fluid: Depression of plasma FSH and LH in the castrated and cryptorchid ram, J. Reprod. Fert., Suppl., 26:97.

Chari, S., Duraiswami, S., and Franchimont, P., 1978, Isolation and characterization of inhibin from bull seminal plasma, Acta Endocrinol., 87:434.

Chari, S., Hopkinson, C. R. N., Daume, E., and Sturm, G., 1979, Purification of "inhibin" from human ovarian follicular fluid, Acta Endocrinol., 90:157.

Davies, R. V., Main, S. J., and Setchell, B. P., 1978, Inhibin: Evidence for its existence, methods of bioassay and nature of active material, Int. J. Androl., Suppl., 2:102.

Demoulin, A., Koulischer, L., Hustin, J., Hazee-Hagelstein, M. T., Lambotte, R., and Franchimont, P., 1979, Organ culture of mammalian testis. III. Inhibin secretion., Hormone Res., 10:177.

Franchimont, P., Chari, S., Hagelstein, M. T., and Duraiswami, S., 1975, Evidence of a follicle-stimulating hormone inhibiting factor, "inhibin" in bull seminal plasma, Nature, 237:402.

Franchimont, P., Demoulin, A., Verstraelen-Proyard, J., Hazee-Hagelstein, M. T., Walton, J. S., and Waites, G. M. H., 1978, Nature and mechanisms of action of inhibin: Perspective in regulation of male fertility, Int. J. Androl., Suppl., 2:69.

Franchimont, P., Demoulin, A., Verstraelen-Proyard, J., Hazee-Hagelstein, M. T., and Tunbridge, W. M. G., 1979, Identification in human seminal fluid of an inhibin-like factor which selectively regulates FSH secretion, J. Reprod. Fert., Suppl., 26:123-133.

Gelsema, W. J., Ligny, C. L., Der Veen, N. G., 1979, Isoelectric points of proteins, determined by isoelectric focusing in the presence of urea and ethanol, J. Chromatography, 171:171.

Godbout, M., 1980, Personal communication, June.

Hermans, W. P., von Leeuwen, E. C. M., Debets, M. H. M., and de Jong, F. H., 1980, Involvement of inhibin in the regulation of follicle stimulating hormone concentrations in prebubertal and adult, male and female rats, J. Endocrinol., 86:79-92.

Hopkinson, C. R. N., Daume, E., Sturn, G., Fritze, E., Kaiser, S., and Hirschhauser, C., 1977, Inhibin-like activity of bovine ovarian extracts in male and female rats, J. Reprod. Fert., 50:93.

de Jong, F. H., and Sharpe, R. M., 1976, Evidence for inhibin-like activity in bovine follicular fluid, Nature, 263:71-72.

de Jong, F. H., Smith, S. D., and van der Molen, H. J., 1979a, Bioassay of inhibin-like activity using pituitary cells in vitro, J. Endocrinol., 80:91.

de Jong, F. H., Welscher, R., Hermans, W. P., Smith, S. D., and van der Molen, H. J., 1979b, Effects of factors from ovarian follicular fluid and Sertoli cell culture medium on in vivo and in vitro release of pituitary gonadotropins in the rat: An evaluation of systems for the assay of inhibin, J. Reprod. Fert., Suppl., 26:47.

Lorenzen, J. R., Channing, C. P., and Schwartz, N. B., 1978, Partial characterization of FSH suppressing activity (Folliculostatin) in porcine follicular fluid using the metestrous rat as an in vivo bioassay model, Biol. of Reprod., 19:635.

Lowry, O. H., Rosebrough, N. J., Farr, A. L., and Randall, R., 1951, Protein measurement with the Folin phenol reagent, J. Biol. Chem., 193:265.

Marder, M. L., Channing, C. P., and Schwartz, N. B., 1977, Suppression of serum follicle stimulating hormone in intact and acutely ovariectomized rats by porcine follicular fluid, Endocrinology, 101:1639.

McCullagh, D. R., 1932, Dual endocrine action of the testes, Science, 76:19-20.

Murthy, H. M. S., Ramasharma, K., and Moudgal, N. R., 1979, Studies on purification and characterization of sheep testicular inhibin, J. Reprod. Fert., Suppl., 26:61.

Nandini, S. G., Lipner, H., and Moudgal, N. R., 1976, A model system for studying inhibin, Endocrinology, 98:1460.

Peek, J. C., and Watkins, W. B., 1980, Synergism between bovine seminal plasma extract and testosterane proprionate in suppressing serum concentrations of gonadotrophins in acutely castrated rats: a role of inhibin, J. Endocrinol., 86:349-355.

Sairam, M. R., Ranganathan, M. R., and Lamothe, P., 1978, Purification and characterization of a follitropin suppressing principle from bull ejaculate, Biol. Reprod. 18 (Suppl. 1): 36A, (Abstract 58).

Sato, E., Miyamoto, H., Ishibashi, T., and Iritani, A., 1978, Identification, purification and immunohistochemical detection of the inhibitor from porcine ovarian follicular fluid to compensatory ovarian hypertrophy in mice, J. Reprod. Fert., 54:263.

Setchell, B. P., and Sirinathsinghji, D. J., 1972, Antigonadotrophic activity in rete testis fluid, a possible "inhibin," J. Endocrinol., 53:lx.

Sheth, A. R., Toshi, L. R., Moodidri, S. B., and Rao, S. S., 1979, Characterization of a gonadal factor involved in the control of FSH secretion, J. Reprod. Fert., Suppl., 26:71.

Schwartz, N. B., and Channing, C. P., 1977, Evidence for ovarian "inhibin": Suppression of the secondary rise in serum follicle stimulating hormone levels in proestrous rats by injection of porcine follicular fluid, Proc. Natl. Acad. Sci. USA, 74:5721.

Scott, R. S., Burgerand, H. G., and Quigg, H., 1980, A simple and rapid in vitro bioassay for inhibin, Endocrinology, 107:1536-1542.

Vaze, A. Y., Thakur, A. N., and Sheth, A. R., 1979, Development of a radioimmunoassay for human seminal plasma inhibin, J. Reprod. Fert., Suppl., 26:135.

Vijayalahshmi, S., Bandivdekar, A. H., Joshi, L. R., Moodbidri, S. B., and Sheth, A. R., 1980, Isolation and characterization of ovine testicular and ovarian inhibin, Arch. Andrology, 5:179.

Ward, D., 1980, Personal communication.

Williams, A. T., Rush, M. E., and Lipner, H., 1979, Isolation and preliminary characterization of inhibin-f, in: "Ovarian Follicular and Corpus Luteum Function, "(Advances in Experimental Medicine and Biology, Vol. 112), C.P. Channing, J. Marsh, and W.A. Sadler, eds., Plenum Press, New York.

Ying, S-Y, Ling, N., Bohlen, P., and Guillemin, R., 1981, Gonadocrinins: Peptides in ovarian fluid stimulating the secretion of pituitary gonadotropins, Endocrinology, 108:1206-1215.

PURIFICATION OF GONADOSTATIN FROM BOVINE SEMINAL PLASMA (BSP)

Shao-Yao Ying, Nicholas C. Ling,
Frederick S. Esch, Roger Guillemin,
and Wayne B. Watkins*

Laboratories for Neuroendocrinology
The Salk Institute for Biological Studies
La Jolla, California 92037, U.S.A.

Introduction

The concept of a gonadal peptide regulating the secretion of gonadotropins was originally proposed by McCullagh in 1932. This hypothetical material, which he called inhibin, was deduced as being a nonsteroidal substance, probably protein or polypeptidic in nature, that is released into general circulation and acts specifically on the pituitary gland, selectively inhibiting the secretion of FSH, but not that of LH. Such preferential regulation of gonadotropin secretion has generated a great deal of attention during the past decade. Inhibin, or inhibin-like activities, have been reported in extracts of testis (Lee et al., 1976; Keogh et al., 1976; Moodbidri et al., 1976; Nandini et al., 1976; Baker et al., 1976), spermatozoa (Lugaro et al., 1974), rete testis fluid (Setchell and Sirianthsinghji, 1972; Setchell and Jacks, 1974), seminal plasma (Franchimont et al., 1975; Chari et al., 1978), and ovarian follicular fluid (Hopkinson et al., 1977; de Jong and Sharpe, 1976; Welschen et al., 1977; Marder et al., 1977; Schwartz and Channing, 1977). While most authors emphasized the selective inhibition of FSH as the most prominent characteristic of inhibin and inhibin-like substances, simultaneous inhibition of the secretion of LH and FSH was also observed (Peek and Watkins, 1979; Chari et al., 1978; Franchimont et al., 1978; Cahoreau et al., 1979; Davies et al., 1976; Chari et al., 1979; de Jong et al., 1979). In a sensitive in vitro bioassay, we have observed inhibi-

*Postgraduate School of Obstetrics and Gynaecology, University of Auckland, Auckland, New Zealand.

tion of the secretion of both LH and FSH by rete testis fluid, seminal plasma, Sertoli cell culture fluid, granulosa cell culture fluid, and follicular fluid (Ying and Guillemin, 1979; Ying et al., 1981b); thus we prefer to refer to the material of gonadal origin that has such inhibitory activity as gonadostatin. Although several investigators have observed inhibin or gonadostatin activity based on the ability of crude or partially purified gonadal extracts to inhibit the secretion of FSH or both FSH and LH in vitro as well as in vivo, none has isolated and characterized the substances that would have been responsible for these observations.

Purification of BSP Gonadostatin

We obtained two large batches of the 86% ethanol precipitate of bovine seminal plasma (BSP) from Wayne Watkins (Peek and Watkins, 1979), dated 11-9-78 and 4-21-80 respectively, and have shown them to be capable of inhibiting the secretion of FSH and LH induced by castration. These two batches of the ethanol precipitate of BSP were used as the starting material in the studies reported here.

Biological Screening Method: The in vitro bioassay used to monitor gonadostatin activity was based on the suppression of the LRF-induced secretion of LH and FSH when purified material was co-incubated in a pituitary monolayer culture system as described previously (Ying et al., 1981a). Pituitaries from immature female rats are employed in the assay because they are highly sensitive to LRF stimulation of FSH and LH release. The LRF challenge dose used was 1.25 ng/ml of culture medium. As an internal control, various doses of [D-Phe^2,D-Trp^6]-LRF, an LRF analog antagonist, were employed. The endpoint was the concentration of LH and FSH in the culture fluids after 4-h incubation with the various treatments, as determined by radioimmunoassay.

G-50 Fine Gel Filtration Chromatography: A 5 × 150 cm column (V_{bed} = 3 L) was packed with Sephadex G-50 Fine in 30% HOAc. After equilibrating with 4 L of 30% HOAc, the column was calibrated with bovine serum albumin (mw 67,000), human ACTH (mw 4536), human [Leu^5]-β-endorphin (mw 3443), γ_3-MSH(1-24) (mw 2670), γ_3-MSH(1-16) (mw 2052), γ_3-MSH(1-12) (mw 1569), γ_3-MSH(1-8) (mw 1094), [D-Trp^2]Met-enkephalin (mw 702) and NaCl. The BSP material was dissolved in 15 ml of 30% HOAc, applied to the column and eluted with the same solvent at a rate of 130 ml/hr. Fractions of 22 ml were collected and monitored by UV absorption at 280 nm. To appropriate aliquots of the column fractions 100 μg of bovine serum albumin (BSA) and 2 ml of water were added and the contents lyophilized and reconstituted for bioassay, usually within 48 h after the end of gel filtration.

As shown in Fig. 1, after Sephadex G-50 Fine chromatography of the 11-9-78 batch of BSP, four biologically active zones with ap-

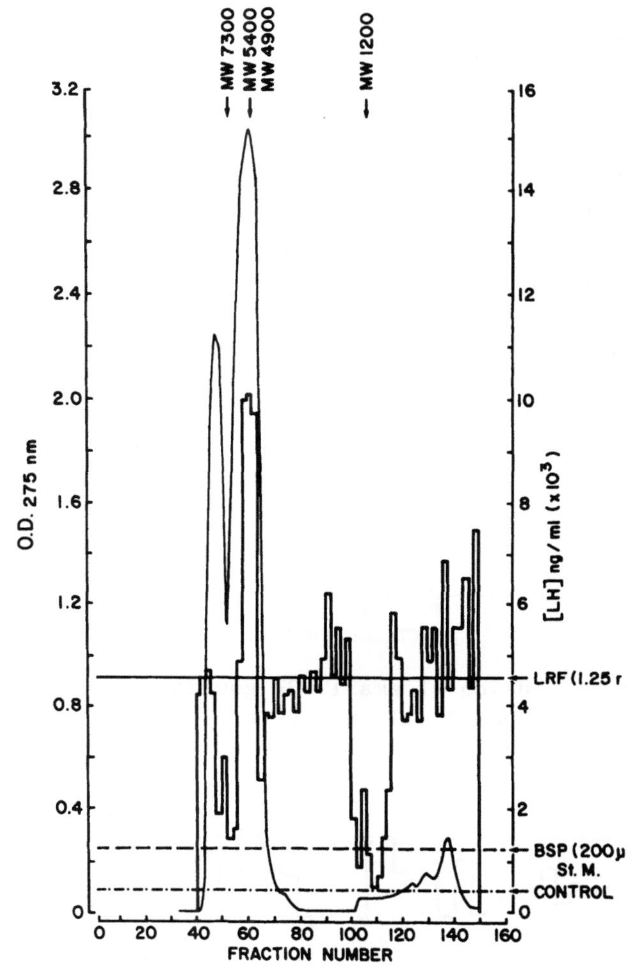

Fig. 1. Sephadex G-50 Fine purification of BSP (11-9-78 batch).

proximate molecular weights of 7300, 5400, 4900, and 1200 respectively were detected in the eluate, with the 7300, 4900, and 1200-dalton materials showing inhibition and the 5400-dalton material showing stimulation of LH release from the pituitary cell culture. A similar FSH profile was observed. During the purification of BSP gonadostatin, we always observed similar LH and FSH profiles; however, for the sake of simplicity, only the LH profile is presented here. Since the 1200-dalton material was the first small molecular weight substance from nature that inhibits gonadotropin secretion, and thus qualifies as a gonadostatin, we decided to purify this substance first. As a result, all of the 11-9-78 batch (13 g) was

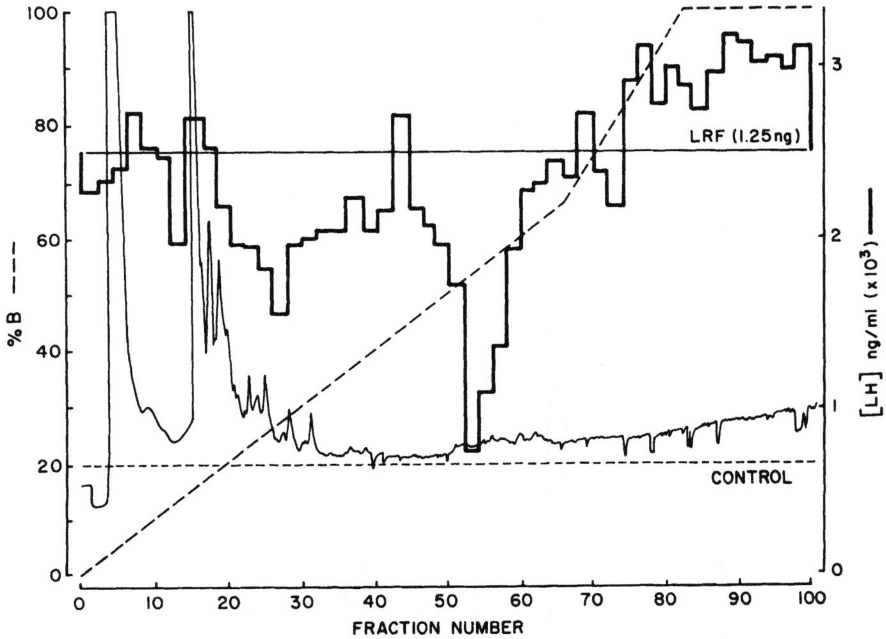

Fig. 2. HPLC purification of small BSP gonadostatin (pyridine formate-1-propanol solvent system).

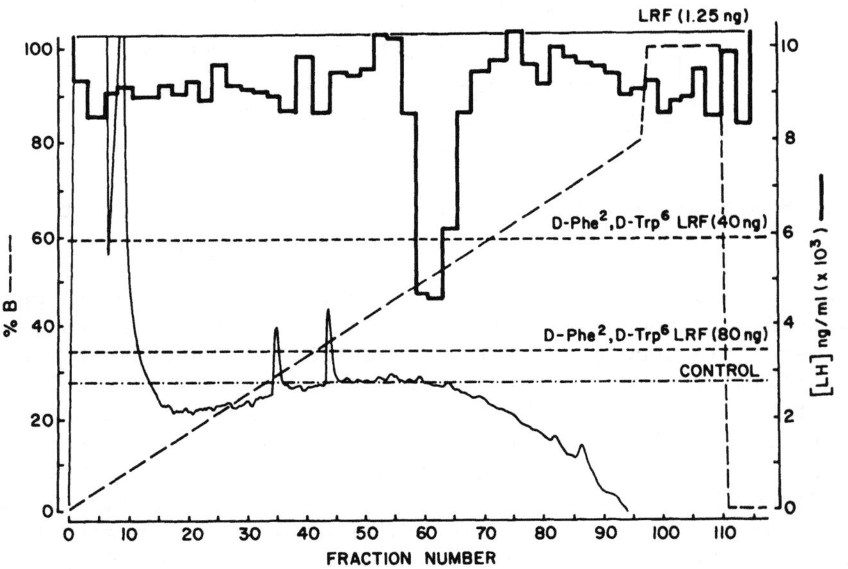

Fig. 3. HPLC purification of small BSP gonadostatin (trifluoroacetic acid-acetonitrile solvent system).

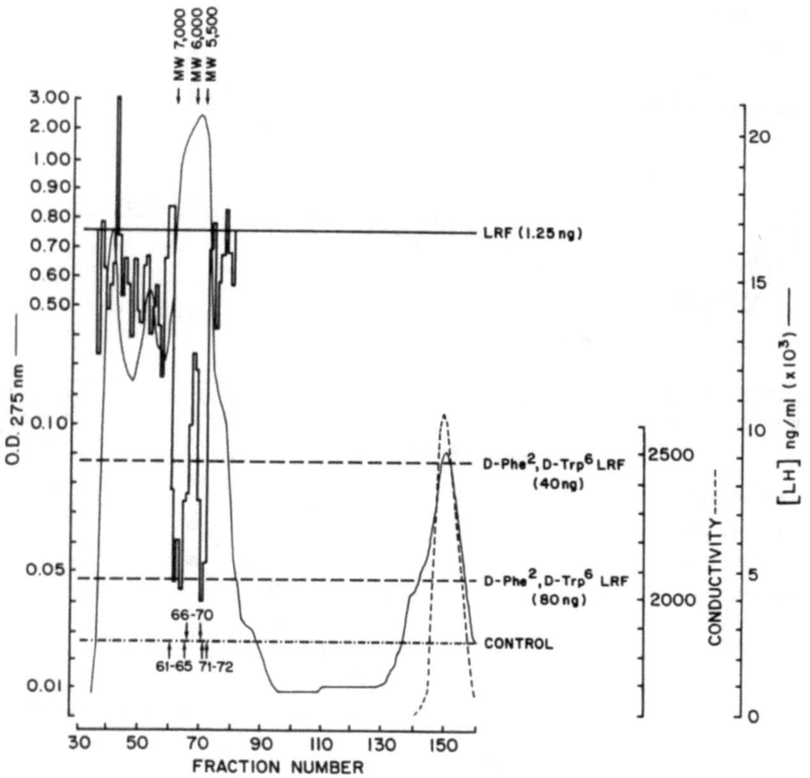

Fig. 4. Sephadex G-75 purification of BSP (4-21-80 batch).

passed through the same G-50 Fine column batchwise and the inhibitory zone in the low molecular weight region was pooled (168 mg).

Five percent of the gel-filtered material was further purified by high-performance liquid chromatography (HPLC) using 1.0% pyridine in 1.5% formic acid-1-propanol as the eluting solvent (Böhlen et al., 1980). Only one major inhibitory zone was located (fractions 53 to 58, Fig. 2) and material in this zone produced no fluorescence with the stream-sampling fluorescamine detection system (Böhlen et al., 1980), indicating that either there were no free amino groups in the compound or they were present in quantities below the level of detection. This material was further purified on another HPLC system using trifluoroacetic acid-acetonitrile buffers (Böhlen et al., 1980). Essentially only one zone of inhibitory activity was found (fractions 58 to 66, Fig. 3), but the material in this zone was present in such minute quantities that no absorbance at 210 nm was detectable. Consequently, we procured another batch of the BSP starting material from Wayne Watkins.

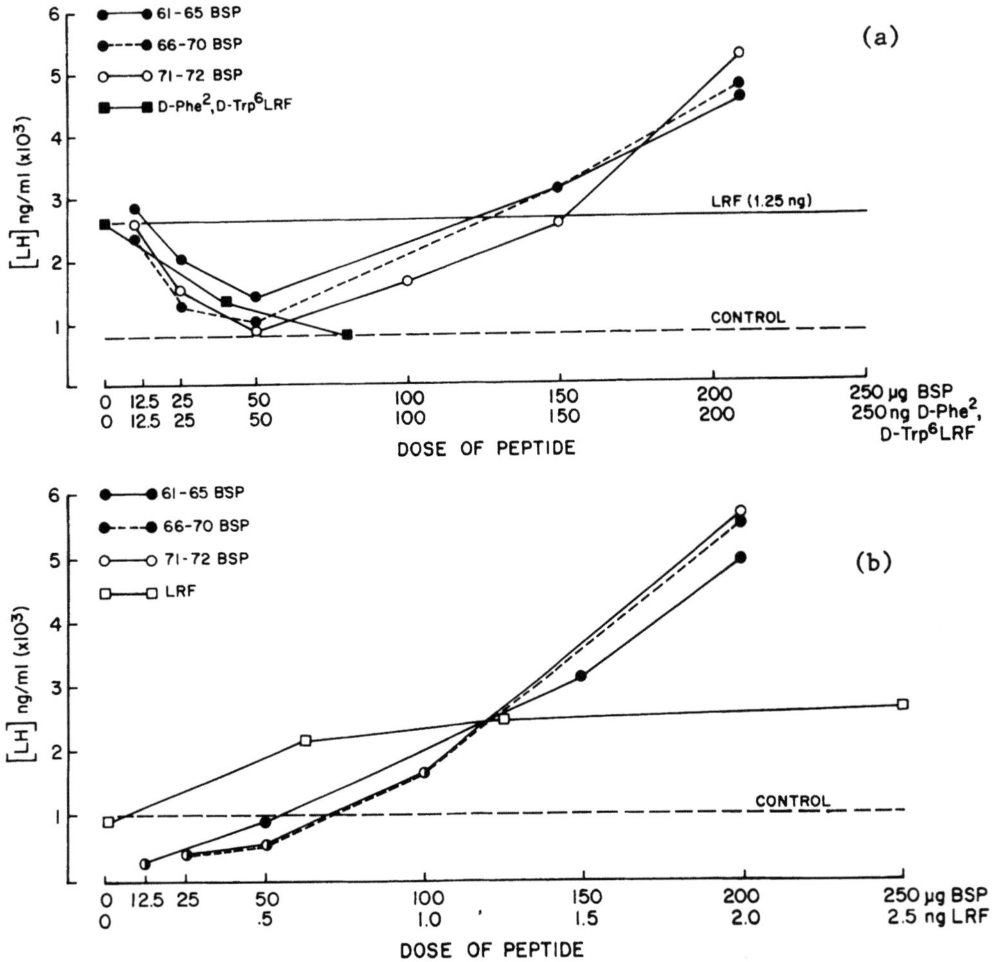

Fig. 5. Inhibition (a) and stimulation (b) of LH release from pituitary cell culture by BSP pooled fractions 61-65, 66-70, and 71-72 from G-75 gel filtration chromatography (coincubated with LRF for (a)).

To our surprise the new batch of BSP, dated 4-21-80, contained only the high molecular weight bioactive substances, but not the 1200-dalton gonadostatin, when fractionated on the G-50 Fine column. Furthermore, this low molecular weight gonadostatin is most probably not a peptide because it is not inactivated by trypsin and chymotrypsin. The non-peptidic nature of this gonadostatin was substantiated by destruction of the activity when placed in 1% NH_4HCO_3 at 100°C for 2 min.

These data indicated to us that isolation and characterization of this small gonadostatin were not a problem that could easily be solved. Therefore we set this aside and pursued the isolation of the high molecular weight active substances in the BSP.

G-75 Gel Filtration Chromatography: To obtain better resolution of the high molecular weight materials, 0.5 g of the ethanol precipitate of BSP from the 4-21-80 batch was fractionated on a 5 × 150 cm (V_{bed} = 3 L) Sephadex G-75 column equilibrated in 30% HOAc. The column was calibrated with the same peptide mixture as above and used in a similar manner to purify the BSP material. Two inhibitory materials with respective molecular weights of 7000 and 5500, as well as one stimulatory material with a molecular weight of 6000, were detected (Fig. 4), as was seen in the previous Sephadex G-50 Fine gel filtration. As a result, three pools from fractions 61-65 (55 mg), 66-70 (111.5 mg) and 71-72 (25 mg), showing inhibition, stimulation and inhibition respectively, were recovered from the gel filtration column. However, when each of the three pooled materials was assayed in a dose-response protocol for inhibitory activity, all showed similar results, with inhibition at the 25-100 µg range but stimulation at the 100-200 µg range (Fig. 5a). When the same assay was repeated without the added challenge dose of LRF, all three pools showed similar stimulatory activity (Fig. 5b). This bi-modal response will be discussed below.

Partition Chromatography on Sephadex G-50 Fine: Since the fractions 66-70 pooled from gel filtration chromatography contained most of the active material, 100 mg of it was further purified by partition chromatography. A 2 × 102 cm Sephadex G-50 Fine column (V_{bed} = 260 ml) was packed in 0.5 M HOAc and then equilibrated with 600 ml of the lower phase of the solvent system 1-butanol:pyridine:acetic acid:water (6:4:1:9). Then the upper phase of the same solvent system was passed through the column until it just emerged from the bottom of the column. The BSP material recovered from gel filtration chromatography was then dissolved in 4 ml of the upper phase of the same solvent and applied to the column. The column was eluted with the same upper phase at a flow rate of 40 ml/hr. Fractions of 10 ml were collected and analyzed by TLC on 0.25 mm-thick pre-coated silica gel 60 plates (EM reagents) using the solvent system 1-butanol:pyridine:acetic acid:water (6:6:1.2:4.8).

Three distinct regions could be discerned from TLC analysis of the effluent, and they were pooled from fractions 23-27 (8 mg), 30-41 (23 mg) and 52-66 (18 mg) respectively. Again, all three partition-chromatography-purified fractions showed similar bi-modal responses when assayed in the range of 12.5-200 µg (see Figs. 6a and 6b).

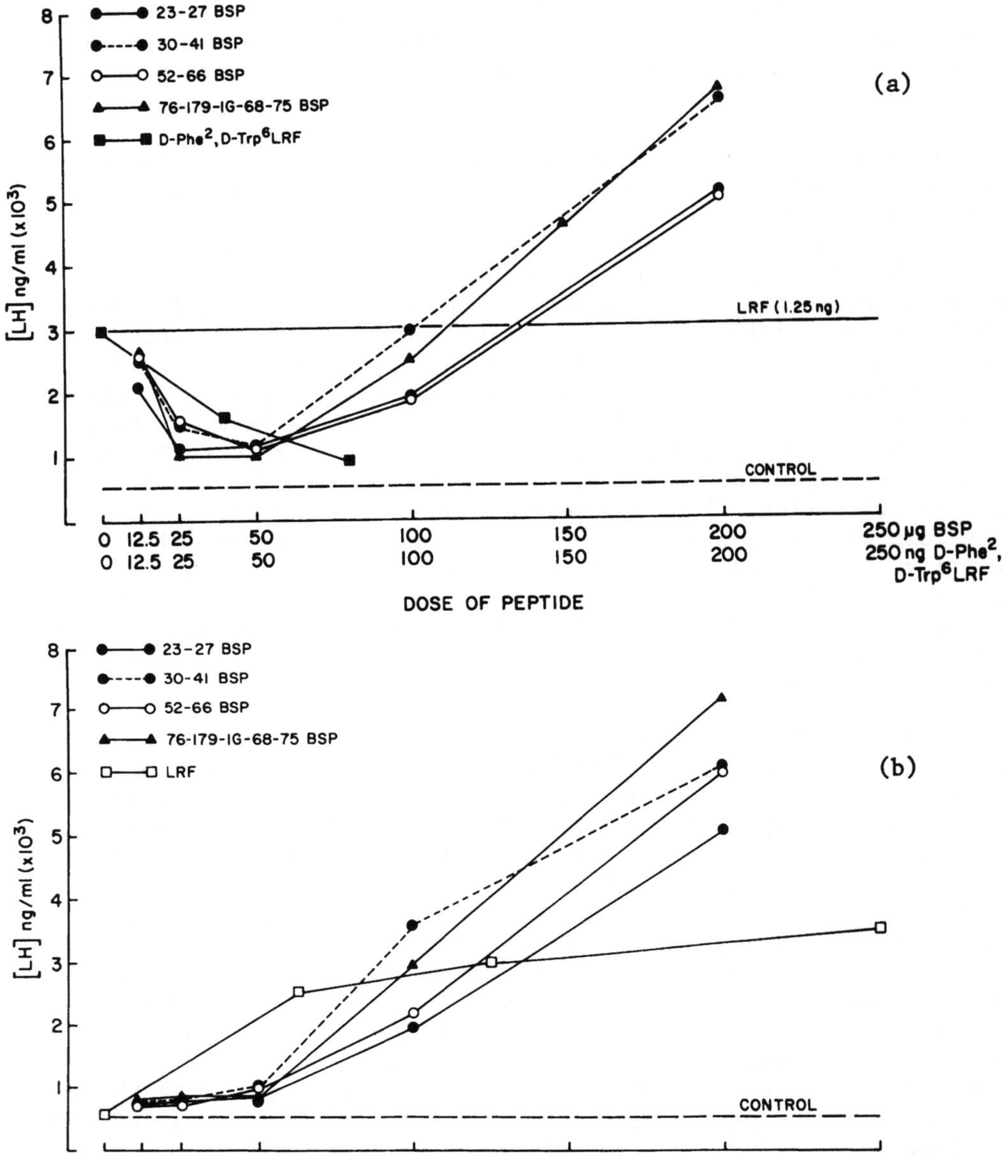

Fig. 6. Inhibition (a) and stimulation (b) of LH release from pituitary cell culture by BSP pooled fractions 23-27, 30-41, and 52-66 from partition chromatography (co-incubated with LRF for (a)). The 76-179-1G-68-75 BSP is a control sample from the pre-partition step.

Table 1. Biological Activity of BSP Gonadostatin after Treatment with Trypsin

Treatment	LH (ng/ml)	FSH (ng/ml)
Control	367 ± 91	163 ± 33*
LRF 1.25 ng	3109 ± 105	1129 ± 43
LRF 1.25 ng + [D-Phe^2,D-Trp^6]LRF 80 ng	704 ± 52*	288 ± 1*
LRF 1.25 ng + [D-Phe^2,D-Trp^6]LRF 40 ng	1417 ± 123*	548 ± 6*
LRF 1.25 ng + Buffer Blank	2844 ± 66	1061 ± 70
LRF 1.25 ng + BSP I Control	990 ± 197*	250 ± 6*
LRF 1.25 ng + BSP II "	919 ± 89*	260 ± 35*
LRF 1.25 ng + BSP III "	1372 ± 157*	433 ± 60*
LRF 1.25 ng + BSP I Digested	2740 ± 232	894 ± 121
LRF 1.25 ng + BSP II "	2385 ± 4	825 ± 3
LRF 1.25 ng + BSP III "	2889 ± 164	1049 ± 141

BSP I, II and III refer to the post-partition chromatography pools 23-27, 30-41 and 52-66 respectively.

* Statistically highly significant from the LRF controls ($P<0.01$).

Trypsin Digestion: Three hundred micrograms of sample were incubated with or without 30 μg trypsin (Calbiochem) in 200 μl 2% NH_4HCO_3 at 37° for 18 h. The enzyme was inactivated with 100 μg soybean trypsin inhibitor (Sigma). After adding 100 μg BSA, the samples were lyophilized for bioassay. Trypsin-digestion of the three post-partition pools 23-27, 30-41 and 52-66 abolished the inhibitory activity (see Table 1), indicating that the activity is associated with a peptide-like substance.

High Performance Liquid Chromatography (Pyridine-Formate-1-Propanol Solvent System): Further purification of the post-partition chromatography material was performed by HPLC using a Beckman Model 322 liquid chromatography system consisting of two model 100A

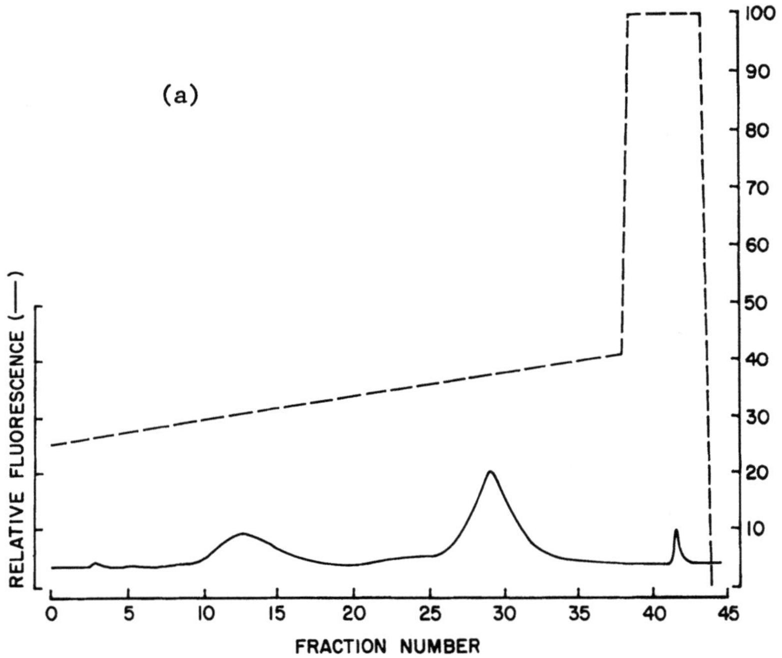

Fig. 7. HPLC of BSP pooled fractions 23-27 (a), 30-41 (b), and 52-66 (c) from partition chromatography (pyridine formate-1-propanol solvent system).

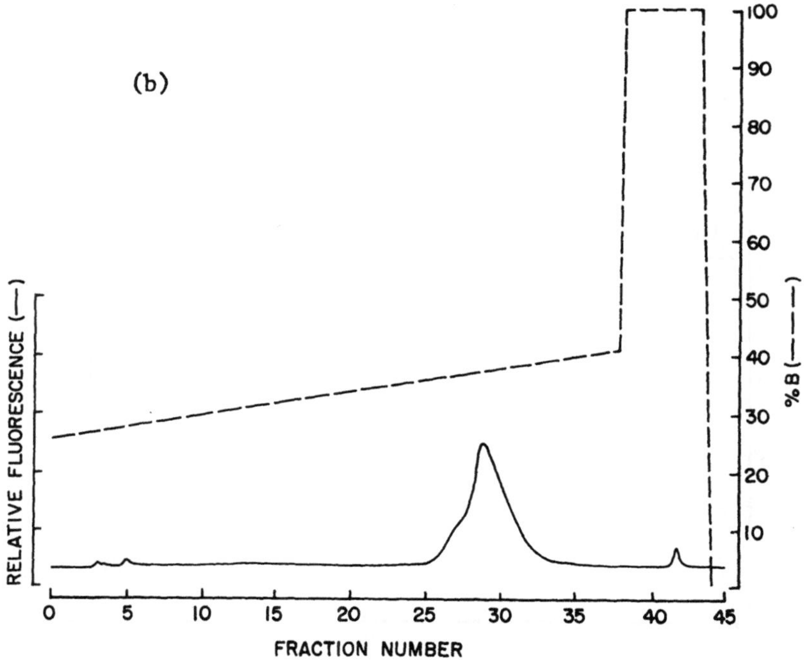
(b)
RELATIVE FLUORESCENCE (—)
% B (- - -)
100
90
80
70
60
50
40
30
20
10
0
5
10
15
20
25
30
35
40
45
FRACTION NUMBER

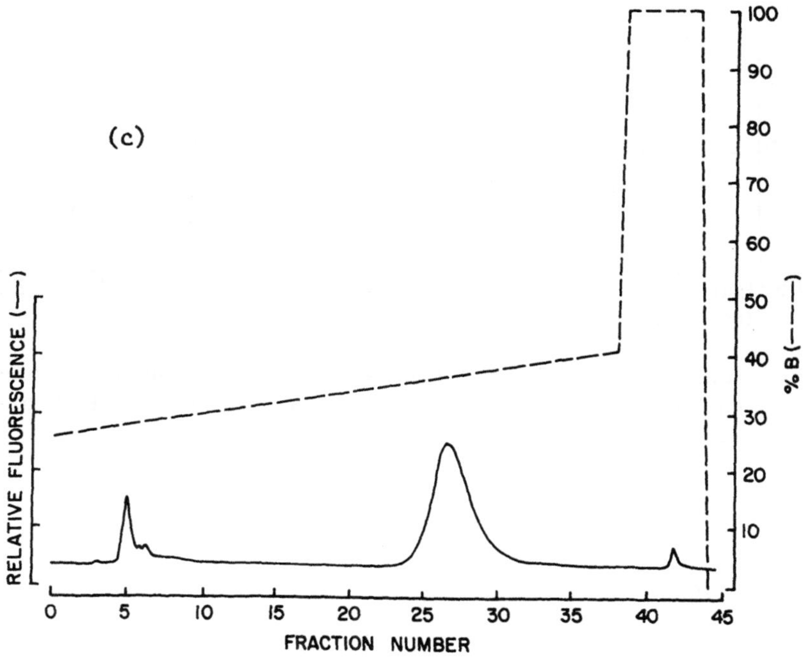
(c)
RELATIVE FLUORESCENCE (—)
% B (- - -)
100
90
80
70
60
50
40
30
20
10
0
5
10
15
20
25
30
35
40
45
FRACTION NUMBER

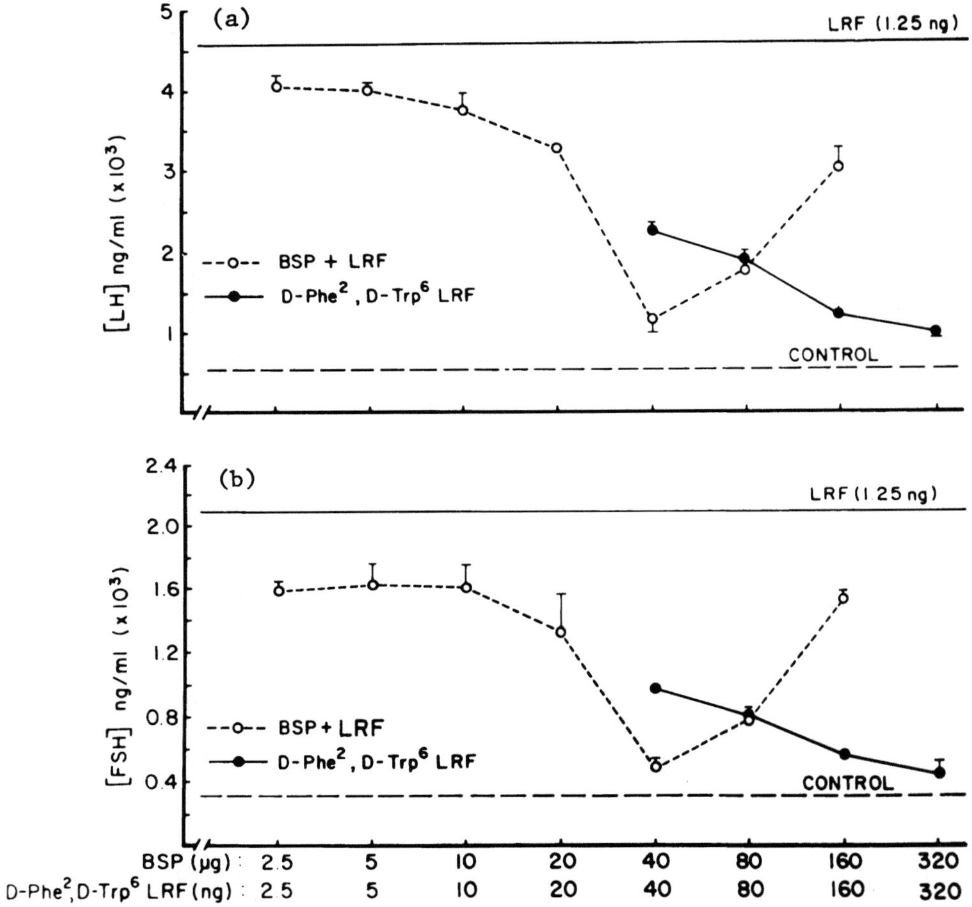

Fig. 8a, b. Inhibition of LH and FSH secretion from anterior pituitary cell culture by 4-h co-incubation of BSP gonadostatin with LRF.

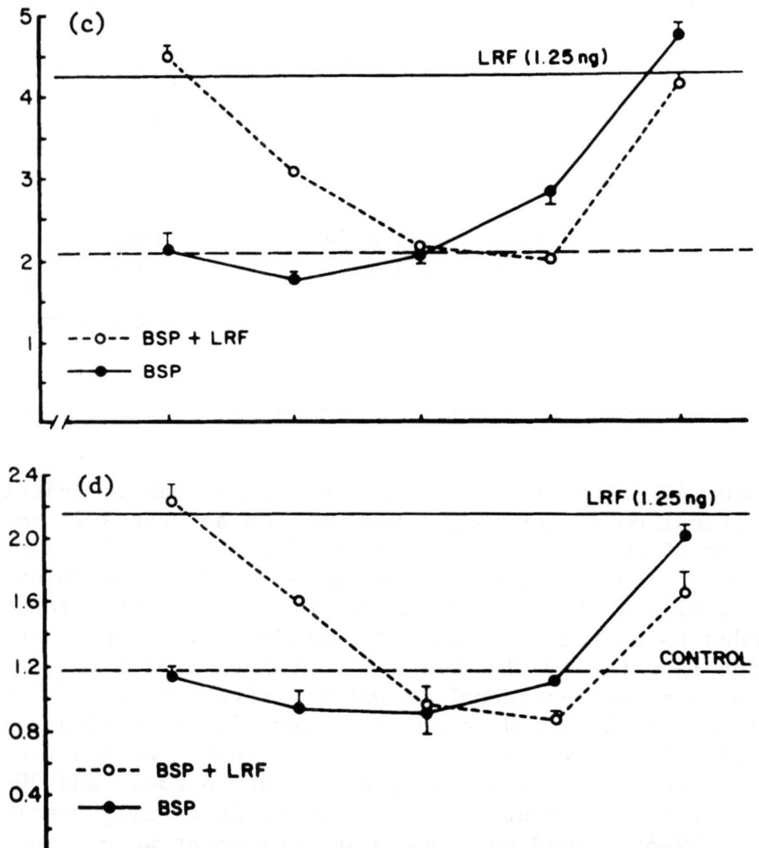

Fig. 8c, d. Inhibition of LH and FSH secretion from anterior pituitary cell culture by 24-h pre-incubation of BSP gonadostatin, followed by a 4-h incubation with or without LRF.

Table 2. No Effect of BSP Gonadostatin on TRF-Mediated Secretion of TSH and PRL

Treatment		TSH (ng/ml)	PRL (ng/ml)
Control		114 ± 19	643 ± 14
TRF 10^{-7}M		474 ± 4	1022 ± 36
TRF 10^{-7}M + BSP	100.0 µg	466 ± 34	857 ± 7
	50.0 µg	382 ± 2	812 ± 30
	25.0 µg	464 ± 14	1001 ± 77
	12.5 µg	533 ± 44	931 ± 80
	6.25 µg	505 ± 44	977 ± 112
	3.1 µg	504 ± 52	872 ± 20

pumps, a model 420 microprocessor and a model 210 sample injection valve. Purification of the sample was done on an 0.46 × 25 cm Spheri-5 Cyano (5 µm) RPLC column protected with an 0.46 × 3.0 cm Spheri-5 Cyano (5 µm) guard column (Brownlee Labs). The column was eluted at 0.6 ml/min with an increasing percentage concentration of Buffer B added to Buffer A. Buffer A contained 1.5% formic acid and 1.0% pyridine in water, while Buffer B contained 1.5% formic acid, 1.0% pyridine and 60% 1-propanol in water. Five percent of the column effluent was diverted to the fluorescamine detection system (Böhlen et al., 1980), while the rest was collected with a fraction collector. As shown in Figs. 7a, b, and c, pools 23-27 and 30-41 contained multiple components which could not be readily resolved. However, pool 52-66 showed only one major component which could be isolated in a highly purified state. As a result, a large amount of the 52-66 pool was processed through HPLC to accumulate enough material for its structural characterization. The substance recovered from HPLC still retained the inhibitory activity.

Biological Actions of BSP Gonadostatin

At the writing of this manuscript we have just obtained sufficiently large amounts of highly purified BSP gonadostatin to permit in vivo studies. These are now in progress. It would be premature, however, to make any statement on the early results obtained. All the present comments on the biology of gonadostatin will therefore pertain to the in vitro methodology described here.

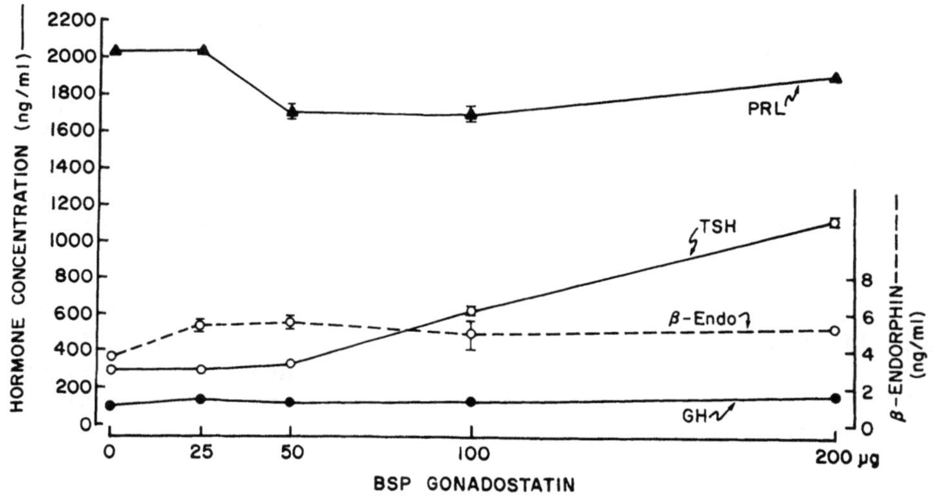

Fig. 9. Effect of BSP gonadostatin on GH, β-endorphin, TSH, and PRL in pituitary culture.

The in vitro biological screening system we have used to monitor the purification of BSP gonadostatin is an acute bioassay, based on the ability of gonadal extracts to inhibit the LRF-mediated secretion of gonadotropins when co-incubated with LRF. Most investigators in the field of inhibin have been using a longer-term incubation of gonadal extracts in the pituitary monolayer system (i.e., gonadal extracts added to the culture system for periods ranging from 16 to 48 h) to test the ability of these substances to inhibit the secretion of gonadotropins, at either the basal levels or LRF-mediated levels. We have also tested our purified BSP gonadostatin in the system used by most investigators to screen inhibin-like biological activity. In one experiment, as shown in Fig. 8, BSP gonadostatin, pre-incubated for 24 h prior to being challenged with LRF, inhibited the LRF-mediated secretion of both LH and FSH, whereas basal levels of LH and FSH were not inhibited. Furthermore, the anti-LRF effect demonstrated by BSP gonadostatin is very specific because no inhibition of TSH or PRL was observed when the cells were challenged with TRF (Table 2).

It is worth noting that in our in vitro biological screening system, as well as in a system designed for testing the biological activity of inhibin, we have consistently observed a bi-modal effect by BSP gonadostatin on the inhibition of gonadotropins; namely, only within a certain narrow range of the dosage was inhibition of the secretion of gonadotropins (i.e., gonadostatin activity) observed. With higher doses of BSP gonadostatin we observed an apparent stimulation of gonadotropin secretion. However, the stimulation by high doses of BSP gonadostatin is not specific for gonadotropins. So far, in tests carried out in our pituitary monolayer culture system, highly significant stimulation of the secretion of TSH has been observed, but there was no effect on pituitary GH, β-endorphin and PRL (see Fig. 9). Based on our Trypan blue exclusion test, these cells, after treatment with 200 μg of purified BSP gonadostatin, are still viable, and morphologically speaking there was no apparent damage to these cells.

To sum up the biological and chemical properties of BSP gonadostatin: 1) It specifically inhibits both LH and FSH secretion induced by LRF in vitro, acting in the manner of synthetic LRF antagonists; 2) At high doses, it releases LH, FSH, and TSH from a pituitary cell culture system.

Since we feel confident that we are now dealing with a highly purified peptide, we believe it would be of major interest to compare gonadostatin with other purified preparations of inhibin in the same bioassays. If inhibin indeed does preferentially inhibit the secretion of FSH, then gonadostatin is not inhibin. If, in the same assay, and at comparable doses, inhibin and gonadostatin do have similar biological activities, then we may well be dealing with the same material under two different names. Resolving the nomenclature should be the least of our problems.

Conclusion

In summary, gonadostatin was purified from bovine seminal plasma, based on its ability to suppress the LRF-stimulated secretion of both gonadotropins, LH and FSH, in a sensitive pituitary monolayer culture system. The purification sequence includes gel filtration, partition chromatography and HPLC. Although other zones of gonadostatin activity were observed in initial gel filtration chromatography experiments, it was decided to concentrate further purification efforts on the larger molecular weight bioactive materials (⩾6000 daltons) for the following reasons: 1) A small gonadostatin (⩾1000 daltons) only appeared in one batch of BSP ethanol precipitate (perhaps due to seasonal factors), and it is most probably not a peptide because it is not inactivated by trypsin and chymotrypsin; 2) The large quantities of the purified high molecular weight gonadostatin obtainable would facilitate its structural characterization.

Using the in vitro method, the highly purified BSP gonadostatin has been found to have the ability to inhibit the secretion of both gonadotropins. Based on these observations, gonadostatin would be a more appropriate term in reflecting the inhibitory activities of peptides of gonadal origin on the secretion of gonadotropins.

Acknowledgement

This work was supported by a program project grant from NIH (HD-09690).

References

Baker, H. W. G., Bremner, W. J., Burger, H. G., de Kretser, D. M., Dulmanis, A., Eddie, L. W., Hudson, B., Keogh, E. J., Lee, V. W. K., and Rennie, G. C., 1976; Testicular control of follicle stimulating hormone secretion, Recent Progr. Horm. Res., 32:429-469.

Böhlen, P., Brazeau, P., Benoit, R., Ling, N., Esch, F., and Guillemin, R., 1980; Isolation and amino acid composition of two somatostatin-like peptides from ovine hypothalamus: somatostatin-28 and somatostatin-25, Biochem. Biophys. Res. Commun., 96:725-734.

Cahoreau, C., Blanc, M. R., Dacheux, J. L., Pisselet, C., and Corot, M., 1979; Inhibition activity in ram rete testis fluid, Depression of plasma FSH and LH in the castrated and cryptorchid ram, J. Reprod. Fertil. Suppl., 26:97-116.

Chari, S., Duraiswami, S., and Franchimont, P., 1978; Isolation and characterization of inhibin from bull seminal plasma, Acta Endocr., 87:434-448.

Chari, S., Hopkinson, C. R. N., Daume, E., and Sturm, G., 1979; Purification of "inhibin" from human ovarian follicular fluid, Acta Endocr., 90:157-166.

Davies, R. V., Main, S. J., Young, M. G. W. L., and Setchell, B. P., 1976; Bioassay of inhibin-like activity in rete testis fluid and its partial purification, J. Endocr., 68:26P (abstr.).

De Jong, F. H., and Sharpe, R. H., 1976; Evidence for inhibin-like activity in bovine follicular fluid, Nature, 263:71-72.

De Jong, F. H., Welschen, R., Hermans, W. P., Smith, S. D., and Van der Molen, H. J., 1979; Effects of factors from ovarian follicular fluid and sertoli cell culture medium on in-vivo and in-vitro release of pituitary gonadotrophins in the rat: An evaluation of system for the assay of inhibin, J. Reprod. Fertil. Suppl., 26:47-59.

Franchimont, P., Chari, S., Hagelstein, M. T., and Duraiswami, S., 1975; Existence of a follicle-stimulating hormone inhibiting factor "inhibin" in bull seminal plasma, Nature, 257:402-404.

Franchimont, P., Demoulin, A., Verstraelen-Proyard, J., Hazee-Hagelstein, M. T., Walton, J. S., and Waites, G. M. H., 1978; Nature and Mechanisms of Action of Inhibin: Perspective in Regulation of Male Fertility, Intern. J. Andrology Suppl., 2:69-80.

Hopkinson, C. R. N., Daume, E., Sturm, G., Fritze, E., Kaiser, S, and Hirschauser C., 1977; Inhibin-like activity of bovine ovarian extracts in male and female rats, J. Reprod. Fertil., 50: 93-96.

Keogh, E. J., Lee, V. W. K., Rennie, G. C., Burger, H. G., Hudson, B., and de Kretser, D. M., 1976; Selective suppression of FSH by testicular extracts, Endocrinology, 98:997-1004.

Lee, V. W. K., Keogh, E. J., Burger, H. G., Hudson, B., and de Kretser, D. M., 1976; Studies on the relationship between FSH and germ cells: Evidence for selective suppression of FSH by testicular extracts, J. Reprod. Fertil. Suppl., 24:1-15.

Lugaro, G., Casellato, M. M., Mazzola, G., Fachini, G., and Carrea, G., 1974; Evidence for the existence in spermatozoa of a factor inhibiting the follicle stimulating hormone releasing hormone synthesis, Neuroendocrinology, 15:62-68.

Marder, M. L., Channing, C. P., and Schwartz, N. B., 1977; Suppression of serum follicle stimulating hormone in intact and acutely ovariectomized rats by porcine follicular fluid, Endocrinology, 101:1639-1642.

Moodbidri, S. B., Joshi, L. R., and Sheth, A. R., 1976; Isolation of an inhibin-like substance from ram testis, I.R.C.S. Med. Sci., 4:217.

McCullagh, D. R., 1932; Dual endocrine activity of the testes, Science, 76:19-20.

Nandini, S. G., Lipner, H., and Moudgal, N. R., 1976; A Model system for studying inhibin, Endocrinology, 98:1460-1465.

Peek, J. C., and Watkins, W. B., 1979; Gonadotropin-inhibiting activity in seminal plasma from intact and vasectomized bulls, J. Reprod. Fertil., 57:281-285.

Schwartz, N. B., and Channing, C. P., 1977; Evidence for ovarian "inhibin": Suppression of the secondary rise in serum follicle stimulating hormone levels in proestrus rats by injection of porcine follicular fluid, Proc. Natl. Acad. Sci. U.S.A., 74: 5721-5724.

Setchell, B. P., and Sirinathsinghji, D. J., 1972; Antigonadotrophic activity in rete testis fluid, a possible "inhibin," J. Endocr., 53:XL-LXI (Abstr.).

Setchell, B. P., and Jacks, F., 1974; Inhibin-like activity in rete testis fluid, J. Endocr., 62:675-676.

Welschen, R., Hermans, W. P., Dullart, J., and de Jong, F. H., 1977; Effect of an inhibin-like factor present in bovine and porcine follicular fluid on gonadotropin levels in ovariectomized rats, J. Reprod. Fertil., 50:129-131.

Ying, S. Y., and Guillemin, R., 1979; Gonadostatine et gonadocrinine, polypeptides d'origine ovarienne à activités hypophysiotropes, C.R. Acad. Sci., Series D (Paris), 289:943-946.

Ying, S. Y., Ling, N., Böhlen, P., and Guillemin, R., 1981a; Gonadocrinins: Peptides in ovarian follicular fluid stimulating the secretion of pituitary gonadotropins, Endocrinology, 108:1206-1215.

Ying, S. Y., Ling, L., Böhlen, P., and Guillemin, R., 1981b; Gonadostatins and gonadocrinin: peptides from the gonads regulating the secretion of gonadotropins, in: "Bioregulators of Reproduction," edited by G. Jagiello and H. J. Vogel, Columbia University Press, New York, N. Y., pp. 389-401.

PROPERTIES OF FOLLICLE STIMULATING HORMONE BINDING INHIBITORS FOUND IN PHYSIOLOGICAL FLUIDS*

Leo E. Reichert, Jr., Michael A. Sanzo,
Paul W. Fletcher, James A. Dias,
and C. Y. Lee

Department of Biochemistry
Albany Medical College of Union University
Albany, New York 12208

Introduction

Our interest in follicle stimulating hormone binding inhibitors (FSH-BI) stemmed from earlier studies on the presence of a low molecular weight testicular factor(s) in buffer or water extracts of mature rat testis which inhibited binding of ^{125}I-FSH to gonadal receptors (Reichert and Abou-Issa, 1977). Shortly thereafter, we described the presence of similar factors in human serum (Reichert et al., 1979) and in bovine follicular fluid (Darga and Reichert, 1978). Most recently, low molecular weight components with FSH-BI activity have been detected in human seminal plasma (Dias et al., 1981). Our initial studies on these factors have been summarized in several reviews (Reichert, 1978; Reichert et al., 1981) and will not be discussed further at this time. Rather, we will present in this report a description of our latest efforts to characterize the FSH binding inhibitors found in human plasma and bovine follicular fluid.

Studies on the Follicle Stimulating Hormone Binding Inhibitor in Human Serum

The starting material for purification and characterization of the serum FSH-BI was derived from human serum by either of two ap-

*This work was supported by United States Department of Agriculture Cooperative Agreement 58-3244-0-158 and a Grant from the Andrew W. Mellon Foundation.

proaches. Initially, human serum was dialyzed against deionized water using a Spectrapor No. 1 membrane (Spectrum Industries, Los Angeles, CA) which passes molecules of molecular weight less than 8000, followed by consecutive gel filtration through columns of Sephadex G-25, G-50, and G-10. In each instance, regions with FSH-BI activity were detected on the basis of the ability of the test fraction to inhibit binding of ^{125}I-hFSH to receptors on rat testes membranes (Reichert et al., 1979). Alternatively, in order to facilitate processing of relatively large volumes of serum, a series of membrane filtrations were utilized rather than the Sephadex filtration series. In this approach, and taking advantage of the stability of the serum FSH-BI to short periods of heat, human serum was heated to 80°C for one hour and the resulting denatured and insoluble proteins removed by filtration through cheesecloth and by centrifugation. The soluble supernatant was then fractionated by pressure filtration through an Amicon Model DC2 Hollow Fiber/Concentration system, equipped with an H1P-5 hollow fiber (Amicon Corporation, Lexington, MA) which passes substances of molecular weight 5000 or less. The passing fraction was then concentrated by pressure filtration through an Amicon UM-2 membrane (retaining substances of molecular weight greater than 1000). The retained fraction was freed of microsolutes isovolumically by dialysis filtration using a six-volume turnover of deionized water, followed by lyophilization. The lyophilized UM-2 retained FSH-BI fraction was then reconstituted with deionized water and fractionated by filtration through a column of Sephadex G-10 also equilibrated and developed with deionized water, to give three fractions having FSH-BI activity. The most active of these, coded G-10-1, emerged from the Sephadex column with a Ve/Vo ratio of 1.10 indicating a molecular weight of about 700. Approximately 1000 mcg of this preparation inhibited binding of ^{125}I-hFSH to receptors from testis of mature and immature rats and from bovine granulosa cells by 50%. Fraction G-10-1 also inhibited binding of ^{125}I-hFSH to receptor solubilized from calf testis with Triton X-100. In addition, however, fraction G-10-1 had the interesting property of augmenting binding of ^{125}I-hLH to membrane receptors from rat testis. A dose of 175 mcg of G-10-1 augmented binding of LH to its receptor by about 50%. A number of experiments were undertaken in an attempt to determine whether the FSH-BI activity and the LH binding augmentor (LH-BA) activity were associated with the same or different molecules. A separation of the two activities was achieved by ion exchange chromatography. Fraction G-10-1 was applied to the strong cation exchanger sulfopropyl-SP-Sephadex C-25 in 0.01 ammonium acetate buffer which was chosen because it was volatile and hence could easily be removed by lyophilization. Under these conditions, the bulk (80%) of recoverable FSH-BI activity emerged unadsorbed. A fraction having both FSH-BI and LH-BA activity was eluted with 0.1 M ammonium acetate, and a fraction having LH-BA activity but not detectable FSH-BI activity was eluted with 0.25 M ammonium acetate.

These results demonstrated that the two activities were separable and not inherent in the same molecule. Characterization of the LH binding augmentor will not be discussed further here, but will be the subject of a separate report.

The FSH-BI fraction emerging unadsorbed from SP-Sephadex was then applied to the strongly basic anion exchanger diethyl-2-hydroxypropyl aminoethyl Sephadex A-25 (QAE-Sephadex) equilibrated with 0.01 M ammonium acetate pH 7.5. A fraction enriched in FSH-BI activity was adsorbed to the exchanger and eluted between 0.2 and 0.3 M ammonium acetate buffer, pH 7.5. Here, as with the fractions described below, the lyophilized powder was extremely hygroscopic, making accurate determination of mass by direct weighing quite difficult. This fraction, after lyophilization and reconstitution with water, inhibited binding of ^{125}I-hFSH to membrane receptor by 50% at a concentration of 25 mcg as determined by fluorescamine assay. The FSH-BI enriched fraction eluted from OAE-Sephadex was further purified by high pressure liquid chromatography on a μBondapack C18 column. This is a reversed phase packing in which long chain alkyl groups are attached to a solid silica matrix backing. Fractions not adsorbed to such material are generally considered highly polar in nature. The FSH-BI fraction was applied in phosphate buffer pH 2.2 at a flow rate of 0.5 ml per min. Fractions were collected every 18 seconds (150 μl per fraction) and the refractive index was recorded to locate eluted components of the applied fraction. The various fractions were than assayed for FSH-BI activity in the usual manner (Reichert et al., 1979). FSH-BI activity eluted as a single peak with a retention time of 6.6 min. No LH-BA activity was detected in this material. The low molecular weight serum precursor fraction having FSH-BI activity can be considered to be the H1P-5 hollow fiber passing fraction, since it presumably represents serum components having a molecular weight of 5000 or less. Lyophilized preparations of that fraction inhibit binding of ^{125}I-hFSH to receptor by 50% at a dose of 6000 μg. The hygroscopic nature of the HPLC purified FSH-BI preparation prevented us from obtaining an accurate estimate of potency based on mass. However, based upon amino acid analysis following acid hydrolysis of the inhibitor fraction, it was calculated that 50% inhibition of ^{125}I-hFSH binding was obtained at a dose of 3.5 μg. This represents an approximate 2000-fold purification of the serum FSH-BI over that present in the H1P-5 passing fraction, and, of course, assumes that the FSH binding inhibitor is a peptide.

Many of the results obtained were similar to those seen when comparable tests were performed on partially purified FSH-BI fractions prepared by slightly different protocols (Reichert et al., 1979; Reichert et al., 1981). Treatment of the purified fractions with N-bromosuccinimide for 1 h at 56°C resulted in a significant (60%) loss in its ability to inhibit binding of ^{125}I-hFSH to receptor. Control experiments indicated that the incubation period and temperature had no effect on FSH-BI activity. N-Bromosuccinimide is a reagent that reacts preferentially with tryptophan and to a lesser degree with tyrosine and histidine. Carboxypeptidase Y is an exopeptidase derived from yeast which has proven of considerable value in structural studies because of its broad specificity. Incubation of FSH-BI with immobilized carboxypeptidase resulted in a 75% decrease in its ability to inhibit binding of FSH to its receptor. Aminopeptidase M is an exopeptidase of similar broad specificity. Incubation of FSH-BI with immobilized aminopeptidase M also caused a significant decrease in its ability to inhibit binding of ^{125}I-hFSH to receptors. One caveat to be noted in interpreting these experiments is that the exopeptidases utilized were coupled covalently to Sepharose. It is conceivable that the loss of FSH-BI activity in the inhibitor solutions could be due to its binding to the Sepharose matrix rather than to hydrolysis of a putative peptide by the attached exopeptidase. This is unlikely, however, in view of the failure of the binding inhibitor to bind to concanavalin A bound to Sepharose (vida infra). Alternatively, it should be recognized that 'immobilized' enzymes may detach ('leak') from the insoluble matrix. If this had occurred then it is possible that detection of residual FSH-BI activity could be due to an effect of the exopeptidases on the ^{125}I-hFSH, rather than on the inhibitor, decreasing binding of the radioligand to the receptor and giving the appearance of binding inhibition.

Extraction of the less purified FSH-BI fraction G-10-1 with diethyl ether did not result in a loss of FSH-BI activity, which was taken to reflect a non-lipid composition. This was confirmed following extraction of the highly purified FSH-BI fraction with chloroform-methanol, which also did not result in a loss of FSH-BI activity. Thus, in addition to not being a simple lipid or steroid, it seemed unlikely that FSH-BI activity was due to a phospholipid component. It was subsequently noted that high doses (one mg) of phosphatidylethanolamine or phosphatidylserine had no effect on the binding of FSH to the receptor. Absorption of FSH-BI solutions with charcoal did not decrease FSH-BI activity, again arguing against it being nucleotide or steroid in nature.

The highly purified FSH-BI fraction derived from HPLC was examined by thin layer chromatography on cellulose coated plates (Brinkman Instruments, Inc., Westbury, N.Y.) in a 7:7:2:4 mixture of N-butanol-acetone-acetic acid-water. The dried plates were scanned for qualitative detection of various classes of compounds. We have previously observed (Abou-Issa and Reichert, 1976) that in our particular systems, certain nucleotides at concentration of 2 mM would inhibit binding of ^{125}I-hFSH to testes receptor. However, no visualization (fluorescence) of nucleotide was apparent on the thin layer plates under ultra-violet light, a method which has a reported detection time of 2 nanomoles. Staining with ninhydrin was positive, indicating the presence of a peptide or primary amines, and staining with aminohippuric acid was negative, suggesting the absence of simple monosaccharides. The purified HPLC fraction was also studied after paper electrophoresis (pH 6.5 in a 900:100:3 mixture of water-pyridine-acetic acid, at 2.5 Kv for 30 min). Staining with ninhydrin was again positive and with aminohippuric acid it was again negative. Color development of FSH-BI solutions with diphenylamine or orcinol-HCl, reagents used to detect the presence of ribose or deoxyribose and therefore of nucleotides, was also negative. FSH-BI was not adsorbed to the lectin concanavalin A bound to Sepharose, indicating that if carbohydrates were present, they were not of the alpha-D-mannopyranose or alpha-D-glucopyranose types, since these would have been adsorbed by the lectin. Finally, some slight color development was seen upon use of the phenol-sulfuric acid reaction, a rather non-specific response that, nevertheless, is suggestive of a carbohydrate presence.

Information on the physical properties of the FSH-BI is sketchy and rather imprecise, given the uncertainties about the purity of even the HPLC-purified FSH-BI serum fraction. However, FSH-BI activity emerges characteristically from a Sephadex G-10 column, packed and developed in distilled water, slightly retarded with a Ve/Vo ratio of 1.10 corresponding to a molecular weight of about 700. The caveats associated with size and molecular weight estimates by the gel filtration technique, especially of relatively small molecules, are well-known and will not be belabored here. Behavior of the FSH-BI during reversed phase column chromatography as well as upon anion and cation exchange chromatography (see previous section) suggest it to be a polar, acidic molecule. Assuming the FSH-BI is a peptide, the latter property would be consistent with the amino acid composition of the HPLC purified fraction, which indicated a predominance of polar and acidic amino acids.

Although much of what has been observed is not inconsistent with the notion that the serum FSH-BI may be peptide, sufficient uncertainties exist to preclude a definitive statement at this time. As always, an unequivocal answer to this central question awaits the availability of rigorously defined homogeneous preparations of the inhibitor.

Studies on the Follicle Stimulating Hormone Binding Inhibitor Present in Bovine Follicular Fluid

Previous studies from our laboratory demonstrated the presence of a factor in bovine follicular fluid which is an inhibitor of FSH binding to bovine granulosa cells (Darga and Reichert, 1978; Darga and Reichert, 1979). Initially, the possibility was considered that binding inhibition might be due to the presence of endogenous FSH in follicular fluid, possibly derived from the serum. However, the cumulative evidence obtained to date indicates rather clearly that all of the FSH-BI activity present in follicular fluid cannot be explained on this basis. It has not been possible to completely abolish FSH-BI activity in follicular fluid by exhaustive dialysis, so that the likelihood that at least a portion of the FSH-BI activity is due to pituitary FSH cannot be completely excluded. Gel filtration studies utilizing Sephadex G-10 indicated FSH-BI activity to be associated primarily with a low molecular weight component, circa 700. Inhibitor activity was not decreased following adsorption with charcoal or extraction with organic solvents, suggesting the active principle was probably not lipid, nucleotide or steroid in nature. FSH-BI activity per ml of follicular fluid was approximately 2-fold greater in fluid from medium sized (3-6 mm) bovine follicles compared to fluid from small (less than 3 mm) follicles, although there was no significant difference in FSH-BI content per ml in fluid from large (greater than 6 mm) follicles compared to fluid from medium sized follicles (Darga and Reichert, 1978). Since we did not have sufficient follicular fluid available to undertake a concerted attack on purification, isolation and chemical identification of the follicular fluid FSH-BI, we decided to characterize the effect of bovine follicular fluid subfractions on ^{125}I-hFSH binding to granulosa cell membranes as well as on postbinding events such as FSH stimulated adenylate cyclase activity or cAMP production by isolated rat granulosa cells.

The starting material for our studies was lyophilized whole bovine follicular fluid, kindly donated by Dr. Jock Findlay, Prince Henry Hospital, Melbourne. The follicular fluid was subfractionated by simple membrane filtration procedures similar to those employed for the processing of human serum. Prior to fractionation, the reconstituted bovine follicular fluid was adsorbed with charcoal to remove interfering steroid, nucleotide or lipid contaminants. Four major fractions were prepared, a charcoal absorbed follicular fluid (FF) fraction used as a control (FF-c); the component of FF that was retained by an Amicon hollow fiber H1P-5 membrane (FF-H1P5-r); a component that passed the H1P-5 membrane but was retained by an Amicon UM-2 membrane (FF-UM2-r) and a component that passed the Amicon UM-2 membrane but was retained by an Amicon UM-05 membrane (FF-UM05-r). The fraction passing the Amicon UM-05 fraction was not studied further in this series of experiments although some of its properties have been reported elsewhere (Darga and Reichert, 1979). Based on information supplied by the manufacturer (Amicon Corporation, Lexington, MA) the molecular weight range for the various subfractions were: for the FF-H1P-r, greater than 5000; for the FF-UM2-r, between 1000 and 5000; and for the FF-UM05-r, between 500 and 1000.

Binding inhibition was assessed on the basis of the ability of the FF subfractions to inhibit binding of ^{125}I-hFSH to its receptors in homogenates of PMSG-primed immature rat ovaries (Lee and Takahashi,

1977). Each of the FF subfractions inhibited ^{125}I-hFSH binding to a significant degree with apparently comparable potency (50% binding inhibition at 1 mg/tube). However, a rigorous assessment of FSH-BI activity was not possible due to limitations on the amount of subfraction available for study. It is possible that significant differences in FSH-BI potency exist amongst the fractions tested and would not be detected when testing is done at only a single point.

The effect of the FF subfractions on FSH stimulated production of cAMP by rat granulosa cells was also examined. Cyclic AMP production was measured in short term whole cell culture media by radioimmunoassay, essentially as described by Steiner et al. (1972) and Jordan et al. (1978). However, control experiments indicated that follicular fluid interfered with the assay, giving spuriously high cAMP values. This difficultly was overcome by addition to the follicular fluid samples to be tested, of an equal volume of 10% trichloroacetic acid (TCA). The TCA was then removed from the supernatant by extraction with diethyl ether. The final aqueous phase was lyophilized and reconstituted with water either to the original sample volume or to a lesser volume in order to concentrate the cAMP and facilitate its detection by radioimmunoassay.

Ovine FSH (NIH-FSH-S9) stimulated cAMP accumulation in media from short term cultures of granulosa cells taken from PMSG-primed immature rat ovaries in a dose-related fashion, over a linear response range of from 10 to 100 ng per tube. At a concentration of 10 ng/tube, ovine FSH stimulated cAMP production in 4×10^6 cells approximately 3-fold over control values. We next studied the ability of the follicular fluid subfractions to affect cAMP produced in response to 10 ng of ovine FSH. Each of the subfractions was tested for its ability to inhibit stimulation of cAMP production in 4×10^6 cells by 10 ng of ovine FSH. At a dose of one mg/tube, fraction FF-H1P5-r had no effect on FSH stimulated cAMP production, but fractions FF-UM2-r and FF-UM05-r inhibited cAMP production by approximately 50%. This same concentration of fractions FF-UM2-r and FF-UM05-r significantly inhibited (by about 50%) binding of ^{125}I-hFSH to receptors in homogenates of rat ovaries. Inhibition of cAMP production by the FF subfractions could have been produced either via a direct inhibitory effect on adenylate cyclase, by activation of phosphodiesterase activity (in the absence of methyl isobutyl xanthine) or, of course, by inhibition of FSH binding to its receptor. In order to differentiate between the former two possibilities, the effect of the FF subfractions on FSH stimulated adenylate cyclase activity in homogenates of granulosa cell membranes was examined. The assay method employed was based on the ability of adenylate cyclase to stimulate conversion of [α ^{32}P]ATP to [^{32}P]cAMP (Birnbaumer et al., 1976).

Fractions FF-UM2-r and UM05-r at a dose of 2 mg/tube strongly (greater than 80%) inhibited FSH (40 to 160 ng/tube) stimulated adenylate cyclase activity. In contrast, fraction FF-H1P5-r had no inhibitor effect on the system tested at the same concentration. The experimental data were analyzed by fitting to the Michaelis-Menton equation using the method described by Hanson et al. (1967). Stimulation of granulosa cell adenylate cyclase activity by ovine FSH indicated an apparent Km = 0.98 μg NIH-FSH-S9/ml, with a Vmax = 190 pMol cAMP/min/mg. Inhibition of the FSH fraction by FF-UM2-r and FF-UM05-r was non-competitive with respect to FSH concentration. The apparent Ki for FF-UM2-r was 0.25 mg/tube and for FF-UM05-r, 0.15 mg/tube. By comparison, the Ki for fraction FF-H1P5-r was 29.9 mg/tube and for charcoal absorbed follicular fluid, 2.9 mg/tube. Further, the inhibitory effects were primarily on NaF stimulatable, rather than basal adenylate cyclase, activity. Thus, the inhibitory effect would presumably be expressed at the catalytic site of the enzyme in addition to an effect at the level of the receptor.

It would appear, therefore, that follicular fluid, in addition to containing low (less than 5000) molecular weight inhibitors of FSH binding to granulosa cells (as well as to testes membrane receptors), also contains low molecular weight components capable of interfering with FSH action at the level of the catalytic (fluoride stimulated) site of adenylate cyclase. Whether this effect is due to single or multiple components is uncertain, since membrane filtration characteristics are affected by many parameters, as referred to previously. It is possible that under appropriate conditions, the UM-2 retained FF fraction would pass the membrane and either be retained by the UM-05 membranes or possibly, pass it as well. In earlier studies, for example, we have shown that FF subfractions passing the UM-05 membrane on ultrafiltration and desalted by gel filtration through Sephadex G-10, will also strongly inhibit FSH binding to bovine granulosa cells, although its effect on cAMP production and adenylate cyclase activity was not studied (Darga and Reichert, 1979). As is usually the situation, final answers to these perplexing problems await purification and chemical identification of the factors involved. Nevertheless, preliminary studies with partially purified precursor fractions are essential in order to allow identification of the various activities that may be present. To proceed directly to purification without intervening studies allows the risk that interesting substances with unique properties may be missed or discarded in the quest for the holy grail.

Conclusions

The nature and significance of serum and follicular fluid fractions that regulate binding of gonadotropins to gonadal receptors and perhaps influence post binding events, remains perplexing. There seems no question that such factors exist, but it has as yet not been possible to demonstrate in vivo effects, possibly due to our incomplete understanding of their biologic and pharmacologic properties as well as due to lack of sufficient quantities of purified material for such studies. As discussed earlier, there are striking similarities between binding regulators from different sources, but there also exist unexplained differences. Although it is tempting and appealing to conclude that certain of these "factors" are peptides, the experimental evidence is as yet far from compelling, mostly due to uncertainties about the purity of the factor being utilized. The term "factors" is used purposefully to reflect this uncertainty. The easy explanation that the presence of FSH-BI activity in a variety of fractions can be explained by problems associated with imperfect size separation (as by gel filtration or membrane sieving) may be an oversimplification. Our results suggest that there may exist a family of binding regulators in serum as well as in follicular fluid, seminal plasma and tissue extracts. We have already demonstrated the presence and separation of FSH binding inhibitors from LH binding augmentors in subfractions of human serum. The appearance of FSH binding stimulatory activity and LH binding inhibitory activity have been observed by us in serum, follicular fluid and testes extracts repeatedly but inconsistently. A mundane but immensely significant practical problem is one of logistics, the need to have available a ready source of large volumes of serum and follicular fluid to allow preparation of quantities sufficient for the performance of definitive biologic and chemical studies. The use of farm animals such as the bovine may solve this difficulty with regard to serum. The problem of acquiring adequate quantities of follicular fluid may be more troublesome, although recent efforts by the Contraceptive Development Branch of the United State National Institutes of Health to make follicular fluid available for basic studies have been of great help. Certainly, further studies of these gonadotropin binding regulators is warranted given their immense potential as tools for understanding and, possibly, for regulation of reproduction.

References

Abou-Issa, H., and Reichert, L. E., Jr., 1976, Properties of the follitropin receptor interaction: Characterization of the interaction of follitropin with receptors in purified membranes isolated from mature rat testes tubules, J. Biol. Chem., 251:3326.

Birnbaumer, L., Yang, P., Hunzicker-Dunn, M., Bockaert, J., and Duran, J. M., 1976, Adenylyl cyclase activities in ovarian tissues. I. Homogenization and conditions of assay in graafian follicles and corpora lutea of rabbits, rats, and pigs: Regulation by ATP, and Some Comparative Properties, Endocrinology, 99:163.

Darga, N. C., and Reichert, L. E., Jr., 1978, Some properties of the interaction of follicle stimulating hormone with bovine granulosa cells and its inhibition by follicular fluid, Biol. Reprod., 19:235.

Darga, N. C.,and Reichert, L. E., Jr., 1979, Evidence for the presence of a low molecular weight binding inhibitor in bovine follicular fluid, in: "Ovarian, Follicular, and Corpus Luteum Function," C. P. Channing, J. M. Marsh, and W. A. Sadler, eds., pp. 338-342, Plenum Press, New York.

Dias, J. A., Treble, D. H. Bennet, A. H., and Reichert, L. E., Jr., 1981, Follitropin receptor binding inhibitors in human seminal plasma, Andrology, 5:259.

Hanson, K. R., Ling, R., and Hovir, E., 1967, A computer program for fitting data to the Michaelis-Menton equation, Biochem. Biophys. Res. Commun., 29:194.

Jordan, A. W., Caffrey, J. L., and Niswender, G. D., 1978, Catecholamine-induced stimulation of progesterone and adenosine 3',5'-monophosphate production by dispersed ovine luteal cells, Endocrinology, 103:385.

Lee, C. Y., and Takahashi, H., 1977, Follicle stimulating hormone receptors in rat ovaries; Decrease in numbers of binding sites associated with luteinization, Endocrinology, 101:869.

Reichert, L. E., Jr., 1978, Some properties of small molecular weight tissue and serum inhibitors of follitropin binding to receptor, in: "Novel Aspects of Reproductive Physiology," C. H. Spilman and J. W. Wilks, eds., pp. 355-368, Academic Press, New York.

Reichert, L. E., Jr., and Abou-Issa, H., 1977, Studies on a low molecular weight testicular factor which inhibits binding of FSH to receptor, Biol. Reprod., 17:614.

Reichert, L. E., Jr., Sanzo, M. A., and Darga, N. S., 1979, Studies on a low molecular weight follicle stimulating hormone binding inhibitor from human serum, J. Clin. Endocrinol. Metab., 49: 866.

Reichert, L. E., Jr., Sanzo, M. A., and Dias, J. A., 1981, Studies on purification and characterization of gonadotropin binding inhibitors and stimulators from human serum and seminal plasma, in: "Intragonadal Regulators of Reproduction," P. P. Franchimont and C. P. Channing, eds., Academic Press, London, in press.

Steiner, A. L., Parker, C. W., Kijnis, D. M., 1972, Radioimmunoassay for cyclic nucleotides. I. Preparation of antibodies and iodinated cyclic nucleotides, J. Biol. Chem., 247:1106.

OOCYTE MATURATION INHIBITOR:

A 1981 PERSPECTIVE

A. Tsafriri and S. Bar-Ami

Department of Hormone Research
The Weizman Institute of Science
Rehovot, 76100, Israel

INTRODUCTION

Meiosis in Mammals

Meiotic division of mammalian oocytes is a protracted process. It begins during embryonic life and is arrested at the diplotene (dictyate) stage of the prophase of first division, just about the time of birth ("first meiotic arrest"). The oocyte, now referred to as "dictyate oocyte" normally resumes meiosis only in large preovulatory Graafian follicles, following the surge of gonadotropins. The meiotic process is arrested again ("second meiotic arrest") at metaphase of the second meiotic division to be reinitiated after interacting with the fertilizing sperm. Though that preovulatory resumption of meiosis includes breakdown of the germinal vesicle (GVB), expulsion of the first polar body and progress to the metaphase of the second meiotic division, I will refer to the first change, namely GVB as "resumption of meiosis" or "maturation" or in short "oocyte maturation."

Mechanism Involved in the Control of Meiosis

Evidence has been presented that initiation of meiosis during embryonic life is triggered by a factor produced by the rete ovarii (Byskov, 1974; O and Baker, 1976). The first meiotic arrest coincides with enclosure of oocytes by granulosa cells and follicle formation (reviewed by Gondos, 1978). It has been suggested that oocytes which do not become enclosed by granulosa cells during this period of follicle formation may proceed beyond diplotene stage to abortive diakinesis and degeneration (Ohno and Smith, 1964). However, whether there is causal relationship between folliculogenesis and meiotic arrest remains obscure. It has been demonstrated that the ability of oocytes to resume meiotic maturation spontaneously in vitro is acquired only at age (days post partum) 15 in mice, (Szybek, 1972; Sorensen and Wassarman, 1976), 23 in hamsters (Iwamatsu and Yanagimachi, 1975) and 26 in rats (Bar-Ami and Tsafrir, submitted). It appears, therefore, that the first meiotic arrest is probably due to an inherent inability of the small oocytes to proceed beyond prophase of the first meiotic division. The biochemical basis for this inability of oocytes to proceed with the meiotic process is not understood as yet. The pattern of protein synthesis in oocytes capable of resuming maturation in culture is remarkably similar to that of oocytes that fail to mature in vitro (Wassarman et al., 1979).

Preovulatory resumption of meiosis is triggered in vivo by the surge of LH (Ayalon et al., 1972; Vermeiden and Zeilmaker, 1974; Tsafriri et al., 1976a). By contrast, Pincus and Enzmann (1935) demonstrated the spontaneous maturation of oocytes explanted into hormone-free medium. Hence they suggested that the resumption of meiosis in fully grown oocytes is prevented by the follicular environment. Here I review the evidence for follicular control of meiotic maturation by a folloicular product, oocyte maturation inhibitor (OMI) and the studies on its characterization and purification.

EVIDENCE FOR FOLLICULAR CONTROL OF OOCYTE MATURATION

The pioneering observation of Pincus and Enzmann (1935) that rabbit oocytes dislodged from their follicles undergo maturation in culture even in hormone-free media, has been confirmed and was extended to many other mammalian species (Biggers, 1972; Donahue, 1972; Schuetz, 1976). By contrast, oocytes explanted within their follicle, prior to the preovulatory surge of gonadotropins into hormone-free medium, remain indefinitely in the dictyate state (reviewed by Tsafriri, 1978; Baker, 1979). In order to test which follicular compartment is responsible for maintaining meiotic arrest, porcine oocytes were cultured within hemisections of follicular wall. Whereas oocytes cultured within theca (Foote and Thibault, 1969; Tsafriri and Channing, 1975) or ovarian bursa (Leibfried and First, 1980) resumed meiosis, hemisections of intact follicular wall exerted a significant inhibition of GVB. That this inhibitory effect is exerted by granulosa cells has been demonstrated by co-culturing oocytes with granulosa cells (Tsafriri and Channing, 1975; Sato and Ishibashi, 1977). Furthermore, this effect of granulosa cells was dose-dependent and cells from small follicles were more potent than cells from medium or large ones. Similarly, when rat oocytes were added to rat granulosa cells previously cultured for 24 hours, spontaneous resumption of meiosis was suppressed, and the degree of inhibition depended on the num-

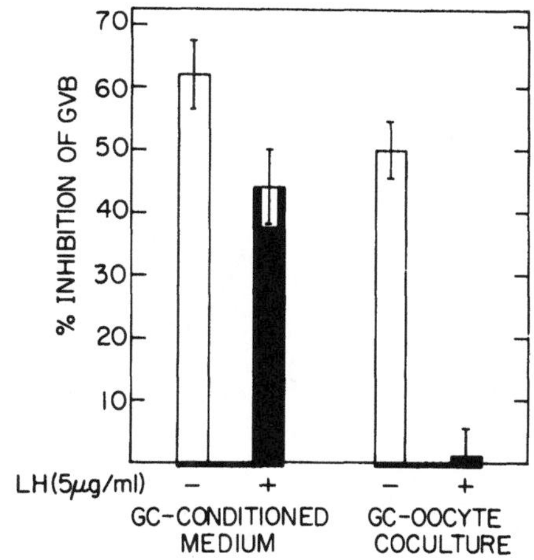

Fig. 1. Inhibition of the resumption of meiotic maturation by co-culture of oocytes with granulosa cells or by culture in a granulosa cell-conditioned medium. Rat oocytes were added either to dishes in which granulosa cells (GC; 10^6/0.2 ml) had been cultured previously for 24 h (GC – oocyte co-culture), or to medium obtained from culture of GC (10^6/ml) for 48 h (GC-conditioned medium). In the latter situation the medium was withdrawn from the culture dish, sterilized by membrane filtration and diluted (1:1) with fresh medium. The oocytes were cultured for 6 h with (black columns) or without (white columns) LH (5 μg/ml).

ber of granulosa cells in the culture (Tsafriri, 1979). This inhibitory effect of granulosa cells was reversed by the addition of LH (5 μg/ml) to the cultures together with the oocytes (Fig. 1).

OOCYTE MATURATION INHIBITOR (OMI)

Meiotic maturation was also inhibited by either porcine granulosa cell extract (Tsafriri et al., 1976b, c; Sato and Ishibashi, 1977) or a "conditioned" medium in which granulosa cells had been cultured previously (Fig. 1; Tsafriri, 1979). Follicular fluid from ovaries of rabbit, ovine, bovine, porcine, hamster and human origin exerted an inhibitory effect on spontaneous maturation of isolated oocytes (Chang, 1955; Tsafriri and Channing, 1975; Gwatkin and Andersen, 1976; Jagiello et al., 1977; Tsafriri et al., 1977; Hillensjö et al., 1978a). This inhibitory effect is not species-specific; thus human FFl inhibitied the maturation of porcine (Hillensjö et al., 1978a) and rat (Hillensjö, personal communication) oocytes, bovine FFl inhibited hamster oocytes (Gwatkin and Andersen, 1976), and porcine FFl inhibited the maturation of mouse

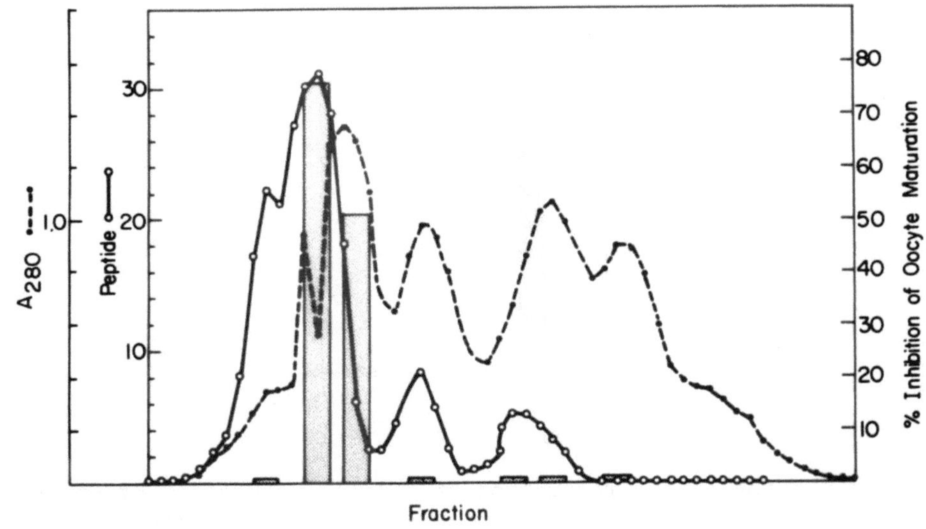

Fig. 2. Purification of OMI by Sephadex G-10 chromatography. Methanol extract of the LMW fraction of pFF1 (320 mg peptide) was applied to a column (1.5 × 65 cm) of Sephadex G-10 and eluted with 0.5 M acetic acid.

and rat oocytes (Tsafriri et al., 1977). The OMI activity in porcine FF1 declines in the course of follicular growth as demonstrated by assaying fresh FF1 collected from small, medium and large follicules (Stone et al., 1978) or fluid collected from pigs on specified days of their reproductive cycle (Van de Wiel, Bar-Ami, and Tsafriri, unpublished observations).

OMI is not extracted from pFF1 by ether or charcoal; it is heat stable (heating of pFF1 at 60°C for 20 min did not destroy OMI activity) and its activity was abolished by treatment of pFF1 by trypsin (Tsafriri et al., 1975, 1976b, c). These results together with evidence that OMI activity is eluted by gel-filtration in fractions corresponding to molecular weight of less than 2000 dalton lead to the conclusion that OMI is probably a peptide (Tsafriri et al., 1976b; Stone et al., 1978; Pomerantz et al., 1979). By consecutive use of Amicon PM-10 membrane filtration, Sephadex G-25 and CM-Sephadex chromatography, an estimated 5000-fold purification of OMI was achieved based on the fluorescamine method described by Udenfriend et al. (1972). We have compared two extraction procedures for the filtrate obtained after filtration of pFF1 in Amicon PM-10 membranes. Whereas both methanol and 2N acetic acid-methanol resulted in the extraction of similar amounts of OMI activity, the acetic methanol extracted 30% more contaminants on the basis of dry weight. Therefore it appears that methanol extraction is preferable for further purification.

Table 1. Effect of Antibody (a) Prepared against the Low Molecular Weight Fraction of Porcine Follicular Fluid (LMW) on Reversing the Action of OMI

	% Inhibition of GVB (Mean ± SEM)
LMW	41.9 ± 7
LMW + a LMW	9.5 ± 11
LMW + purified a LMW	10.8 ± 8
LMW + LMW adsorbed serum	70.3 ± 12

The methanol extract of the PM-10 filtrate was applied to a column of Sephadex G-10 and eluted with 0.5 M acetic acid. The elution pattern obtained is depicted in Fig. 2. One peak of OMI activity was obtained and it did not correspond exactly with the peptide peak (as measured by fluorescamine assay) nor with absorbance at 280. Sephadex G-10 appears to be somewhat advantageous as compared to G-25 in giving one peak of activity, while offering similar purification. Furthermore, this procedure supports the notation that OMI may be of smaller MW than we originally thought.

PHYSIOLOGICAL CONSIDERATIONS OF OMI ACTION

Reversibility of OMI

The OMI activity of pFFl and its partially purified fractions can be reversed by transferring the oocytes after 24 h-culture into a medium without OMI (Stone et al., 1978; Hillensjö et al., 1979). An antiserum was prepared by immunization of male rabbits (Tsafriri et al., 1979) or rats with the low molecular weight (LMW) fraction of pFFl conjugated to BSA with soluble carbodiimide. Addition of the antiserum to rat oocyte cultures prevented the inhibitory action of OMI. Furthermore, the LMW fraction of pFFl was conjugated to rabbit serum albumin (RSA) and the conjugate to sepharose. The antibody was passed through the LMW-RSA-sepharose column and absorbed to the column bed. The effluent (LMW-adsorbed serum) did not prevent the OMI activity. The purified antibody (a LMW) was eluted with $MgCl_2$ (5 M), and this material was effective in neutralizing the action of OMI (Table 1). Thus it appears that the activity of pFFl and its purified fraction is not due to an unspecific action on the cumulus-oocyte complex.

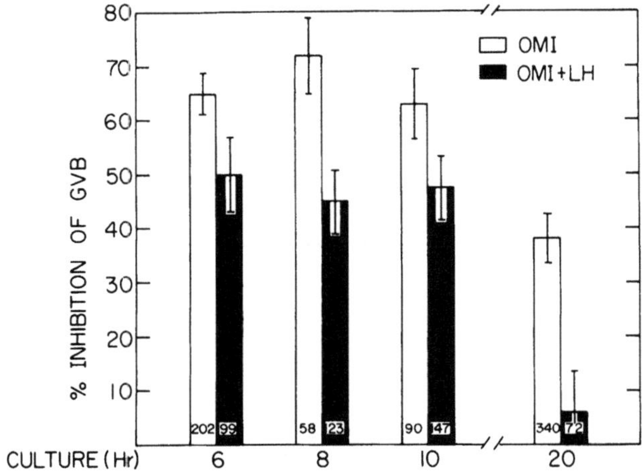

Fig. 3. Effects of duration of culture and LH on GVB of rat oocytes. Oocytes were cultured with LMW fraction of pFFl (0.170 mg protein/ml) with (black columns) or without 5 μg/ml LH (white columns).

Action of LH

Further support for the physiological role of OMI is derived from the finding that addition of LH to cultures of porcine oocytes in the presence of pFFl or its purified fractions overcomes the inhibition of meiosis (Tsafriri et al., 1975). Similarly the action of bFFl or hamster FFl on maturation of hamster oocytes was overcome by addition of LH (Gwatkin and Andersen, 1976).

The inhibitory effect of pFFl, the LMW fraction of pFFl and of granulosa cell-conditioned medium were reversed by the addition of LH. However, the maturational changes were delayed even in the presence of LH; more than 10 h of incubation were required before GVB comparable to control cultures occurred in the presence of the LMW fraction of pFFl (Fig. 3). Similarly, addition of LH to oocytes cultured in granulosa cell-conditioned medium failed to induce GVB in the majority of oocytes within 6 h culture-period, whereas LH was more effective in co-cultures of granulosa cells with oocytes (Fig. 1). Our previous studies demonstrated that GVB was completed within 2 h in isolated rat oocytes cultured without the inhibitor, a culture period of 4 h was adequate for this change to occur in follicle-enclosed oocytes exposed to LH (Tsafriri, 1979).

Recently, specific LH/hCG receptors were demonstrated also on cumulus cells, although their number was much lower on cumulus cells than on mural granulosa cells (Amsterdam and Tsafriri, 1979; Channing et al., 1979; Lawrence et al., 1980). The presence of LH re-

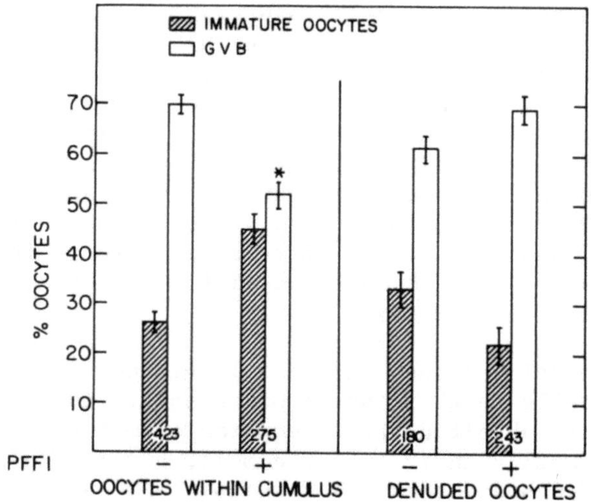

Fig. 4. The role of cumulus cells in OMI action on the spontaneous maturation of rat oocytes. The oocytes were cultured within their cumulus or after removal of adherent cells by repeated transfers through thin bore glass capillaries (denuded oocytes). Culture was for 6 h in the presence (+) or absence (—) of the LMW fraction of pFF1 (PFF1). The asterisk indicates a significant reduction ($p < 0.005$) relative to oocyte cultured without pFF1.

ceptors on cumulus cells was further supported by their response to gonadotropins involving enhanced steroidogenesis (Nicosia and Mikhail, 1975; Hillensjö and Channing, 1980) and expansion (Hillensjö et al., 1976; Dekel and Kraicer, 1978; Hillensjö and Channing, 1980). By contrast, no specific binding of LH/hCG on the oocyte has been observed. It appears, therefore, that the effect of LH on oocytes maturation is exerted through the granulosa and cumulus cells. The delay in GVB caused by OMI even in the presence of LH may be due to the inability of the cumulus cells included in the culture to act as substitutes for the granulosa cells. This inability might be due to the low number of cumulus cells included in the culture or to the reduced ability of cumulus cells, compared to mural granulosa cells, to bind LH and hence to respond to it.

Role of Cumulus-Oocyte Communication

Gap junctions have been described between adjacent granulosa cells and also between the cytoplasmic extension of the corona radiata cells that transverse the zona pellucida and the oolemma (Amsterdam et al., 1976; Anderson and Albertini, 1976). Such junctional association between cells may facilitate transfer of small molecules between neighboring cells and bring about their metabolic coupling (Azarnia et al., 1974; Gilula et al., 1972; Sheridan, 1971). Bidirectional electrical coupling between cumulus cells, as well as transfer of injected fluorescein dye from oocyte to cumulus cells was demonstrated (Gilula et al., 1978). This ionic coupling was maximal prior to hCG stimulation and it decreased as ovulation approached. By using ^{3}H-labeled choline, uridine and inositol, Moor et al. (1980) measured cumulus-oocyte coupling in the sheep. They found that gonadotropins administered in vivo or in vitro reduced the coupling between these two cell types within 12-15 h, but did not eliminate it completely.

Studies with OMI provided additional evidence for the essential role of cumulus cells in the control of meiosis. The LMW fraction obtained from pFFl inhibited the resumption of meiosis only in oocytes cultured within their intact cumuli, but it did not affect the maturation of denuded oocytes of the pig (Hillensjö et al., 1979) or rat (Fig. 4).

The close association between cumulus cells and the oocyte resulting from gap junctions may allow the propagation of the meiosis-inducing stimulus as well as the inhibitory effect of OMI from mural granulosa and cumulus cells to the oocyte. The finding that cumulus-oocyte junctions are disrupted following the ovulatory stimulus led to the hypothesis that release of the oocyte from inhibition of meiosis may result from dismantling or functional disengagement of the oocyte/cumulus cell junctions, possibly induced by the ovulatory hormone (Lindner et al., 1974; Gilula et al., 1978; Moor et al., 1980). Thus, LH may induce meiotic maturation secondarily, following functional seqestration of the oocyte. However, the main difficultly in this hypothesis is in the temporal association of the meiotic changes and disruption of oocyte/cumulus junctions. The morphological dissociation of cumulus/oocyte junctions appears to follow rather than precede GVB in rat (Dekel and Kraicer, 1978) and rabbit oocytes (Szöllösi et al., 1978). Similarly, the reduction of intercellular transport of choline from sheep cumulus cells to the oocyte occurred only after that meiotic maturation had progressed to prometaphase or even first metaphase (Moor et al., 1980). Furthermore, termination of cumulus/oocyte ionic coupling in vivo (Gilula et al., 1978) and in vitro (Dekel et al., 1981) follows GVB. Thus, whereas GVB in explanted follicle-enclosed oocytes was completed in 90% of the oocytes within 4 h after LH addition all the complexes were still coupled by that time; uncoupling to the extent of 50% occurred after 9 h and only after 20 h of culture were 100% of complexes uncoupled (Dekel et al., 1981). Since in this study the absolute lack of channels through which small ions may pass was assayed, it is still possible that the flow of substances larger than the ions measured may be blocked before detectable breakdown of ionic communication. Thus a substantial

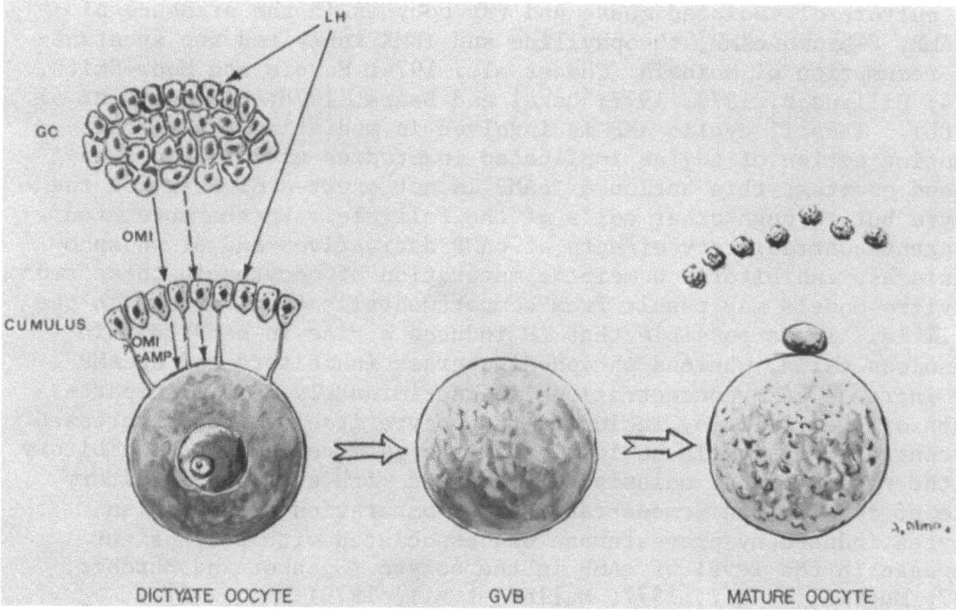

Fig. 5. Schematic representation of the control of meiotic maturation. See text for explanation.

reduction in the transmission of larger molecules (dyes) occurred in the presence of ionic communication in mouse perimplantation embryos (Lo and Gilula, 1979). While it appears that junctional association between the oocyte and cumulus cells is essential for maintaining meiotic arrest, evidence is lacking to substantiate the hypothesis that the meiosis-inducing action of LH is expressed solely by breakdown of oocyte/cumulus communication.

Role of Cyclic AMP

The involvement of cAMP in the regulation of meiosis in mammals is much debated, but not adequately understood. The feature common to most of the agents that induce maturation of follicle-enclosed oocytes in vitro is their ability to stimulate the production of cAMP (and hence to activate protein kinase) in the follicle (Tsafriri et al., 1972; Lindner et al., 1974). Injection of dbcAMP into the follicular antrum (Tsafriri et al., 1972) or short-term exposure of follicles to a medium containing 8-bromo-cAMP Hillensjö et al., 1978b) or dbcAMP or IBMX (Dekel et al., 1981) triggered GVB. It should be emphasized though, that continuous presence of phosphodiesterase inhibitors or cAMP derivatives in the medium prevented the LH-induced maturation of follicle-enclosed oocytes (Lindner et al., 1974; Hillensjö et al., 1978b). Similarly, the culture of isolated mouse and rat oocytes in the presence of dbcAMP, 8-bromo-cAMP, theophylline and IBMX inhibited the spontaneous resumption of meiosis (Cho et al., 1974; Mekola and Moor-Smith, 1974; Hillensjö, 1976, 1977; Dekel and Beers, 1978; Hillensjö et al., 1978b). Thus if cyclic AMP is involved in mediating the meiosis-inducing action of LH, as implicated in studies with follicle-enclosed oocytes, this action of cAMP is not exerted directly on the oocyte but through other cells of the follicle. Furthermore, the apparent contradictory effects of cAMP derivatives and of phosphodiesterase inhibitors on meiotic maturation of oocytes in these two in vitro models may result from compartmentalization of cAMP in the follicle. It is possible that LH induces a rise in cAMP only in granulosa cells, whereas phosphodiesterase inhibitors and dbcAMP may increase cAMP concentrations indiscriminately in all compartments of the follicle, including the oocyte itself. This increased concentration of cyclic nucleotide in the germ cell may be inhibitory to the resumption of meiosis. In concert with such an inhibitory role of cAMP is the demonstration that maturation of amphibian oocytes induced by progesterone was associated with a transient decrease in the level of cAMP in the oocyte (Speaker and Butcher, 1977; Morrill et al., 1977; Maller et al., 1979).

These observations regarding inhibition of meiosis by cAMP and the junctional association between the oocyte and cumulus cells led to the hypothesis that cAMP serves as the physiological inhibitor of meiosis which is transmitted from cumulus cells to the oocyte via gap junctions. According to this hypothesis LH induces the resumption of meiosis by disrupting oocyte/cumulus communication (Lindner et al., 1974; Dekel and Beers, 1978; Moor et al., 1980; Dekel et al., 1981). Dekel et al. (1981) demonstrated that cAMP derivatives and IBMX bring about the interruption of oocyte/cumulus communication. Thus it is possible that the meiosis-inducing effect resulting from brief exposure of follicle-enclosed oocytes to cAMP derivatives and to IBMX is related to their effect on cumulus/oocyte communication. Whereas cAMP may be a physiological inhibitor of meiosis in the oocyte the temporal relationship between GVB and oocyte/cumulus uncoupling, discussed above, is still a weak point of this hypothesis.

CONCLUDING REMARKS

The control of meiotic maturation in mammals appears to be most complicated. The intricate communication system between mural granulosa cumulus oophorus cells and the oocyte seem to be essential in this process. Furthermore, OMI and cAMP, apparently produced by the somatic cells of the follicle, as well as the preovulatory surge of LH all participate in the controlling of the meiotic process. In Fig. 5, theputative mechanisms involved in the control of meiotic maturation are summarized: in antral follicles, meiosis is prevented by OMI produced by granulosa cells. The resulting inhibitory signal is conveyed through cumulus/oocyte junctions. Whether OMI or cAMP serve as the inhibitor transferred into the oocyte remains to be elucidated. LH induces the resumption of meiosis, possibly through the mediation of mural granulosa cells. Whether the meiosis-inducing action of LH is exerted by breakdown of cumulus/oocyte communication, blockage of OMI production, or by the generation of some unspecified positive signal, remains unknowns as yet.

The hypothesis that meiotic maturation is controlled within the follicle, first put forward by Pincus and Enzmann (1935) has been recently supported by experimental evidence. While OMI is clearly involved, it appears to be only one component of the intricate mechanisms controlling the meiotic process. Purification of OMI to homogeneity will allow assessment of the precise physiological role of this factor in the control of meiosis.

ACKNOWLEDGEMENTS

The studies reported here were supported by generous grants from the United-States—Israel Binational Agricultural Research and Development Fund (BARD), by the Ford Foundation and the Rockefeller Foundation. The generous supply of part of the pFFl by the Center for Population Research, NICHD, NIH, and of LH by NIAMDD, is gratefully acknowledged. We thank Mrs. R. Slager and Mrs. A. Tsafriri for skilful technical assistance, Dr. D. J. Dierschke and Dr. N. Dekel for helpful discussion, and Mrs. M. Kopelowitz for excellent secretarial assistance.

REFERENCES

Amsterdam, A., Josephs, R., Lieberman, M. E., and Lidner, H. R., 1976, Organization of intramembrane particles in freeze-cleaved gap junctions of rat Graafian follicles: optical-diffraction analysis, J. Cell. Sci., 21:93.

Amsterdam, A. and Tsafriri, A., 1979, In vitro binding of ^{125}I-human chorionic gonadotrophin in the preovulatory follicle: absence of receptor sites on oocyte, J. Cell Biol., 83:255A.

Anderson, A. and Albertini, D. F., 1976, Gap junction between the oocyte and companion follicle cells in the mammalian ovary, J. Cell Biol., 71:680.

Ayalon, D., Tsafriri, A., Lidner, H. R., Cordova, T., and Harell, A., 1972, Serum gonadotrophin levels in pro-ooestrous rats in relation to the resumption of meiosis by the oocytes, J. Reprod. Fert., 31:51.

Azarnia, R., Larsen, W. J., and Loewenstein, W. R., 1974, The membrane junctions in communicating and noncommunicating cells, their hybrids and segregants, Proc. Nat. Acad. Sci. U.S.A., 71:880.

Baker, T. G., 1979, The control of oogenesis in mammals, in: "Ovarian Follicular Development and Function," A. R. Midgley and W. A. Sadler, eds., Raven Press, New York

Biggers, J. D., 1972, Metabolism of the oocyte, in: "Oogenesis," J. D. Biggers and A. W. Schuetz, eds., pp. 241-251, University Park Press, Baltimore.

Byskov, A. G., 1974, Does the rete ovarii act as a trigger for the onset of meiosis? Nature, 252:396.

Chang, M. C., 1955, The maturation of rabbit oocytes in culture and their maturation, activation, fertilization, and subsequent development in Fallopian tube, J. Exp. Zool., 128:378.

Channing, C. P., Stone, S. L., Fowler, S., Edelson, S., and Schwartz-Kripner, A., 1979, Comparison of porcine granulosa and cumulus cell properties: LH/hCG receptors, ability to secrete progesterone and ability to respond to LH, Biol. Reprod., (Suppl.), 20:74A (abstract 125).

Cho, W. K., Stern, S., and Biggers, J. D., 1974, Inhibitory effect of dibutytyl c-AMP on mouse oocyte maturation in vitro, J. Exp. Zool., 187:383.

Dekel, N. and Beers, W. H., 1978, Rat oocyte maturation in vitro: Relief of cyclic AMP inhibition by gonadotropins, Proc. Natl. Acad. Sci., 75:4369.

Dekel, N. and Kraicer, P. F., 1978, Induction in vitro of mucification of rat cumulus oophorus by gonadotropins and adenosine 3',5'-monophosphate, Endocrinology, 102:1797.

Dekel, N., Lawrence, T. S., Gilula, N. B., and Beers, W. H., 1981, Modulation of cell-to-cell communication in the cumulus-oocyte complex and the regulation of oocyte maturatin by LH, Develop. Biol., in press.

Donahue, R. P., 1972, The relation of oocyte maturation to ovulation in mammals, in: "Oogenesis," J. D. Biggers and A. W. Schuetz, eds., Chap. 22, pp. 413-438, University Park Press, Baltimore.

Foote, W. D. and Thibault, C., 1969, Recherches experimentales sur la maturation in vitro des ovocytes de truie et de veau, Annls. Biol. anim. Biochem. Biophys., 9:329.

Gilula, N. B., Reeves, O. R., and Stienback, A., 1972, Metabolic coupling, ionic coupling and cell contacts, Nature (Lond.), 325:262.

Gilula, N. B., Epstein, M. L., and Beers, W. H., 1978, Cell-to-cell communication and ovulation; a study of the cumulus-oocyte complex, J. Cell. Biol., 78:58.

Gondos, B., 1978, Oogonia and oocytes in mammals, in: "The Vertebrate Ovary," R. E. Jones, ed., pp. 83-120, Plenum Press, New York.

Gwatkin, R. B. L. and Andersen, O. F., 1976, Hamster oocyte maturation in vitro inhibition by follicular components, Life Sciences, 19:527.

Hillensjö, T., 1976, Oocyte maturation and glycolysis in isolated preovulatory follicles of PMS-injected immature rats, Acta Endocrinol., 82:809.

Hillensjö, T., 1977, Dissociation of preovulatory maturational events in rat oocytes and cumuli in the presence of dibutyryl cAMP, Acta Physiol. Scand., 100:261.

Hillensjö, T., Dekel, N., and Ahrén, K., 1976, Effect of gonadotrophins on the cumulus oophorus of isolated preovulatory rat follicles, Acta Physiol. Scand., 96:558.

Hillensjö, T., Batta, S. K., Schwartz-Kripner, A., Wentz, A. C., Sulewski, J., and Channing, C. P., 1978a, Inhibitory effect of human follicular fluid upon the maturation of porcine oocytes in culture, J. Clin. Endocr. Metab., 47:1332.

Hillensjö, T., Ekholm, C., and Ahrén, K., 1978b, Role of cyclic AMP in oocyte maturation and glycolysis in the preovulatory rat follicle, Acta Endocr., 87:377.

Hillensjö, T., Kripner, A., Pomerantz, S. H., and Channing, C. P., 1979, Action of porcine follicular fluid oocyte maturation inhibitor in vitro: Possible role of cumulus cells, in: "Ovarian Follicular and Corpus Luteum Function," C. P. Channing, J. M. Marsh, and W. J. Sadler, eds., pp. 283-290, Plenum Press, New York.

Hillensjö, T. and Channing, C. P., 1980, Gonadotropin stimulation of steroidogeneis and cellular dispersion in cultured porcine cumuli oophori, Gamete Res., 3:233.

Iwamatsu, T. and Yanagimachi, R., 1975, Maturation in vitro of ovarian oocytes of prepubertal and adult hamsters, J. Reprod. Fert., 45:83.

Jagiello, G., Graffeo, J. Ducayen, M., and Prosser, R., 1977, Further studies of inhibitors of in vitro mammalian oocyte maturation, Fertil. Steril., 28:476.

Lawrence, T. S., Dekel, N., and Beers, W. H., 1980, Binding of human chorionic gonadotropin by rat cumuli oophori and granulosa cells: A comparative study, Endocrinol., 106:1114.

Leibfried, L. and First, N. L., 1980, Follicular control of meiosis in the porcine oocyte, Biol. Reprod., 23:705.

Lindner, H. R., Tsafriri, A., Lieberman, M. E., Zor, U., Koch, Y., Bauminger, S., and Barnea, A., 1974, Gonadotrophin action on cultured Graafian follicles: induction of maturation division of the mammalian oocyte and differentiation of the luteal cell, in: "Recent Progress in Hormone Research, Vol. 30," R. O. Greep, ed., pp. 79-138, Academic Press, New York.

Lo, W. C. and Gilula, N. B., 1979, Gap junctional communication in the perimplantation mouse embryo, Cell, 18:399.

Maller, J. L., Butcher, F. R., and Krebs, E. G., 1979, Early effect of progesterone on levels of cyclic adenosine 3':5'-monophosphate in Xenopus oocytes, J. Biol. Chem., 254:579.

Moor, R. M., Smith, M. W., and Dawson, R. M. C., 1980, Measurement of intercellular coupling between oocytes and cumulus cells using intracellular markers, Exp. Cell Res., 126:15.

Morrill, G. A., Schatz, F., Kostellow, A. B., and Poupko, J. M., 1977, Changes in cyclic AMP levels in the amphibian ovarian follicle following progesterone induction of meiotic maturation. Differentiation, 8:97.

Nekola, M. V. and Moor-Smith, D. M., 1975, Failure of gonadotropins to induce in vitro maturation of mouse oocytes treated with dibutyryl cyclic AMP, J. Exp. Zool., 194:529.

Nicosia, S. V. and Mikhail, G., 1975, Cumuli oophori in tissue culture: hormone production, ultrastructure and morphometry of early luteinization. Fertil. Steril., 26:427.

O., W. S. and Baker, T. G., 1976, Initiation and control of meiosis in hamster gonads in vitro. J. Reprod. Fert., 48:399.

Ohno, S. and Smith, J. B., 1964, Role of fetal follicular cells in meiosis of mammalian oocytes, Cytogenetics, 3:324.

Pincus, G. and Enzmann, E. V., 1935, The comparative behavior of mammalian eggs in vivo and in vitro, J. Exp. Med., 62:655.

Pomerantz, S. H., Channing, C. P., and Tsafriri, A., 1979, Studies on the purification and action of an oocyte maturation inhibitor isolated from porcine follicular fluid, in: "Peptides: Structure and Biological Function," E. Gross and J. Meienhofer, eds., Pierce Chemical Company, Rockford, Il.

Sato, E. and Ishibashi, T., 1977, Meiotic arresting action of the substance obtained from cell surface of porcine ovarian granulosa cells, Jap. J. Zootech. Sci., 48:22.

Schuetz, A. W., 1974, Role of hormones in oocyte maturation, Biol. Reprod., 10:150.

Sheridan, J. D., 1971, Dye movement and low resistance junctions between reaggregated embryonic cells, Devl. Biol., 26:627.

Sorensen, R. A. and Wassarman, P. M., 1976, Relationship between growth and Molecular events during oocyte maturation, in: "Biochemistry of Animal Development, Vol. III," R. Weber, pp. 1-46, Academic Press, San Francisco.

Speaker, M. G. and Butcher, F. R., 1977, Cyclic nucleotide fluctuations during steroid induced meiotic maturation of frog oocytes, Nature, 267:848.

Stone, S. L., Pomerantz, S. H., Schwartz-Kipner, A., and Channing, C. P., 1978, Inhibition of oocyte maturation from porcine follicular fluid: further purification and evidence for reversible action, Biol. Reprod., 19:585.

Szöllösi, D. Gérard, M. Ménézo, Y., and Thibault, C., 1978, Permeability of ovarian follicle: corona cell-oocyte relationship in mammals, Ann. Biol. anim. Bioch. Biophys., 18:511.

Szybek, K., 1972, In vitro maturation of oocytes from sexually immature mice, J. Endocrinol., 54:527.

Tsafriri, A., Lindner, H. R., Zor, U., and Lamprecht, S. A., 1972, In vitro induction of meiotic division in follicle-enclosed rat oocytes by LH, cyclic AMP and prostaglandin E_2, J. Reprod. Fertil., 31:39.

Tsafriri, A. and Channing, C. P., 1975, An inhibitory influence of granulosa cells and follicular fluid upon porcine oocyte meiosis in vitro. Endocrinology, 96:922.

Tsafriri, A., Lieberman, M. E., Koch, Y., Bauminger, S., Chobsieng, P., Zor, U., and Lindner, H. R., 1976a, Capacity of immunollogically purified FSH to stimulate cyclic AMP accumulation and steroidogenesis in Graafian follicles and to induce ovum maturation and ovulation in the rat, Endocrinology, 98:655.

Tsafriri, A., Pomerantz, S. H., and Channing, C. P., 1976b, Follicular control of oocyte maturation, in: "Ovulation in the Human, Vol. 8," P. G. Grosignani and D. R. Mishel, eds., pp. 31-39, Academic Press, New York.

Tsafriri, A., Pomerantz, S. H., and Channing, C. P., 1976c, Inhibition of oocyte maturation by porcine follicular fluid: partial characterization of the inhibitor, Biol. Reprod., 14:511.

Tsafriri, A., Channing, C. P., Pomerantz, S. H., and Lindner, H. R., 1977, Inhibition of maturation of isolated rat oocytes by porcine follicular fluid, J. Endocr., 75:285.

Tsafriri, A., 1978, Oocyte maturation in mammals, in: "The Vertebrate Ovary," R. E. Jones, ed., pp. 409-442, Plenum Press, New York.

Tsafriri, A., 1979, Mammalian oocyte maturation: Model systems and their physiological relevance, in: "Ovarian Follicular and Corpus Luteum Function," C. P. Channing, J. Marsh, and W. J. Sadler, eds., pp. 269-281, Plenum Press, New York.

Tsafriri, A., Weinstein, Y., Bar-Ami, S., Channing, C. P., Pomerantz, S. H., and Lindner, H. R., 1979, The control of meiotic maturation of the rat oocyte, in: "Research on Steroids, Vol. 8," E. Conti, ed., pp. 193-198, Academic Press, London.

Udenfriend, S., Stein, S., Bohlen, P., Dairman, W., Leimgruber, W., and Weigle, M., 1972, Fluorescamine: a reagent for assay of aminoacids, peptides, proteins and primary amines in the picomole range, Science, 178:871.

Vermeiden, J. P. W. and Zeilmaker, G. M., 1974, Relationship between maturation division, ovulation and luteinization in the female rat, Endocrinology, 95:341.

Wassarman, P. M., Schultz, R. M. Letourneau, G. E., LaMarca, M. J., Josefowicz, W. J., and Bleil, J. D., 1979, Meiotic maturation of mouse oocytes in vitro, in: "Ovarian Follicular and Corpus Luteum Function," C. P. Channing, J. M. Marsh, and W. A. Sadler, eds., pp. 251-268, Plenum Press, New York.

MEIOTIC ARRESTING SUBSTANCE SEPARATED FROM PORCINE OVARIAN GRANULOSA CELLS AND HYPOTHETICAL ARRESTING MECHANISM OF MEIOSIS

Eimei Sato, Takehiko Ishibashi,
and Akira Iritani

Department of Animal Science
College of Agriculture
Kyoto University, Kyoto 606, Japan

Introduction

In mammals, oogonia enter prophase of the first meiotic division in prenatal life, and shortly after birth, all oocytes are arrested in a late prophase known as the dictyate stage, and when an oocyte in an ovary reaches the end of its growth period, it contains a single large nucleus called the germinal vesicle (Franchi et al., 1962). It is now generally accepted that the resumption of the arrested first meiotic division occurs as a result of the surge of luteinizing hormone (LH) at each ovarian cycle (Shuetz, 1969; Tsafriri, 1978a), although the ability to complete the first meiotic division is related as well to the stage of follicular development (Tsafriri and Channing, 1975a; Sorensen and Wassarman, 1976).

Follicle-enclosed oocytes remain in the dictyate stage when cultured (Edwards, 1962; Foote and Thibault, 1969). On the other hand, isolated oocytes will resume meiosis spontaneously in culture even in hormone-free media (Pincus and Enzmann, 1935; Biggers, 1977). The usual interpretation of the resumption of meiosis in mammalian oocytes placed in vitro is that some hypothetical inhibitory agents which hold the oocyte in the dictyate condition are removed by the isolation of the oocyte from its normal environment. It has been observed that oocyte contact with the granulosa cell layer prevents maturation of the oocytes beyond the dictyate stage (Foote and Thibault, 1969), and also that follicular fluid (FFl) inhibits the resumption of meiosis (Tsafriri and Channing, 1975a, b; Tsafriri et al., 1976; Tsafriri et al., 1977; Hillensjö et al., 1978; Stone et al., 1978; Tsafriri, 1978b; Channing et al., 1980).

Therefore, it is assumed that the granulosa cells appear to exert their inhibitory influence upon meiosis by the secretion of a chemical message.

We also have been engaged in research work on the resumption of meiosis in vitro (Sato and Ishibashi, 1977; Sato et al., 1978; Sato et al., 1980), and our work on the meiotic arresting substances separated from porcine granulosa cells and on the arresting mechanism of meiosis is described.

Separation of Meiotic Arresting Substance from Porcine Granulosa Cells

Porcine ovaries were obtained from non-cycling, prepuberal gilts of approximately 100 kg body weight at a local slaughterhouse. The ovaries were excised within 30 min after sacrifice and placed in physiological saline (0.154 M NaCl) containing penicillin (100 IU/ml) and streptomycin (50 μg) at 4°C. The granulosa cells or granulosa cell layers were obtained by needle aspiration from medium-sized (2-5 mm) follicles, and the granulosa cells obtained were separated from the FFl by centrifugation and washed with 50 mM Tris-HCl buffer (pH 7.4) three times, and then a 1 ml suspension of granulosa cells was incubated with 10 ml of 10 mM Tris-HCl buffer containing 1 M urea and 5 mM EDTA (pH 8.3) at 37°C for 30 min. After removing the cells from the incubation mixture by centrifugation at 4°C, 10,000 g, the supernatant was loaded on a column of Sephadex G-25 to remove urea and EDTA, as well as to fractionate the substances from the granulosa cells. Elution was carried out with distilled water.

A cell aggregation factor has been extracted by dissociating sea urchin embryo with 1 M urea-5 mM EDTA solution (Kondo and Sakai, 1971), and the substances from the cell surfaces of cultured fibroblasts have been extracted with the same solution (Igarashi and Yaoi, 1975). Investigation has proven that 1 M urea-5 mM EDTA solution is also useful in the extraction of substances from cell surfaces and in the extraction of extracellular materials in higher animals with minimal destruction of cells, because 1 M urea-5 mM EDTA solution and the cells are nearly isotonic. Therefore, it may be considered that the substances extracted here from granulosa cells with 1 M urea-5 mM EDTA solution are from the cell surface.

The substances extracted from the cell surface were eluted from the column of Sephadex G-25 (3.5 × 95 cm) into 120 fractions of 4 ml each. The elution curve of these fractions examined with 280 nm ultraviolet absorption revealed two peaks. Material absorbing at 280 nm was evaluated for its ability to arrest meiosis of cultured cumulus-enclosed porcine oocytes (Fig. 1, Sato et al., 1978).

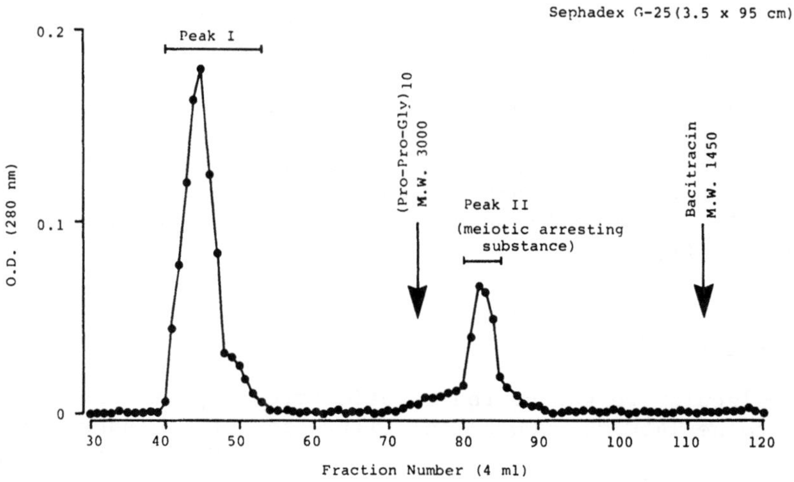

Fig. 1. Gel filtration of the substances obtained from cell surfaces of porcine granulosa cells on a column of Sephadex G-25. The substances separated from the granulosa cells from medium (2-5 mm) follicles by incubation with 10 mM Tris-HCl buffer containing 1 M urea and 5 mM EDTA (pH 8.3) were loaded on a column of Sephadex G-25 (3.5 × 95 cm). Elution was carried out with distilled water. Fractions of 4 ml were collected and used to determine the meiotic arresting activity. The arrows indicate the positions of $(Pro\text{-}Pro\text{-}Gly)_{10}$ and bacitracin used as standards (Sato et al., 1978).

To examine the effect of the substances obtained from cell surface of granulosa cells on oocyte maturation, immature oocytes at the germinal vesicle stage were collected by puncturing medium-sized (2-5 mm) follicles with a needle. These oocytes surrounded by cumulus cells were cultured with each test substance for 25 h in a modified Krebs Ringer bicarbonate solution (NaCl, 114.97 mM; KCl, 4.81 mM; $CaCl_2 \cdot 2H_2O$, 1.72 mM; KH_2PO_4, 1.20 mM; $MgSO_4 \cdot 7H_2O$, 1.20 mM; $NaHCO_3$, 24.64 mM) supplemented with crystalline BSA fraction V (1 mg/ml), penicillin G potassium (100 IU/ml) and streptomycin sulfate (50 µg/ml) in an atmosphere of 5% CO_2 and 95% air. Examinations of cultured oocytes at hourly intervals revealed that GVB occurred after about 14 h of culture and the percentage of oocytes which resumed meiosis (inhibited GVB) became constant within 25 h of culture (Sato et al., 1978b). The extent of oocyte maturation was assessed by mounting the oocytes on slide and staining them with aceto-orcein.

The presence of 100, 300, or 600 µg dry weight of peak II per milliliter in the culture medium prevented germinal vesicle breakdown (GVB) significantly, whereas the occurrence of GVB was not prevented in the medium containing 100, 300, and 600 μg dry weight of peak I. Urea contamination was not found in the fractions containing meiotic arresting substance, and urea and EDTA in 100, 300, or 600 μg of themselves did not possess the meiotic arresting action. As mentioned above, the substance obtained from the surfaces of granulosa cells was fractionated into two peaks on a column of Sephadex G-25, and the substance in the second peak possessed meiotic arresting activity.

Physico-Chemical Characteristics of the Meiotic Arresting Substance from Porcine Granulosa Cells

The molecular weight of the meiotic arresting substance was estimated to be 1450 to 3000 by gel filtration of Sephadex G-25 using $(Pro-Pro-Gly)_{10}$ and bacitracin (Fig. 1). The meiotic arresting substance was insoluble in acetone and ether, and could not be dialyzed with a cellulose tube (VT 801, Nakari Chemicals, Kyoto, Japan). Colorations of the meiotic arresting substance by ninhydrine, phenol and biuret reactions were very strong, and photometric analysis of the meiotic arresting substance dissolved in distilled water showed a peculiear absorption maximum near 260 nm. Biological activity of the meiotic arresting substance disappeared when digested by pronase, but did not disappear when treated with heat (56°C, 15 min). Heparin, chondroitin sulfate and hyaluronic acid which were considered the intracellular materials did not possess the meiotic arresting capability. From these results, it is concluded that the meiotic arresting substance is a peptide.

The oocyte maturation inhibitor (OMI) has been separated from porcine FFl, and it has been shown that the inhibitor is a peptide with a molecular weight of less than 2000 daltons. Its action is abolished by trypsin digestion, but unaffected by heating to 60°C (Tsafriri et al., 1976; Stone et al., 1978; Channing et al., 1980). The physico-chemical characteristics of the OMI seem very similar to those of the substance separated here. Additional studies are required to establish whether these are indeed similar.

Necessary Condition to Reveal Meiotic Arresting Activity of Granulosa Cells

Foote and Thibault (1969) have demonstrated the inhibitory effect of porcine granulosa cells on the resumption of meiosis in the porcine oocyte; oocytes grafted on theca cells invariably resumed meiosis as free oocytes whereas oocytes grafted on granulosa cells reamined in the dictyate stage. Tsafriri and Channing (1975b) went further to show that this effect of granulosa cells is dose-dependent and that cells from small, medium, and large Graafian follicles have the inhibitory effect, though cells from small follicles are more potent in this respect.

Table 1. Effects of Different Adhesion Patterns of Oocytes to Granulosa Cell Layers (Composed of about 10^7 Granulosa Cells from Medium-Sized Follicles) on the Resumption of Meiosis in Porcine Oocytes

	Pattern of adhesion	Sketch	No. of oocytes examined	No. of oocytes which resumed meiosis	% of oocytes which resumed meiosis
1	Cumulus-oocyte complex alone		51	38	74.5
2	Co-culture of cumulus-oocyte complex with granulosa cell layer (separated from each other)		32	23	71.9
3	The denuded oocyte adhering to a granulosa cell layer		54	32	59.3
4	The denuded oocyte completely covered by a granulosa cell layer		56	1	1.8*
5	Cumulus-oocyte complex adhering to a granulosa cell layer		61	0	0.0*
6	Cumulus-oocyte complex adhering to the fibrous materials extending from granulosa cell layer		39	28	71.8

 Oocyte

 Granulosa cell layer

 Cumulus cells

 Fibrous materials extending from granulosa cell layer

* Significantly different from the control value ($P < 0.01$)

As shown in Table 1, GVB was not inhibited in the co-culture of an oocyte with a granulosa cell layer (composed of about 10^7 granulosa cells) from medium-sized (2-5 mm) follicles if separated from each other. GVB was inhibited only when oocytes were kept in contact with a granulosa cell layer, and moreover, the inhibitory effect of granulosa cells upon GVB was dependent on the pattern of adhesion of the oocytes to the granulosa cell layer; GVB of the oocyte surrounded by cumulus cells was inhibited even when a portion of the granulosa cell layer adhered to the cumulus-oocyte com-

Table 2. Effect of Granulosa Cell Layer on the Resumption of Meiosis in Porcine Oocyte Surrounded by an Expanded Cumulus

	Duration of culture (hrs)	No. of oocytes examined	No. of oocytes which resumed meiosis	% of oocytes which resumed meiosis
1 Oocyte surrounded by an expanded cumulus	0	30	5	16.7
2 Oocyte surrounded by an expanded cumulus	25	30	27	90.0
3 Oocyte surrounded by an expanded cumulus adhering to a granulosa cell layer from medium-sized(2-5 mm) follicle	25	30	25	83.3

plex, but in the oocyte without cumulus cells, the inhibition occurred only when the entire surface of the oocyte was covered by the granulosa cell layer.

It has been pointed out that OMI appears to exert its inhibitory action on the resumption of meiosis through the mediation of cumulus cells, seeing that oocytes denuded of their cumulus cells mature in the presence of the OMI (Channing et al., 1980). Our data also suggested that the cumulus cells have an important role in mediating the effect of the inhibitory activity of the granulosa cell layer. Fibrous materials extending from the granulosa cell layer after culturing did not have any inhibitory activity.

As shown in Table 2, in the oocytes surrounded by expanded cumulus cells recovered from large Graafian follicles, a portion of oocytes had already resumed meiosis. When cultured, a high proportion of oocytes resumed meiosis, and the resumption of meiosis was not inhibited when the granulosa cell layer from medium-sized (2-5 mm) follicles adhered to the cumulus-oocyte complexes. As these results show, the inhibitory activity of granulosa cells layer was unavailable to the oocytes surrounded by expanded cumulus cells. Alternately the oocytes harvested from large follicles could be refractory to the inhibitory action of the granulosa cell layer because they have already started to mature and had passed beyond the point of being inhibitable.

The granulosa cell layer from large Graafian follicles also possessed the meiotic arresting activity when the granulosa cell layer adhered to the cumulus-surrounded oocytes.

Hypothetical Arresting Mechanism of Meiosis

Follicle-enclosed oocytes remain in the dictyate stage when cultured (Edwards, 1962; Foote and Thibault, 1969). On the other hand, dictyate oocytes, released from the follicles, mature spontaneously in culture (Pincus and Enzmann, 1935; Biggers, 1977), therefore it has been considered that some inhibitory agents which hold the oocyte in the dictyate condition are acting in the follicle.

Recent studies suggest that OMI activity is present in the granulosa cells, in the cumulus oophorus and in the FFl (Chang, 1955; Tsafriri and Channing, 1975b; Gwatkin and Andersen, 1976; Tsafriri et al., 1976; Jagiello et al., 1977; Tsafriri et al., 1977; Hillensjö et al., 1978; Tsafriri, 1978b). LH acts by neutralizing the OMI secreted by the granulosa cells and/or the cumulus oophorus and accumulated in the FFl (Tsafriri et al., 1976). Physiological studies have revealed that LH partially or completely neutralize the inhibitory action of FFl, and normally, the preovulatory surge of gonadotropins induces the resumption of oocyte meiosis (Tsafriri et al., 1976). However, the mechanism that LH acts to neutralize OMI is still obscure. A comparison of the inhibitory effects of granulosa cells and of FFl upon the spontaneous resumption of meiosis by isolated oocytes revealed two major differences: a) granulosa cells inhibit the spontaneous maturation of 80-90% of isolated oocytes whereas FFl is only 45-60% effective; and b) the inhibitory action of FFl, but not that of granulosa cells, can be overcome by the addition of LH (Tsafriri et al., 1976).

Now we would like to propose another hypothesis concerning the arresting mechanism of meiosis. Prior to oocyte maturation, the cells in the periphery of the cumulus oophorus separate and begin to bleb as mucopolysaccharide filled the intercellular interstices (Dekel et al., 1978; Dekel et al., 1978; Dekel et al., 1979; Dekel et al., 1979), and the dispersion of the cumulus cells induces the expansion of the cumulus. As shown in Table 1 and 2, the inhibitory effect of the granulosa cell layer upon GVB was dependent on the pattern of adhesion of the cumulus-oocyte complex to the granulosa cell layer. The GVB of the oocytes surrounded by an expanded cumulus was not inhibited even if the cumulus-oocyte complex was in contact with the granulosa cell layer, but that of the oocyte surrounded by a non-expanded cumulus was strongly inhibited if at all in contact. Therefore it may be considered that the resumption of meiosis is controlled by the adhesion pattern of the cumulus-oocyte complex to the granulosa cell layer. The most important point of this interpretation is whether or not LH induces the dispersion of the cumulus cells and the resultant expansion of the cumulus.

Biological grade preparations of follicle stimulating hormone (FSH) and LH stimulate hyaluronic acid synthesis by isolated mouse oocyte-cumulus cell complexes and lead to the expansion of the cumulus (Eppig, 1979; Eppig, 1980). In addition, the complexes are much more sensitive to FSH than to LH, suggesting that the response to biological grade LH is due to contamination with FSH (Eppig, 1979a, b; Eppig, 1980). However, as Eppig (1979a) has pointed out, an indirect role for LH in regulating cumulus expansion in vivo cannot be excluded. Eppig (1979a) has suggested that the FFl of preovulatory follicles contains significant amounts of FSH before the LH surge, but this FSH does not stimulate cumulus expansion before the gonadotropin surge. Perhaps LH affects the preovulatory follicle in some way that permits the oocyte-cumulus cell complex to respond to FSH and synthesize hyaluronic acid. It, therefore, might be thought that LH may have an important role in expanding the cumulus cells, although more information is required.

Although inhibition of meiosis by FFl has been demonstrated previously in various species, the inhibitory activity of FFl becomes weaker with follicular development (Channing et al., 1980). Consequently, only very slight inhibitory activity is seen in the fluid from large follicles (Stone et al., 1978) and immature cumulus-oocyte complexes are able to resume meiosis in FFl from large follicles (Sato et al., unpublished data).

Gwatkin and Andersen (1975) have reported evidence that the arrest of meiosis is due to the cumulus oophorus in the hamster, however, our experiments in the pig did not succeed in showing any such meiotic arresting action by the cumulus oophorus (Sato et al., 1977b). The cumulus cells metabolize substrates which would not be available for oocytes, and they presumably provide substances necessary for the maturation of oocytes, thus the cumulus cells have a metabolically active role in the maturation of oocytes (Sato et al., 1977a, b). In addition oocytes denuded of cumulus cells degenerate more easily, therefore it is considered that the cumulus cells also have a defensive role against the degeneration of oocytes (Sato et al., 1977b; Sato et al., 1979).

Biochemical studies on the hours requires for GVB in the porcine cumulus-oocyte complex using metabolic inhibitors, i.e., actinomycin D, puromycin and 2,4-dinitrophenol, demonstrated that mRNA is produced in the oocyte immediately after placing the oocyte in culture (Sato et al., 1982).

These data suggest that the mechanism of GVB is as follows (Fig. 2): the metiotic arresting substance located at the surface of granulosa cells acts through adhesion to the cumulus-oocyte complex to inhibit the production of mRNA necessary for GVB, this inhibitory effect is mediated by cumulus cells. Gonadotropins stimu-

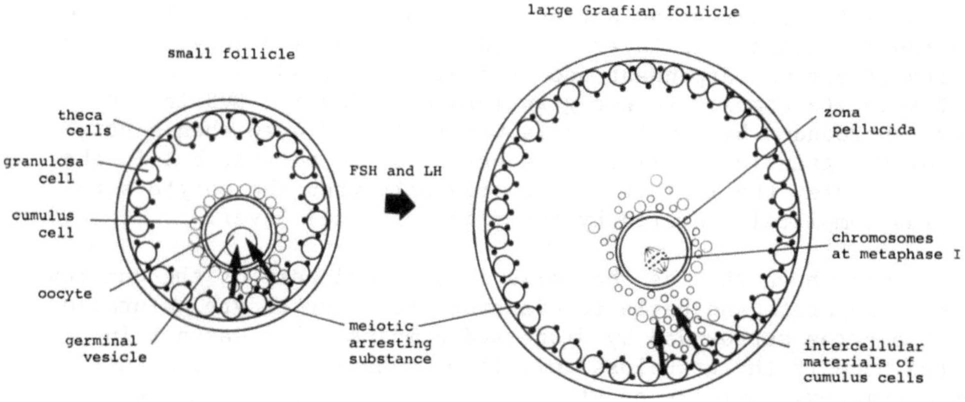

Fig. 2. Diagrammatic sketch of hypothetical mechanism of the resumption of meiosis; the meiotic arresting substance located at the surface of granulosa cells is active through adhesion to the cumulus-oocyte complex, and the arresting activity is mediated by cumulus cells. Germinal vesicle breakdown (GVB) occurs through the separation of the oocyte from the granulosa cell layer through cumulus expansion induced by gonadotropins and release from the arresting activity of the meiotic arresting substance.

late cumulus expansion through increased hyaluronic acid synthesis by the cumulus oophorus; consequently, contact between the cumulus-oocyte complex and the granulosa cell layer becomes loose. Thus the gonadotropins induce separation of cumulus-oocyte complex from the granylosa cell layer thereby releasing the cumulus-oocyte complex from the meiotic arresting activity of the substance in the granulosa cells. FFl and cumulus cells, too, have active roles in oocyte maturation: FFl is important as a culture medium and cumulus cells metabolize substrates for oocytes. In turn, gonadotropins induce GVB and oocytes extrude the first polar body and enter the second meiotic metaphase.

Summary

It has been reported that the OMI is present in granulosa cells and in FFl. We isolated the meiotic arresting substance from porcine granulosa cells. The elution curve of fractionated substances which had been obtained from granulosa cell surfaces with 10 mM Tris-HCl buffer containing 1 M urea-5 mM EDTA (pH 8.3) showed two peaks. The meiotic arresting activity was found in the substance represented by the second peak, and it was shown that the meiotic arresting substance is a peptide of which the molecular weight was estimated to be 1450 to 3000.

The meiotic arresting action of granulosa cells was effective when the granulosa cell layer adhered to the cumulus-oocyte complex, and its effect was dependent upon the pattern of adhesion of the cumulus-oocyte complex to the granulosa cell layer. GVB in the oocyte surrounded by cumulus cells was inhibited even when a portion of the granulosa cell layer adhered to the oocyte, but in the denuded oocyte, the inhibition occurred only when the oocyte surface was completely covered by the granulosa cell layer.

It was shown that as the cells in the periphery of the cumulus oophorus separate and begin to disperse preceding oocyte maturation, GVB of oocytes surrounded by dispersed cumulus cells was not inhibited even if the granulosa cell layer adhered to the cumulus-oocyte complex. Cumulus-oocyte complexes were able to resume meiosis in FFl from large follicles. The cumulus cells have a metabolically active role in the maturation of oocytes and have a defensive role against their degeneration.

From these results, it may be concluded that the mechanism of the resumption of meiosis is as follows: the meiotic arresting substance located at the surface of granulosa cells acts through adhesion to the cumulus-oocyte complex, the arresting activity being mediated by cumulus cells. GVB occurs as a result of the separation of the oocytes from the granulosa cell layer through cumulus dispersion and the resultant release from the arresting activity of the meiotic arresting substance.

Acknowledgements

This work was supported in part by Grant 704-0404 from the Ford Foundation and by Grants 276164 and 576165 from the Ministry of Education, Japan.

References

Biggers, J. D., 1973, Oogenesis and ovum maturation, in: "The Regulation of Mammalian Reproduction," S. J. Segal, R. Crozier, P. A. Corfman, and P. G. Condliffe, eds., pp. 273-283, Charles C. Thomas, Springfield, Illinois.

Chang, M. C., 1955, The maturation of rabbit oocytes in culture and their maturation, activation, fertilization, and subsequent development in fallopian tube, J. Exp. Zool., 128:378.

Channing, C. P., Schaerf, F. W., Anderson, L. D., and Tsafriri, A., 1980, Ovarian follicular and luteal physiology, in: "International Review of Physiology, Vol. 22, Reproductive Physiology III," R. O. Greep, eds., pp. 117-201, University Park Press, Baltimore.

Dekel, N., Hillensjö, T., and Kraicer, P. F., 1979, Maturational effects of gonadotropins on the cumulus-oocyte complex of the rat, Biol. Reprod., 20:191.

Dekel, N., and Kraicer, P. F., 1978, Induction in vitro of mucification of rat cumulus oophorus by gonadotropins and c-AMP, Endocrinology, 102:1797.

Dekel, N., Kraicer, P. F., Phillips, D. M., Sanchez, R., and Segal, S. J., 1978, Cellular association in the rat oocyte-cumulus cell complex: morphology and ovulatory changes, Gamate Res., 1:47.

Dekel, N., and Phillips, D. M., 1979, Maturation of the rat cumulus oophorus: A scanning electron microscopic study, Biol. Reprod., 21:9.

Edwards, R. G., 1962, Meiosis in ovarian oocytes of adult mammals, Nature, 196:446.

Eppig, J. J., 1979a, FSH stimulates hyaluronic acid synthesis by oocyte-cumulus cell complexes from mouse preovulatory follicles, Nature, 281:483.

Eppig, J. J., 1979b, Gonadotropin stimulation of the expansion of cumulus oophori isolated from mice: general conditions for expansion in vitro, J. Exp. Zool., 208:111.

Eppig, J. J., 1980, Role of serum in FSH stimulated cumulus expansion by mouse oocyte-cumulus cell complexes in vitro, Biol. Reprod., 22:629.

Foote, W. D., and Thibault, C., 1969, Recherches experimentales sur la maturation in vitro des ovecytes de truie et de veau, Ann. Biol. Anim. Biochim. Biophys., 9:329.

Franchi, L. L., Mandl, A. M., and Zuckermann, S., 1962, The development of the ovary and the process of oogenesis, in: "The Ovary, Vol. 1," S. Zuckermann, ed., pp. 1-88, Academic Press, New York.

Gwatkin, R. B. L., and Andersen, O. F., 1976, Hamster oocyte maturation in vitro: Inhibition by follicular components, Life Sci., 19:527.

Hillensjö, T., Batta, S. K., Schwartz-Kripner, A., Wentz, A. C., Sulewski, J., and Channing, C. P., 1978, Inhibitory effect of human follicular fluid upon the maturation of porcine oocytes in culture, J. Clin. Endocrinol. Metabol., 47:1332.

Igarashi, Y., and Yaoi, Y., 1975, Growth-enhancing protein obtained from cell surface of cultured fibroblasts, Nature, 254:248.

Jagiello, G., Graffeo, J., Ducayen, M., and Prosser, R., 1977, Further studies of inhibitors of in vitro mammalian oocyte maturation, Fertil. Steril., 28:476.

Kondo, K., and Sakai, H., 1971, Demonstration and preliminary characterization of reaggregation-promoting substances from embryonic sea urchin cells, Devel. Growth Differ., 13:1.

Pincus, G., and Enzmann, E. V., 1935, The comparative behavior of mammalian eggs in vivo and in vitro, I. The activation of ovarian eggs, J. Exp. Med., 62:665.

Sato, E., Furubayashi, R., and Ishibashi, T., 1978, Physicochemical characteristics of the meiotic arresting substance obtained from cell surface of porcine granulosa cells, Jpn. J. Anim. Reprod., 24:59.

Sato, E., Iritani, A., and Nishikawa, Y., 1977a, Factors involving in maturation of pig and cattle follicular oocytes cultured in vitro, Jpn. J. Anim. Reprod., 23:12.

Sato, E., Iritani, A., and Nishikawa, Y., 1977b, Effects of energy sources on oocyte maturation in pig and cattle, Jpn. J. Zootech. Sci., 48:333.

Sato, E., Iritani, A., and Nishikawa, Y., 1978a, Maturation and activation of cattle follicular oocytes cultured in vitro, Jpn. J. Zootech. Sci., 49:236.

Sato, E., Iritani, A., and Nishikawa, Y., 1978b, Rate of maturation division of pig follicular oocytes cultured in vitro, Jpn. J. Zootech. Sci., 49:400.

Sato, E., Iritani, A., and Nishikawa, Y., 1979, "Formation of nucleus" and "cleavage" of pig follicular oocytes cultured in vitro, Jpn. J. Anim. Reprod., 25:95.

Sato, E., Iritani, A., and Nishikawa, Y., 1980, Mechanism of inhibition of germinal vesicle breakdown in pig follicular oocytes with special reference to the role of granulosa cell layer, Jpn. J. Fertil. Steril., 25:229.

Sato, E., Iritani, A., and Nishikawa, Y., 1982, Analysis of the hours required for germinal vesicle breakdown in pig and cattle oocytes, Jpn. J. Fertil. Steril., 27:112.

Sato, E., and Ishibashi, T., 1977, Meiotic arresting action of the substance obtained from cell surface of porcine ovarian granulosa cells, Jpn. J. Zootech. Sci., 48:22.

Schuetz, A. W., 1969, Oogensis: processes and their regulation, Advances in Reprod. Physiol., 4:99.

Sorensen, R. A., and Wassarman, P. M., 1976, Relationship between growth and meiotic maturation of the mouse oocyte, Dev. Biol., 50:531.

Stone, S. L., Pomerantz, S. H., Schwartz-Kripner, A., and Channing, C. P., 1978, Inhibition of oocyte maturation from porcine follicular fluid: Further purification and evidence for reversible action, Biol. Reprod., 19:585.

Tsafriri, A., 1978a, Oocyte maturation in mammals, in: "The Vertebrate Ovary," R. E. Jones, ed., pp. 409-442, Plenum Press, New York.

Tsafriri, A., 1978b, Inhibition of nuclear maturation of isolated rat oocytes by follicular consituents, Ann. Biol. Anim. Biochim. Biophys., 18:523.

Tsafriri, A., 1979, Mammalian oocyte maturation: model systems and their physiological relevance, in: "Ovarian Follicular and Corpus Luteum Function," Advances in Experimental Medicine and Biology, Vol. 112, C. P. Channing, J. M. Marsh, and W. A. Sadler, eds., p. 269, Plenum Press, New York.

Tsafriri, A., and Channing, C. P., 1975a, Influence of follicular maturation and culture conditions on the meiosis of pig oocytes in vitro, J. Reprod. Fert., 43:149.

Tsafriri, A., and Channing, C. P., 1975b, An inhibitory influence of granulosa cells and follicular fluid upon porcine oocyte meiosis in vitro, Endocrinology, 96:922.
Tsafriri, A., Channing, C. P., Pomerantz, S. H., and Lindner, H. R., 1977, Inhibition of maturation of isolated rat oocytes by porcine follicular fluid, J. Endocrinol., 75:285.
Tsafriri, A., Pomerantz, S. H., and Channing, C. P., 1976, Inhibition of oocyte maturation by porcine follicular fluid: Partial characterization of the inhibitor, Biol. Reprod., 14:511.

ROLE OF CUMULUS CELLS IN OOCYTE MATURATION

Torbjörn Hillensjö, Claes Magnusson,
Carl Ekholm, Håkan Billig,
and Lars Hedin

Department of Physiology
University of Göteborg
Box 33031, S-400 33 Göteborg
Sweden

Introduction

Resumption of meiosis (oocyte maturation) is normally triggered by the preovulatory LH surge, and the first meiotic division is completed at the time of follicle rupture. The mature follicle has receptors for LH both in the theca and granulosa cells but no specific LH receptors have been demonstrated on the oocyte itself. Therefore, the triggering effect of LH on meiosis is likely to be mediated indirectly by an effect on the somatic cell types, probably the granulosa cells. We have studied the mechanism of LH action on meiosis with particular emphasis on the relation to the cumulus cells. In this review we will summarize some of our previous findings as well as report on some new results. We focus on studies where oocyte and cumulus maturation have been studied in relation to metabolic changes (1) of the whole isolated follicle and (2) of the isolated oocyte-cumulus complex.

Studies on Whole Preovulatory Follicles

Tsafriri and co-workers (1972) developed an organ culture system which enabled the study of meiosis in isolated preovulatory rat follicles. They have used this system in a number of elegant studies (Lindner et al., 1974). For instance they have demonstrated that not only LH but also a number of other hormones and pharmacological agents can induce meiosis in vitro. Several of these compounds, such as prostaglandins and cholera toxin, are able to stimulate cAMP formation. Furthermore, DBcAMP, when injected into the follicles but not when present in the culture medium, also induced meiosis.

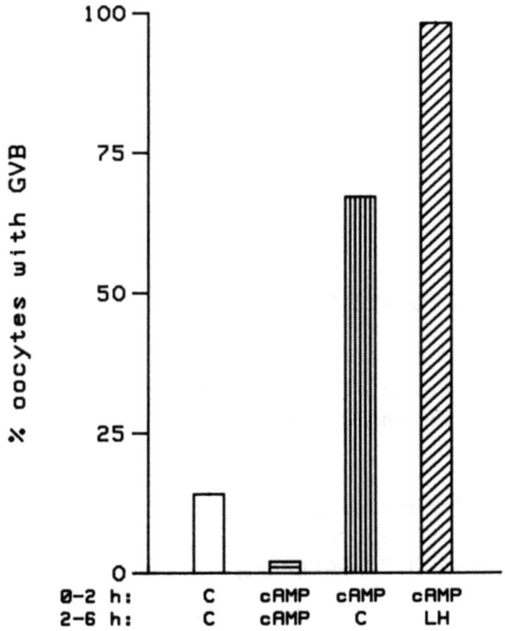

Fig. 1. Germinal vesicle breakdown (GVB) after incubation of follicle-enclosed oocytes in control medium (C), or in the presence of 10 mM DBcAMP (cAMP) or in 10 µg/ml oLH. After an initial 2 h incubation in C or DBcAMP, respectively, the follicles were transferred to new flasks for an additional 4 h incubation in C, DBcAMP or LH. There were 40-50 oocytes in each group examined (data adapted from Hillensjö et al., 1978).

We have used a slightly different incubation system with defined medium to study isolated follicles of PMSG-injected immature rats (Nilsson, 1974). When the follicles are incubated in the absence of hormone the oocyte-cumulus complex remains morphologically unchanged throughout a 10 h incubation period. In the presence of LH or FSH an orderly sequence of maturational changes occurs in the oocyte as well as the cumulus oophorus (Hillensjö et al., 1976). Under these conditions oocyte meiosis is resumed within 4 h as evidenced by germinal vesicle breakdown (GVB) and completed within 10 h as shown by polar body (PB) formation. Characteristic dispersion and mucification develop in the cumulus complex but these changes become apparent only after GVB. Drug-induced inhibition of follicular steroidogenesis did not affect the morphological response to gonadotrophin in the oocyte or cumulus complex.

In agreement with the results of Tsafriri and colleagues

(1972) we could not induce meiosis by addition of DBcAMP to the incubation medium, although 8-Br-cAMP caused a significant induction of meiosis (Hillensjö et al., 1978). Continuous exposure of the follicles to DBcAMP or phosphodiesterase inhibitors prevented LH from inducing meiosis. Cumulus mucification, however, was stimulated by these compounds (Hillensjö, 1977). It is known that LH causes an acute increase in endogenous follicular cAMP levels and that these levels decline within a few hours (Marsh et al., 1972; Nilsson et al., 1974). In an attempt to mimic this pattern we incubated the follicles for 2 h in a high concentration of DBcAMP whereafter the follicles were transferred to fresh medium without DBcAMP. This treatment led to a high rate of oocyte meiosis (Fig. 1), suggesting that although continuous high levels of cAMP are inhibitory, cAMP may be involved in the early changes leading to resumption of meiosis. A similar suggestion was made by Lindner and co-workers (1974) to explain the blocking effect of theophylline on LH-induced meiosis. Thus the exact role of cAMP in the resumption of meiosis is yet not clear and one has to consider also other possible mediators for the LH effect.

Recently we found that GnRH and agonists of GnRH are potent inducers of meiosis when added to incubated follicles (Hillensjö and LeMaire, 1980) or when injected into rats shortly after hypophysectomy (Ekholm et al., 1981). Interestingly, GnRH also causes mucification of the cumulus cells (Fig. 2b). GnRH in vitro stimulated prostaglandin accumulation in isolated rat granulosa cells without any apparent stimulation of cAMP formation (Clark et al., 1980). Also stimulation of meiosis by GnRH occurs without any apparent changes in follicular cAMP levels (Hillensjö, LeMaire, and Ahrén, to be published). In the pituitary gland GnRH action has been suggested to involve calcium (Naor et al., 1980). Tsafriri and Bar-Ami (1978) showed that follicle-enclosed oocytes resumed meiosis after addition of the divalent cation ionophor A 23187 to the culture medium. We have also seen this effect in our system and, furthermore, we have found that the ionophor causes cumulus mucification (Fig. 2). The importance of ion fluxes in the oocyte for meiosis has been demonstrated earlier. During the meiotic maturation depolarization of the mouse oocyte plasma membrane occurs (Powers and Tupper, 1974, 1975) and meiosis can be inhibited by membrane potential hyperpolarization (Powers and Biggers, 1976). We have found that ouabain, an inhibitor of Na/K ATPase, when added to incubated follicles, regularly induces meiosis (Fig. 2a). Interestingly this agent also causes mucification in most of the cumulus complexes (Fig. 2b). It can therefore be concluded that a number of compounds, which act primarily on other cellular systems than cAMP does, can induce meiosis in follicle-enclosed oocytes. It is not known whether GnRH, ionophor A 23187 and ouabain act directly on the oocyte or indirectly via the cumulus or other follicular cells.

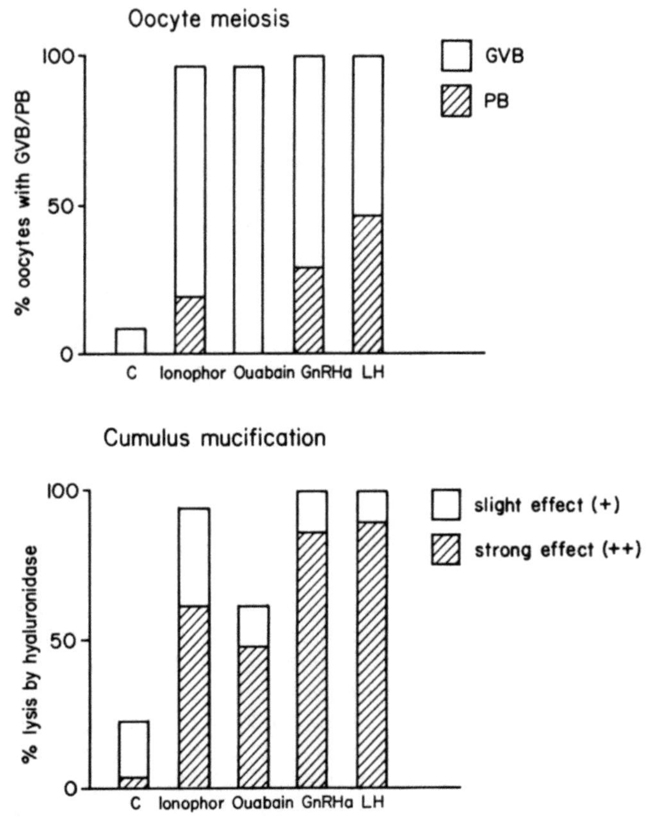

Fig. 2. Isolated preovulatory follicles were incubated for 10 h in modified KRB medium (Hillensjö 1976) under control conditions (C) or in the presence of 10^{-5} M Ionophor A23187, 10^{-3} M Ouabain, 1 µg/ml (D-Ala^{6}, des-gly^{10}-NH_2) LHRH ethylamide (GnRHa) or 1 µg/ml oLH S19. There were 35-40 follicles in each group.

The upper panel demonstrates oocyte meiosis (GVB = germinal vesicle breakdown, PB = polar body).

The lower panel demonstrates cumulus mucification as determined semiquantitatively by the sensitivity to testicular hyaluronidase (800 IU/ml PBS) of the isolated cumulus-oocyte complexes immediately after recovery.

Beers and colleagues have demonstrated intercellular communication via gap junctions between the oocyte and the cumulus cells (Gilula et al., 1978). This communication ceases in relation to meiosis but so far these changes have not been demonstrated until after GVB. The significance of the cell-to-cell communication for the induction of meiosis remains an open question, but transport of cAMP and/or ions through gap junctions may be important early events in resumption of meiosis.

Studies on Isolated Oocyte-Cumulus Complexes

As discussed above the functional relationships between the cumulus cells and the oocyte may be crucial for resumption of meiosis. Furthermore, the interaction between these cell types may be important also for the events after ovulation. Apart from morphological observations surprisingly few studies have been devoted to the cumulus cells until recently. It has become evident that the granulosa cell population of the follicle may be heterogeneous with respect to morphology, hormone receptors, and enzyme activities (Amsterdam et al., 1975; Zoller and Weisz, 1979). In the case of the oocyte-cumulus complex, biochemical studies have been hampered by the small amount of tissue available. Studies on mouse oocytes have revealed specific demands for energy substrates. It has been found that pyruvate/lactate are necessary for maturation in vitro but that glucose cannot be utilized (Biggers et al., 1967; Brinster, 1971; Zeilmaker et al., 1972). Using sensitive microrespirometric techniques we have studied energy metabolism in the rat oocyte. Addition in vitro of pyruvate but not glucose stimulated oocyte respiration (Hillensjö et al., 1975). During meiotic maturation oocyte oxygen consumption was increased whether meiosis was induced by hormone in vivo prior to isolation or occurring spontaneously in culture (Magnusson et al., 1977). Similarly, GnRH-induced meiosis was accompanied by an increase in oocyte respiration (Magnusson and LeMaire, 1981). Recently, it was found that also rabbit oocytes, induced to mature in vivo by hCG, increased their oxygen consumption (Magnusson et al., 1981).

In the isolated rat cumulus complex LH decreased oxygen consumption, whether administered in vivo or in vitro (Hillensjö et al., 1976). LH added in vitro rapidly decreased respiration whereas a lag period of 4 h was seen after LH in vivo (Magnusson and Hillensjö, 1981). One possible explanation for this inhibitory action of LH on oxygen consumption is that it represents a Crabtree effect (Crabtree, 1929), due to an increased glycolysis. Ovarian tissue is known to have a high rate of aerobic glycolysis which can be stimulated by LH (Ahrén et al., 1969). Recent experiments have shown that also isolated cumulus complexes can produce lactate and that gonadotrophins markedly stimulate lactate accumulation (Billig et al., 1981; Fig. 3). Thus in the physiological situation the rat cumulus cells may increase their lactate production at a time when the oocyte energy metabolism is increased during meiosis. It is likely that this represents a permissive rather than a triggering factor for meiosis.

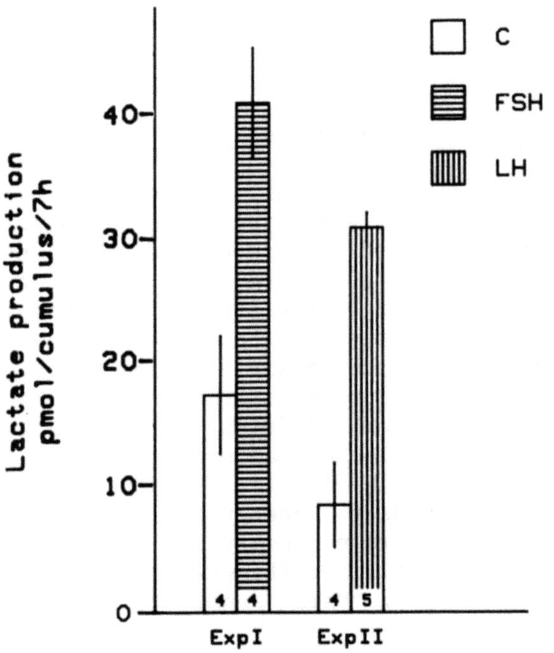

Fig. 3. Lactate production by isolated cumulus complexes cultured in modified BMOC (Magnusson et al., 1977) containing 5.5 mM glucose and 0.25 mM pyruvate as substrates. Groups of 20-30 cumulus complexes were incubated in 200 µl medium with and without oFSH-S13 (10 µg/ml) or oLH-S19 (10 µg/ml) for 7 h. Lactate accumulation in the media was analyzed by a highly sensitive microfluorometric assay (Billig et al., 1981). The values shown are the means +/− SE of 4-5 determinations per group. Stimulatory effects by FSH and LH are highly significant ($p < 0.01$).

Another biochemical parameter we have studied is progesterone synthesis in the cumulus cells during short-term culture. Cumulus complexes explanted into culture on the morning of proestrus had low basal progesterone synthesis but addition of oLH or oFSH markedly stimulated the synthesis within 12 h. Cumulus complexes isolated from the follicles shortly after the endogenous LH/FSH peak, or from the oviduct after ovulation, produced large amounts of progesterone in culture (Magnusson et al., 1979; Hillensjö et al., 1981). High progesterone formation in vitro by the periovulatory and postovulatory rat cumulus has also been reported by Sherizly and Kraicer (1980) and Schuetz and Dubin (1981). Gonadotrophin stimulation of cumulus progesterone secretion in vitro is paralleled by a morphological response suggestive of mucification (Fig. 4). In the isolated mouse cumulus complex mucification and hyal-

TABLE 1. Maturation and Respiration of Oocytes Cultured with Intact Cumulus or Denuded

Cumulus present during culture	Meiosis (% GVB of total (n)) yes	Meiosis (% GVB of total (n)) no	Respiration nl/h/oocyte (mean±SEM, n=no of samples) yes	Respiration nl/h/oocyte (mean±SEM, n=no of samples) no
Culture time (min)				
0	-		0.121 ± 0.003 (50)	
30	9 (32)	7 (29)	-	-
45	13 (68)	9 (58)	-	-
60	50 (142)	31 (133)	-	-
75	66 (70)	62 (60)	-	-
120	84 (135)	86 (101)	0.122 ± 0.007 (20)	0.118 ± 0.006 (19)
240	83 (72)	91 (88)	0.166 ± 0.006 (14)**	0.095 ± 0.007 (13)**
360	94 (31)	86 (35)	0.175 ± 0.013 (7)**	0.111 ± 0.005 (15)
600	82^x(120)	60^x(156)	-	-

^{x}PB formation

** $p < 0.01$ vs non-cultured control

Data adapted from Magnusson 1980b.

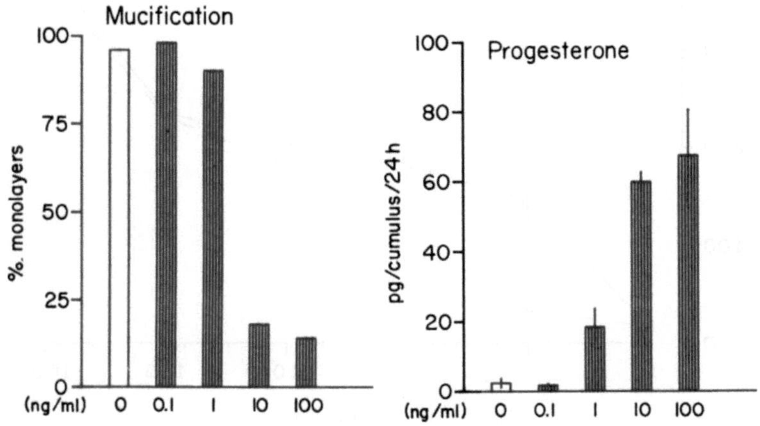

Fig. 4. Effect of various concentrations of oFSH-S13 on isolated cumulus complexes (5 × 10 cumuli per group) cultured in modified Eagle's medium (Hillensjö et al., 1981) for 24 h. The left panel shows mucification response (inversely proportional to the number of cumuli forming monolayers) and the right panel shows the progesterone accumulation (mean +/− SE).

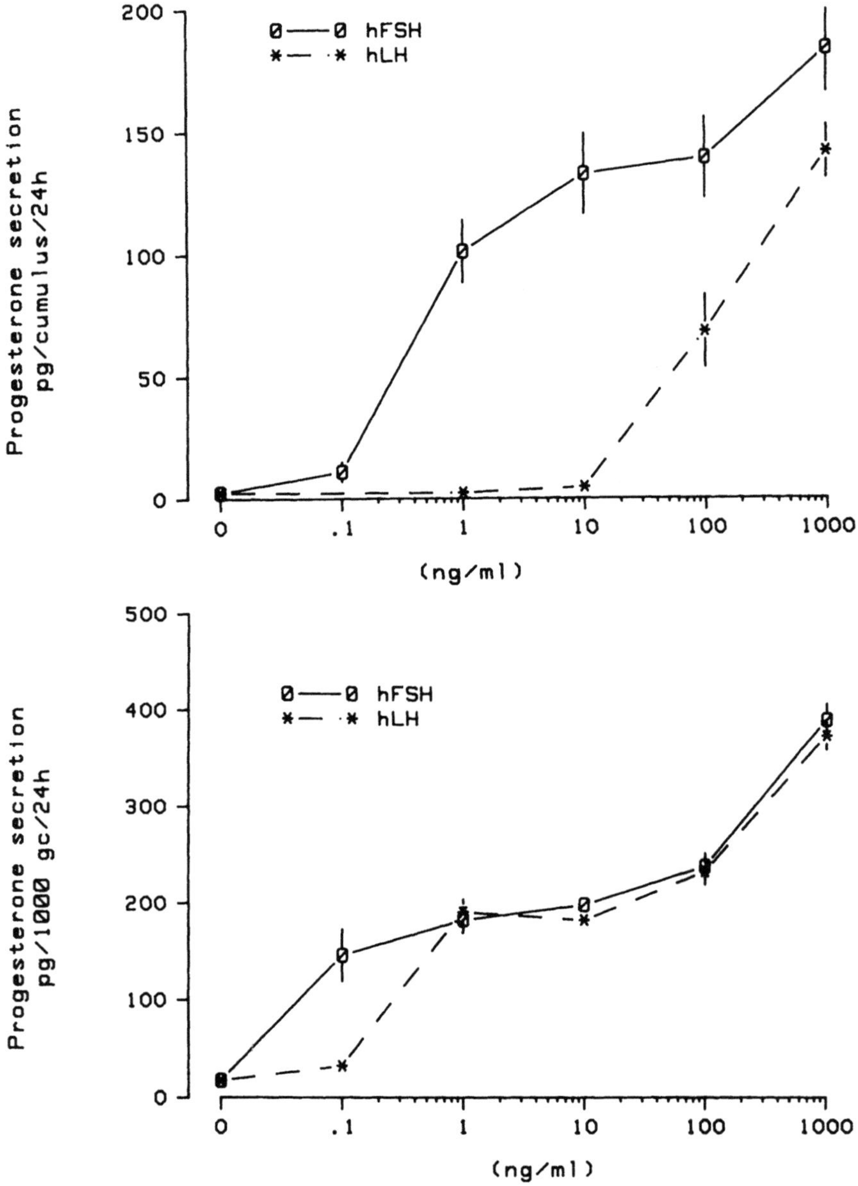

Fig. 5. Effect of various concentrations of purified hLH (Kabi, 11600 IU/mg) or purified hFSH (Kabi, 14000 IU/mg) on progesterone secretion by isolated cumulus complexes (upper panel) or mural granulosa cells (lower panel). Each value is the mean +/− SE of 5 determinations.

uronic acid synthesis is a specific FSH response that cannot be elicited by highly purified LH (Eppig, 1979a, b).

This hormone specificity may exist also in the rat, since highly purified hFSH was clearly more potent than purified hLH in the cumulus cells (Fig. 5). On the other hand, mural granulosa cells responded readily to both gonadotrophins. These differences in hormone specificity may, however, not be functionally important. The normal trigger of meiosis, LH, may act on cells located more peripherally in the follicle and produce the physiologically response in the cumulus and oocyte via intercellular connections (gap junctions).

Effect of Removal of the Cumulus Cells

The cumulus cells may have a role both in maintaining meiotic arrest and in resumption of meiosis. For instance, in the hamster (Gwatkin and Andersen, 1976) and the rat (Dekel and Beers, 1980) removal of the cumulus was reported to accelerate meiosis in vitro. In the mouse removal of the cumulus increased the proportion of oocytes reaching metaphase II stage (Cross and Brinster, 1970). In our studies we did not find any difference in the time-course for spontaneous GVB after removal of the cumulus (Table 1). However, the increase in oxygen consumption which normally is seen during meiosis was absent in denuded oocytes (Table 1). As discussed more extensively by Magnusson (1980a) the interpretation of these findings is not clearcut, but it may be suggested that the cumulus cells are required for maturational changes in the cytoplasm. A lack of energy substrates in the oocyte cannot account for this effect (Magnusson, 1980b). Neither does it seem to be due to a deficiency of steroid hormones after removal of the cumulus cells, since culture of cumulus-enclosed oocytes in the presence of aminoglutethimide does not affect oocyte respiration (Magnusson, to be published).

The partially purified OMI from pig follicular fluid appears to require the presence of viable cumulus cells to exert its inhibition of meiosis as shown both for pig, rat, and mouse oocytes (for ref. see Channing et al., 1980) and direct effects of the partially purified OMI on cumulus differentiation in vitro have been reported (Hillensjö et al., 1980). A low molecular weight fraction of human follicular fluid, which may or may not be identical to pig OMI, inhibits rat oocyte meiosis and cumulus cell progesterone secretion in vitro (Hillensjö et al., 1981b). Another substance that may inhibit meiosis is cAMP. Its effect is direct on the oocyte since removal of the cumulus cells did not diminish the inhibition produced by DBcAMP (Fig. 6). The inhibition of meiosis that under certain conditions can be demonstrated with puromycin is present also in the denuded oocyte (Fig. 6).

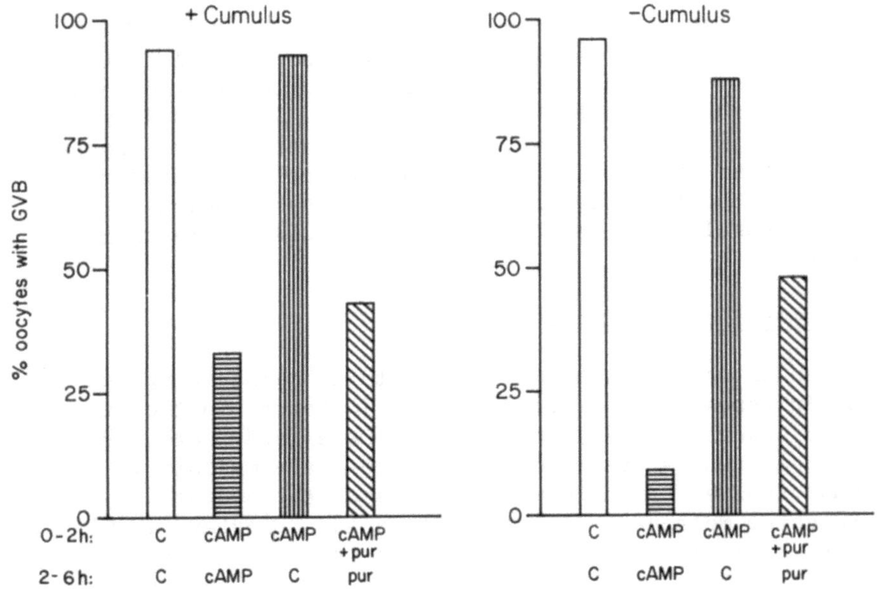

Fig. 6. Germinal vesicle breakdown (GVB) in cumulus-enclosed versus mechanically denuded oocytes cultured for 2 h in media containing various additions followed by transfer to new medium as indicated for an additional 4 h culture. The medium was modified BMOC without (C) or with the addition of DBcAMP (cAMP, 1 mM), or puromycin (pur, 10 µg/ml) or the combination of DBcAMP and puromycin. Adapted from Ekholm and Magnusson (1979).

Concluding Remarks

It is obvious that the cumulus cells undergo morphological and biochemical changes at a time when the oocyte is completing its meiotic maturation. The precise changes in the cumulus cells that are important for the oocyte have not yet been identified and some of the changes discussed above may only represent parallel phenomena. It is likely that some of them, such as the enhanced cumulus cell glycolysis, represent permissive factors for meiosis. One aspect of oocyte differentiation not discussed above is the cytoplasmic maturation necessary for fertilization (Thibault and Gérard, 1973). It is possible that stimulation of cumulus steroid secretion, as observed in this study, may play a role in this process.

Acknowledgements

We thank professor Kurt Ahrén for constant support and interest in our studies, and Mrs. Harriet Thelander for expert technical assistance. Kabi Läkemedel, Stockholm, Sweden generously supplied the purified hLH and hFSH. We also thank the hormone distribution officer of the NIAMDD for the gift of oLH and oFHS. The study was supported by grants from the Swedish Medical Research Council (5650 and 27), Magnus Bergvalls Stiftelse, Harald Jeanssons Stiftelse, and the Medical Faculty, University of Göteborg, Sweden.

References

Ahrén, K., Hamberger, L., and Rubinstein, L., 1969, Acute in vivo and in vitro effects of gonadotrophins on the metabolism of the rat ovary, in: "The Gonads," K. W. McKerns, ed., pp. 327-352, Appleton-Century-Crofts, New York.

Amsterdam, A., Koch, Y., Lieberman, M. E., and Lindner, H. R., 1975, Distribution of binding sites for human chorionic gonadotropin in the preovulatory follicle of the rat, J. Cell Biol., 67: 894.

Biggers, J. D., Whittingham, D. F., and Donahue, R. P., 1967, The pattern of energy metabolism in the mouse oocyte and zygote, Proc. Natl. Acad. Sci., U.S.A., 58:560.

Billig, H., Hedin, L., Magnusson, C., and Hillensjö, T., 1981, Gonadotrophin stimulation of rat cumulus and mural granulosa cell lactate production, Acta Endocrinol. (Suppl. 243) (Copenh), 97:A56.

Brinster, R. L., 1971, Oxidation of pyruvate and glucose by oocytes of the mouse and rhesus monkey, J. Reprod. Fertil., 24:287.

Channing, C. P., Schaerf, F. W., Anderson, L. D., and Tsafriri, A., 1980, Ovarian follicular and luteal physiology, Int. Rev. Physiol., 22:117.

Clark, M. R., Thibier, C., Marsh, J. M., and LeMaire, W. J., 1980, Stimulation of prostaglandin accumulation by luteinizing hormone-releasing hormone (LHRH) and LHRH analogs in rat granulosa cells in vitro, Endocrinology, 107:17.

Crabtree, H. G., 1929, Observations on the carbohydrate metabolism of tumours, Biochem. J., 23:536.

Cross, P. C., and Brinster, R. L., 1970, In vitro development of mouse oocytes, Biol. Reprod., 3:298.

Dekel, N., and Beers, W. H., 1980, Development of the rat oocyte in vitro: Inhibition and induction of maturation in the presence or absence of the cumulus oophorus, Dev. Biol., 75:247.

Dekel, N., Hultborn, R., Hillensjö, T., Hamberger, L., and Kraicer, P., 1976, Effect of luteinizing hormone on respiration of the preovulatory cumulus oophorus of the rat, Endocrinology, 98: 498.

Ekholm, C., Hillensjö, T., and Isaksson, O., 1981, Gonadotropin releasing hormone agonist stimulates oocyte meiosis and ovulation in hypophysectomized rats, Endocrinology, 108:2022.

Eppig, J. J., 1979, Gonadotropin stimulation of the expansion of cumulus oophori isolated from mice: general conditions for expansion in vitro, J. Exp. Zool., 208:111.

Eppig, J. J., 1979, FSH stimulates hyaluronic acid synthesis by oocyte-cumulus cell complexes from mouse preovulatory follicles, Nature, 281:483.

Gilula, N. B., Epstein, M. L., and Beers, W. H., 1978, Cell-to-cell communication and ovulation. A study of the cumulus-oocyte complex, J. Cell Biol., 78:58.

Gwatkin, R. B. L., and Andersen, O. F., 1976, Hamster oocyte maturation in vitro: inhibition by follicular components, Life Sci., 19:527.

Hillensjö, T., 1977, Dissociation of preovulatory maturational events in rat oocytes and cumuli in the presence of dibutyryl cyclic AMP, Acta Physiol. Scand., 100:261.

Hillensjö, T., and LeMaire, W. J., 1980, Gonadotropin releasing hormone agonists stimulate meiotic maturation of follicle-enclosed rat oocytes in vitro, Nature, 287:145.

Hillensjö, T., Hamberger, L., and Ahrén, K., 1975, Respiratory activity of oocytes isolated from ovarian follicles of the rat, Acta Endocrinol. (Copenh), 78:751.

Hillensjö, T., Dekel, N., and Ahrén, K., 1976, Effects of gonadotrophins on the cumulus oophorus of isolated rat Graafian follicles, Acta Physiol. Scand., 96:558.

Hillensjö, T., Ekholm, C., and Ahrén, K., 1978, Role of cyclic AMP in oocyte maturation and glycolysis in the pre-ovulatory rat follicle, Acta Endocrinol. (Copenh), 87:377.

Hillensjö, T., Pomerantz, S. H., Kripner, A. S., Anderson, L. D., and Channing C. P., 1980, Inhibition of cumulus cell progesterone secretion by low molecular weight fractions of porcine follicular fluid which also inhibit oocyte maturation, Endocrinology, 106:584.

Hillensjö, T., Chari, S., Magnusson, C., Daume, E., and Sturm, G., 1981b, Inhibitory effects of low molecular weight fractions of human follicular fluid upon rat granulosa cells and oocytes in vitro, Excerpta Medica 1981, in press.

Hillensjö, T., Magnusson, C., Svensson, U., and Thelander, H., 1981a Effect of LH and FSH on progesterone synthesis in cultured rat cumulus cells, Endocrinology, 108:1920.

Lindner, H. R., Tsafriri, A., Lieberman, M. E., Zor, U., Koch, Y., Bauminger, S., and Barnea, A., 1974, Gonadotropin action on cultured Graafian follicles: Induction of maturation division of the mammalian oocyte and differentiation of the luteal cell, Recent Prog. Horm. Res., 30:79.

Magnusson, C., 1980a, Mammalian oocyte maturation. Studies on the hormonal regulation and the role of the cumulus cells, Thesis, Göteborg, ISBN 91-7222-330-8.

Magnusson, C., 1980b, Role of cumulus cells for rat oocyte maturation and metabolism, Gamete Res., 3:133.

Magnusson, C., and Hillensjö, T., 1981, Further studies on the gonadotrophin-induced inhibition of respiration in the preovulatory rat cumulus oophorus, Acta Physiol. Scand., 113:17.
Magnusson, C., and LeMaire, W. J., 1981, A gonadotrophin-releasing hormone agonist stimulates oxygen consumption and maturation of follicle-enclosed rat oocytes in vitro, Acta Physiol. Scand., 111:377.
Magnusson, C., Hillensjö, T., Tsafriri, A., Hultborn, R., and Ahrén, K., 1977, Oxygen consumption of maturing rat oocytes, Biol. Reprod., 17:9.
Magnusson, C., Svensson, U., Thelander, H., and Hillensjö, T., 1979, Gonadotrophin stimulation of progesterone secretion by cultured rat cumulus oophorus, Acta Endocrinol. (Suppl. 225) (Copenh)., 91:156.
Magnusson, C., LeMaire, W. J., and Hillensjö, T., 1981, Stimulation by hCG in vivo of oxygen consumption by rabbit oocytes in vitro, J. Reprod. Fertil., 61:185.
Marsh, J. M., Mills, T. M., and LeMaire, W. J., 1972, Cyclic AMP synthesis in rabbit graafian follicles and the effect of luteinizing hormone, Biochim. Biophys. Acta, 273:389.
Naor, Z., Leifer, A. M., and Catt, K. J., 1980, Calcium-dependent action of gonadotropin-releasing hormone on pituitary guanosine-3'5'-monophosphate production and gonadotropin release, Endocrinology, 107:1438.
Nilsson, L., 1974, Acute effects of gonadotrophins and prostaglandins on the metabolism of isolated ovarian follicles from PMSG-treated immature rats, Acta Endocrinol. (Copenh), 77:540.
Nilsson, L., and Rosberg, S., and Ahrén, K., 1974, Characteristics of the cyclic 3'5'-AMP formation in isolated ovarian follicles from PMSG-treated immature rats after stimulation in vitro with gonadotrophins and prostaglandins, Acta Endocrinol. (Copenh), 77:559.
Powers, R. D., and Biggers, J. D., 1976, Inhibition of mouse oocyte maturation by cell membrane potential hyperpolarization, J. Cell Biol., 70:352.
Powers, R. D., and Tupper, J. T., 1974, Some electrophysiological and permeability properties of the mouse egg, Dev. Biol., 38:320.
Powers, R. D., and Tupper, J. T., 1975, Ion transport and permeability in the mouse egg, Exp. Cell Res., 91:413.
Schuetz, A. W., and Dubin, N. H., 1981, Progesterone and prostaglandin secretion by ovulated rat cumulus cell-oocyte complexes, Endocrinology, 108:457.
Sherizly, I., and Kraicer, P. F., 1980, Progesterone secretion by the postovulatory rat cumulus oophorus, Gamete Res., 3:115.
Thibault, C., and Gérard, M., 1973, Cytoplasmic and nuclear maturation of rabbit oocytes in vitro, Ann. Biol. Anim. Biochim. Biophys., 13:145.

Tsafriri, A., and Bar-Ami, S., 1978, Role of divalent cations in the resumption of meiosis of rat oocytes, J. Exp. Zool., 205: 293.

Tsafriri, A., Lindner, H. R., Zor, U., and Lamprecht, S. A., 1972, In vitro induction of meiotic division in follicle-enclosed rat oocytes by LH, cyclic AMP and prostaglandin E2, J. Reprod. Fertil., 31:39.

Zeilmaker, G. H., Hulsmann, W. C., Wensinck, F., and Verhamme, C., 1972, Oxygen-triggered mouse oocyte maturation in vitro and lactate utilization by mouse oocytes and zygotes, J. Reprod. Fertil., 29:151.

Zoller, L. C., and Weisz, J., 1979, A quantitative cytochemical study of glucose-6-phosphate dehydrogenase and Δ5-3β-hydroxysteroid dehydrogenase activity in the membrana granulosa of the ovulable type of follicle of the rat, Histochemistry, 62:125.

ACTIONS OF HORMONES AND OTHER FACTORS UPON OOCYTE MATURATION

Cornelia P. Channing*, Seymour H. Pomerantz,**
In-Ha Bae*, Valerie W. Evans*
and Susan J. Atlas*

University of Maryland School of Medicine
660 West Redwood Street
Baltimore, Maryland 21201

INTRODUCTION

The mammalian oocyte is arrested in the dictyate state of meiosis from shortly before or at the time of birth until 1-2 days prior to ovulation. Under the influence of the preovulatory LH surge the oocyte in the large preovulatory follicle destined to ovulate undergoes maturation. The germinal vesicle breaks down and the first polar body is extruded. This first meiotic maturation must be completed in order to render the oocyte fertilizable. The physiological mechanism responsible for keeping the oocyte in the state of meiotic arrest is unknown. It is well documented that if the oocytes are removed from medium or large-sized follicles and cultured they resume meiotic maturation before they normally would in vivo. (Reviewed by Tsafriri, 1978; Channing et al., 1980; Schuetz, 1974.) Addition of a low molecular weight fraction of follicular fluid (FF1) (porcine, human, bovine) can inhibit the resumption of meiosis of cultured cumulus-enclosed porcine (Tsafriri and Channing, 1975; Tsafriri et al., 1976; and Stone et al., 1978), rat (Tsafriri et al., 1977; Pomerantz, Tsafriri, and Channing, 1979), and hamster (Gwatkin and Andersen, 1976) oocytes suggesting that a meiotic inhibitor is present in follicular fluid.

An oocyte maturation inhibitor (OMI) has been partially purified by us from porcine follicular fluid (Tsafriri, Pomerantz, and Channing, 1976; Stone, et al., 1978; Pomerantz, Tsafriri, and

*Departments of Physiology and **Biological Chemistry.

Table 1. Purification of Oocyte Maturation Inhibitor from Porcine Follicular Fluid

Fraction	Volume (ml)	Peptide* (mg/ml)	Units/mg	Total Units	Fold* Purification
FF1	900	---	---	1,800**	---
Amicon PM-10 Filtrate	24.5	105	0.48	1,225	1
Sephadex G-25 Peak A	20	43.4	1.38	1,100	2.9
CM-Sephadex	15	29	5.17	2,200***	10.8
DEAE-Sephadex (on 1,100 units of CM-Sephadex fraction)	2	1.67	60	200	125

* Based on peptide determination by the fluorescamine method (Udenfriend et al., 1972) using β-MSH as a standard.
** Estimated, 1:1 dilution inhibited maturation by approx. 50%.
*** The observed increase in activity probably due to removal of a stimulator by purification.
(Taken from Pomerantz, Channing, and Tsafriri (1979) with permission.)

Channing, 1979) and found to be a polypeptide of less than 2000 daltons. The inhibitor appears to be secreted by the granulosa cells (rat: Tsafriri et al., 1979; pig: Centola et al., 1981; present observations (see below), Sato and Ishibashi, 1977).

Several questions arise: Is OMI the principal factor in controlling oocyte maturation and what is its chemical nature? What is the mechanism of OMI action upon mammalian oocytes and what (if any) role is there for Ca^{2+} in this process? Is the secretion of OMI under hormonal control and is the hormonal control of its synthesis by the follicle appropriate for regulation of oocyte maturation in vivo? For example, when a follicle becomes atretic and the oocyte resumes maturation prematurely (Matson et al., 1981; Hirshfield and Midgley, 1978) is this due to diminished granulosa cell secretion of OMI? Is the resumption of oocyte maturation prior to ovulation the result of (a) decreased granulosa cell secretion of OMI brought about by luteinization or (b) the direct antagonizing effect of LH on the action of OMI in the follicle? We can only begin to answer some of these questions at present.

EVIDENCE FOR A MEIOTIC INHIBITOR AND STIMULATOR IN PORCINE FOLLICULAR FLUID

A procedure for the partial purification of OMI is shown in Table 1 and Fig. 1. The fold purification is about 15,000 relative to raw follicular fluid. An example of the elution pattern of OMI using Sephadex G-25 and CM-Sephadex chromatography is shown in Figs.

Follicular fluid → PM-10 filtration → PM-10 → Lyophilize → Dry solid

→ Extract with 2 $\underline{N}$ acetic acid-methanol → Acetic acid-methanol → Evaporate Lyophilize → Dry solid → Extract with methanol →

Methanol supernatant → Evaporate lyophilize → Dry solid → Dissolve in 0.1 $\underline{N}$ acetic acid → Solution Extract with ether (Discard ether)

→ Lyophilize aqueous phase →

Dry solid → Dissolve in 0.1 $\underline{N}$ acetic acid → acetic acid fraction → Sephadex G-25 in 0.1 $\underline{N}$ acetic acid → OMI Peak

→ CM-Sephadex elute with NH_4OAC, pH 5 in a step gradient of 0.01 $\underline{M}$ 0.05 $\underline{M}$, 0.1 $\underline{M}$, and 0.2 $\underline{M}$. → Three peaks of OMI activity

Fig. 1. Current purification scheme for OMI from pig follicular fluid.

2 and 3. The three peaks obtained from CM-Sephadex at pH 5 will be investigated by high pressure liquid chromatography. Since Stone et al. (1978) observed that fluid from small- (1-2 mm) and medium- (3-5 mm) sized porcine follicles contained more OMI activity than did fluid harvested from large-sized (6-12 mm) porcine follicles most subsequent studies on purification of OMI employed pools of FFl from small- and medium-sized follicles.

The observation of a decline in OMI in FFl as the follicle matures was confirmed in staged pig ovarian FFl by Tsafriri (unpublished observations). He observed that fluid harvested from large-sized follicles of pigs in estrus contained non-detectable levels of OMI using the rat oocyte bioassay. In contrast, fluid harvested from medium-sized follicles prior to estrus contained significant amounts of OMI activity. This observation suggests that the concentration of OMI has physiological and regulatory significance since a low level at the time of ovulation would allow oocyte maturation to occur.

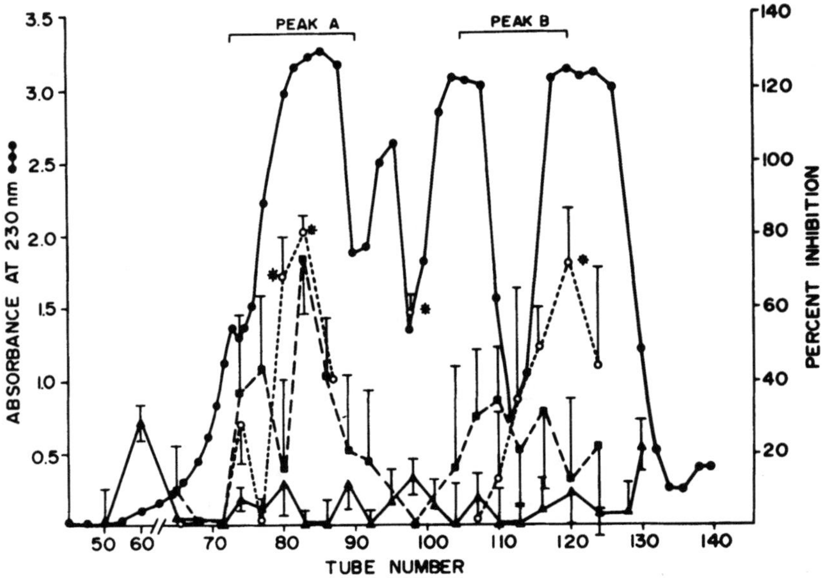

Fig. 2. Sephadex G-25 chromatography of methanol-acetic acid treated PM-10 filtrate of follicular fluid (preparation number AL 37). Pooled follicular fluid (1670 ml) was filtered through an Amicon PM-10 filter and extracted with methanol-acetic acid according to the procedure in Fig. 1. The final samples (containing 4.16 g of material by dry weight, 2.31 g peptide by fluorescamine) was dissolved in 0.1 N acetic acid, applied to a Sephadex G-25 column (5 × 70 cm), and eluted with 0.1 N acetic acid. The flow rate of the column was 50 ml/hr, and 15 ml fractions were collected. Individual fractions were lyophilized and resuspended in 2.0 ml of distilled water. Aliquots of this solution were used to test inhibition of oocyte maturation (■ — ■) and progesterone secretion by the oocyte-cumulus complex (O — O) as well as the ability to inhibit the binding of ^{125}I-hCG to granulosa cell LH/hCG receptors (▲ — ▲) compared to control samples. The points are graphed with the bars indicating the standard error. Significance of oocyte maturation inhibition compared to control was determined by chi-square analysis, while inhibition of progesterone and ^{125}I-hCG binding was determined by Student's t-test. Those data points marked with an asterisk are significantly different from the control ($p < 0.05$). Absorbency at 230 nm is indicated by ● — ●. (Atlas et al., 1980).

Table 2. Recovery of OMI Activity after Chromatography on CM-Sephadex

Experiment	Activity Applied (Units)	Activity Recovered (Units)	Recovery (Per Cent)
1	590	1300	220
2	510	1000	196
3	475	665	140
4	450	470	104

In each case the activity applied was an OMI fraction from a Sephadex G-25 column. In experiments 1 and 2 the CM-Sephadex (2 × 27 cm) was suspended in 0.01 M NH_4HCO_3, pH 8, and the sample was eluted with that buffer after about (116 ml) had been passed through the column. In experiments 3 and 4, the column (2 × 50 cm) was suspended in 0.01 M NH_4OAC, pH 5 and the column was eluted in succession with 0.01 M, 0.05 M, 0.10 M, and 0.20 M NH_4OAC. The activity recovered is the sum of three areas of activity.

Table 3. Observations by H. A. Kent on Inhibition of Ovulation by Synthetic Samples of Tetrapeptide (H-thr-pro-arg-Lys-OH)

		Ova/pair of oviducts (hamster)		
Dose in µg/100g/day	Control	Initial CBC Material	Two peaks eluted from Sephadex G10: Peak #2	Peak #1
0.00	11.3	--	--	--
0.03		7.6	7.5	11.1
0.10		2.1	--	13.0
0.3		0	0.0	14.0
3.0		3.1	0.0	15.0
30.0		7.6	0.0	15.0
		(Mixture)	(Inhibitor)	(Stimulator)

The tetrapeptide was synthesized and the original material (CBC) assayed for its inhibitory effects upon ovulation in the golden hamster (given sc for 4 days). The CBC was then chromatographed on Sephadex G-10. Two peaks were eluted and assayed. (Kent, 1978).

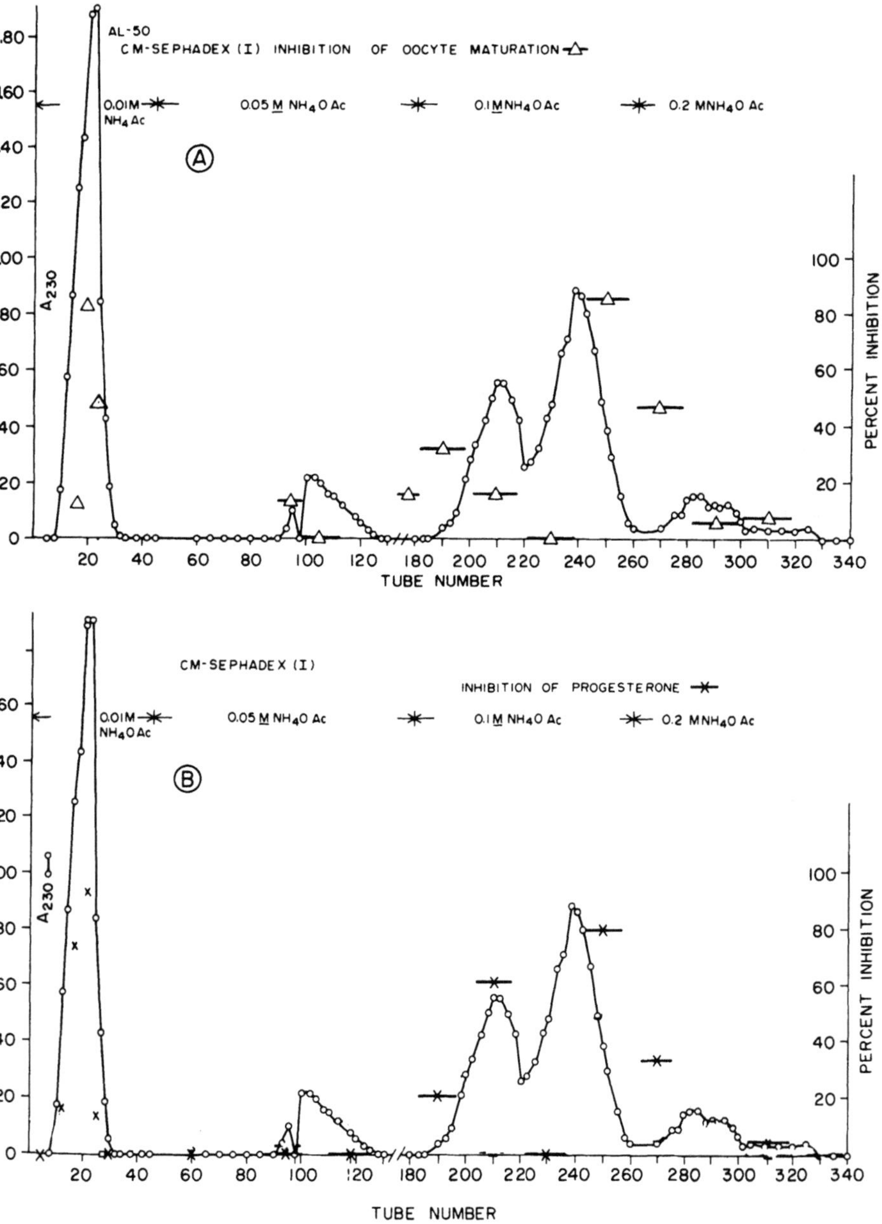
AL-50
CM-SEPHADEX (I) INHIBITION OF OOCYTE MATURATION
0.01M NH4 Ac
0.05 M NH4 O Ac
0.1 M NH4 O Ac
0.2 MNH4 O Ac
Ⓐ
A230
PERCENT INHIBITION
TUBE NUMBER
CM-SEPHADEX (I)
INHIBITION OF PROGESTERONE
0.01M NH4 O Ac
0.05 M NH4 O Ac
0.1 M NH4 O Ac
0.2 MNH4 O Ac
Ⓑ
A230
PERCENT INHIBITION
TUBE NUMBER

Fig. 3. Chromatography of the Sephadex G-25 peak on CM-Sephadex. The OMI fraction from a column of Sephadex G-25 (585 units; 0.57 units/mg peptide) was adjusted to pH 5.0 with acetic acid. It was then applied to a column of CM-Sephadex (2 × 50 cm) suspended in 0.01 M NH_4OAc, pH 5 and the column eluted as shown in the figures with 0.01 M, 0.05 M, 0.10 M and 0.20 M NH_4OAc, pH 5. Tube volume was 7 ml. Both A and B show the elution pattern observed at 230 nm. For assay purposes the tubes were combined into 28 fractions, lyophilized to dryness, and dissolved in 2 or 4 ml of water. For OMI (A), 60 µl of each fraction was taken for assay at a final dilution of (1:25). (B) shows the inhibition of progesterone formation by the granulosa cell clumps for which 30 µl of each fraction was taken for assay at a final dilution of 1:100. Granulosa cells were pooled from many small (<2 mm) porcine follicles and washed several times. A large number of clumps were selected, pooled and stirred in medium 199A, and then dispensed (100 µl) into culture wells containing test fractions, standards, and controls. The results are graphed as percent inhibition of oocyte maturation and progesterone formation. The span of each bar indicates the tubes within a given fraction. (Pomerantz and Channing, unpublished observations.)

Table 4. Ability of Anti-rProlactin Antiserum to Neutralize OMI

Treatment of Oocytes	Number of Oocytes	Percent Maturation	Percent Inhibition	Standard Error	P vs Control
199 A Control	32	53.1	---	---	---
1/60 PM-10 AL 48 OMI	51	27.4	48.4	12	0.05
1/100 Dilution Anti PR1 antiserum	43	41.8	21.1	15	NS
1/100 Dilution Anti PR1 antiserum + 1/60 PM-10 AL 48	47	57.4	0	0	NS
1/1000 Dilution Anti PRL antiserum	51	39	26	14	NS
1/1000 Dilution Anti PR1 antiserum + 1/60 PM-10 AL 48 OMI	48	39.5	25.6	14.8	NS
1/25 Sephadex Peak A AL 50 OMI	45	46.6	12.2	16	NS
1/25 Sephadex Peak A AL-50 OMI + 1/100 Dilution Anti PR1 antiserum	34	73.5	(Stimulation over Control) + 37% ± 18% p=0.02		
1/25 Sephadex Peak A AL 50 OMI + 1/1000 Dilution Anti PR1 antiserum	29	55.2	0	19	NS

Oocytes were harvested from small (1-2 mm) porcine follicles and cultured for two days in the presence of various mixtures of anti PR1 and OMI. At the end of the 2 day period the oocytes were evaluated for maturation. Two doses of anti PR1 antiserum (NIH-S7) (1:100 and 1:1000 dilution) were incubated either alone or with either 1:60 PM-10 A148 OMI or 1/25 Sephadex Peak A A150 OMI for 1 hour at 37° in Medium 199 A prior to addition of the oocytes. Note that 1:100 dilution of anti PR1 completely neutralizes the inhibitory effect of 1/60 PM-10 A148 whereas 1:1000 dilution partially neutralizes the inhibitory effect. Data were analyzed by Chi Square Test. (Channing and Bae, unpublished observation).

This hypothesis may however require revision in the light of other observations. Liebfried and First (1980), Tsafriri (unpublished) and we (Pomerantz and Channing, unpublished) have found that some samples of pooled porcine follicular fluid or low molecular weight fraction contain little or no OMI activity. However when we carried out further purification by chromatography on Sephadex G-25 followed by CM Sephadex there was an increase in OMI activity with a yield >100% (Table 2). These data suggested that further purification of OMI led to the separation of a stimulator from the inhibitor. Using CM-Sephadex column chromatography Atlas and Channing (unpublished experiments) have recovered a stimulator fraction which was separated from the inhibitor present in conditioned porcine granulosa cell culture medium. These findings may mean that even fluid from large-sized follicles may have OMI but that its action is masked by a stimulator of oocyte maturation.

Kent (1978) has made a similar observation in samples of synthetic tetrapeptide H-thr-pro-arg-Lys-OH prepared by Calbiochem (CBC) (Table 3). He found that the tetrapeptide produced a variable dose response curve when tested for ability to inhibit ovulation in the cycling hamster. However, if the material was passed over a column of Sephadex G-10 two peaks were recovered, one of which was an inhibitor (Peak 2) and the other a stimulator (Peak 1) of ovulation (Table 3). The confusing dose response curve of the original fraction could then be explained. Poor detection of OMI in crude porcine follicular fluid or even the low molecular weight fraction could also be due to inactivation due to binding to a carrier protein or peptide as was the case in somatomedin (Knauer and Smith, 1980). The existence of carrier proteins which modify actions of luteinization inhibitor, oocyte maturation inhibitor and perhaps inhibin could account for the disparate data obtained in different laboratories.

Recently Nolin and Bogdanove (1980; Nolin, 1980a) observed that incubation of antiserum to rat prolactin with slices of rat ovaries led to immunohistochemical reaction product localized over oocytes, follicles, and corpus luteum tissue. This could be explained if follicular tissue contained either prolactin or a closely related component which cross-reacted with prolactin antiserum. Since it was found by Baker and Hunter (1978) that addition of prolactin antiserum to follicle-enclosed porcine oocytes hastened maturation of the oocyte, it is possible that the follicular factor cross reacting with the anti PRL antisera might be OMI. To test this hypothesis we preincubated antiserum to rat prolactin (NIH S7) for 1 h at 37° with two crude OMI preparations and subsequently assayed the mixtures and the parent OMI fraction in the cultured porcine oocyte assay for ability to inhibit oocyte maturation (Table 4). (The experiment was repeated two more times with similar results.) Incubation of the OMI AL 48 with 1/100 dilution of antiserum completely reversed, while incubation with 1/1000 dilution, partially reversed its inhibitory action. If the antiserum was added to a less potent OMI preparation (1/25 Sephadex Peak A of AL 50) the antiserum reversed its small 12% inhibitory action and resulted in a significant 38% stimulatory action upon oocyte maturation. The data can be explained by the existence of an inhibitor and a stimulator in the Sephadex G-25 fraction and when the inhibitor is neutralized the stimulator exerts its effect. Addition of prolactin alone to cultured oocytes failed to inhibit oocyte maturation (data not shown). Nolin (1980b) observed that incubation of a PM-10 fraction of porcine follicular fluid with prolactin antiserum before adding it to rat ovarian slices inhibited the reaction of antiserum with ovarian slices. These data are consistent with the hypothesis that OMI

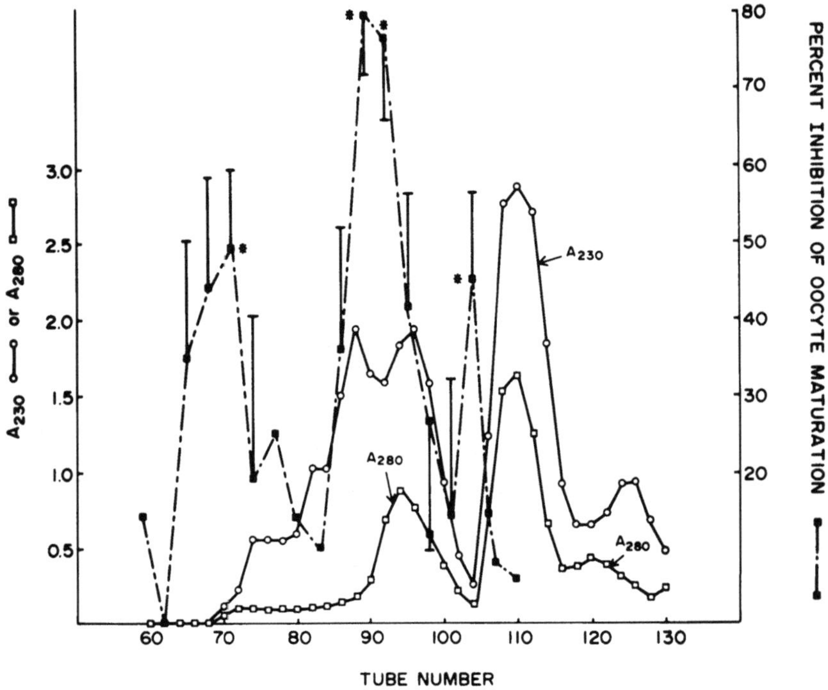

Fig. 4. Chromatography of methanol-acetic acid fractionated conditioned medium from culture of pig granulosa cells. Granulosa cells were incubated (10^6 cells/ml) without agitation in 50 ml lots in siliconized bottles at 37° for two days in 5% CO_2 — 95% air in medium 199 containing 10% pig serum. The cells were centrifuged at 20,000 rpm for 30 min. A total of 2,040 ml of culture medium was pooled and filtered through an Amicon PM-10 filter. The PM-10 filtrate (1510 ml) was lyophilized to give 26.6 g dry powder. The powder (23.9 g) was treated with acetic acid-methanol, methanol, and acetic acid as given in Fig. 1. The volume of the final 0.1 N acetic acid solution was 20 ml and the OMI activity was estimated as 70 units/ml by dose-response assay. A Sephadex G-25 column (5 × 70 cm) suspended in 0.1 N acetic acid was loaded with sample (9.4 ml, 1.49 g peptide by fluorescamine, 660 OMI units) and eluted with 0.1 N acetic acid. The elution diagram shows A_{230} O———O, A_{280} □———□, and OMI in percent inhibition ■---■. For OMI measurement the individual fractions were lyophilized, and the residues dissolved in 2 ml water. These solutions were assayed at a dilution of (1:25). Three areas of OMI activity emerged: tubes (64-73), (84-99), and (102-106). The major peak of activity, (84-99), corresponds to the position of Peak A of follicular fluid. The bars show the S.E. The

cross reacts with antiserum to rat prolactin. Whether OMI represents a fragment of prolactin or whether OMI cross reacts with another site on the antiserum having little to do with prolactin cannot be answered yet. Further studies with different antisera and examination of other cross reactions will be needed to resolve this issue. Nevertheless, this observation may enable us to use antiprolactin antisera to advantage in the purification of OMI.

HORMONAL CONTROL OF GRANULOSA CELL SECRETION OF OOCYTE MATURATION INHIBITOR

I has been shown by Tsafriri and Channing (1975), and Centola et al. (1981) that medium conditioned by culture with porcine granulosa cells contains OMI activity. The more recent studies by Centola et al. (1981) showed that low molecular weight extracts of porcine granulosa cells but not those of liver or lung also contain OMI activity. These investigators demonstrated that the OMI activity in medium conditioned by granulosa cells from small follicles was greater than that in medium conditioned by granulosa cells harvested from large follicles. The extract of granulosa cells harvested from small follicles had more OMI activity than extracts of granulosa cells harvested from large follicles. These data indicate that the potential of granulosa cells to secrete OMI parallels the level of OMI in follicular fluid and perhaps the physiological state of the follicle. It is appropriate that a large follicle containing a maturing oocyte should have a low level of OMI in its fluid and that there be a poor potential for its production by large follicle granulosa cells.

Recently a preliminary chemical characterization of the OMI activity in medium conditioned by porcine granulosa cells was made. Sephadex G-25 chromatography of the low molecular weight fraction of conditioned porcine granulosa medium revealed that the principle peak of OMI activity eluted in a region comparable to that of follicle fluid OMI (Figs. 2 and 4). An additional 2 peaks of OMI activity are also eluted, one prior to and one after the major one. The late eluting peak corresponds to Peak B sometimes found in porcine follicular fluid and the earlier eluting peak corresponds to a higher molecular weight form of OMI. Addition of 1 μg/ml ovine LH

points marked with an asterisk indicate that those values of percent inhibition of maturation were significantly different from zero (number 71, $p < 0.005$; 89, $p \ll 0.005$; 92, $p \ll 0.005$; 104, $p \ll 0.05$). The calculation of percent inhibition of oocyte maturation and the method of statistical evluation were described in Stone et al. (1978). (Pomerantz and Channing, unpublished observations.)

Table 5. Lack of Inhibitory Effect of Ovine Prolactin upon Maturation of Cumulus-Enclosed Porcine Oocytes

Addition to Culture	Number of Oocytes	Percent Maturation
Control	31	90%
0.1 μg/ml oPRl	29	93%
1.0 μg/ml oPRl	34	88%
10 μg/ml oPRl	30	93%

Porcine oocytes were harvested from medium sized follicles and cultured for 2 days in the presence of various doses of ovine prolactin (NIH oPRl S-13) and the percent maturation evaluated. (Bae and Channing, unpublished observations.)

to oocytes cultured in the presence of the main Sephadex peak recovered from conditioned medium reversed its inhibitory action, as does LH when added to OMI from follicular fluid. Fractionation of a comparable volume of blank medium gave no OMI activity.

Hormonal control of OMI secretion can be examined advantageously in suspension cultures of porcine granulosa cells harvested from medium-sized follicles. Anderson and Stone (1980) observed that addition of human FSH (1 μg/ml) stimulated OMI secretion by the cultured cells. In contrast addition of 1.4 μg/ml testosterone or dihydrotestosterone (a nonaromatizable androgen) led to a decrease in OMI secretion. Addition of either androgen or estrogen alone to cultured porcine oocytes failed to inhibit oocyte maturation. These data indicate that the elevated follicular androgens associated with atresia (McNatty et al., 1979a, b) or with the LH/FSH surge (Ainsworth et al., 1980) may play a role in triggering oocyte maturation by causing a decrease in OMI secretion by granulosa cells. Addition of either 1.4 μg/ml estradiol or LH (1 μg/ml) to the cultured granulosa cells had no consistent effect upon OMI secretion. Since the granulosa cell is the source of OMI and since atresia is associated with oocyte maturation and death of granulosa cells (McNatty et al., 1979a, b; Hirschfield and Midgley, 1978), the decline of oocyte maturation inhibitor in these follicles could be accounted for merely by death of the granulosa cells. Studies by Channing et al. (1981a) demonstrated that atretic mare follicles contain less inhibin than viable follicles. Cystic human follicles which contain fewer granulosa cells also contain less inhibin and OMI compared to viable follicles (Channing et al., 1981b).

Table 6. Effect of Ovine Prolactin upon Porcine Granulosa Cell Secretion of Oocyte Maturation Inhibitory Activity

Group	Treatment (Addition to oocyte culture)	Number of Oocytes	% Maturation	% Inhibition of Maturation	Standard Error of % Inhibition	P vs Control (I)	P vs Cells Only (IV)
I	Control 199 A	113	63	---	---	---	---
II	1/50 PM-10 Filtrate OMI	28	34	49	15	0.009	---
III	Suspension Medium Blank	95	58	5.1	9.1	NS	NS
IV	Medium Conditioned By Granulosa cells for 2 1/2 days	83	52	16.5	9.6	0.05	---
V	Medium + 0.1 μg/ml oPRL Conditioned by Granulosa Cells for 2 1/2 days	75	57	7.5	10.0	NS	NS
VI	Medium + 1.0 μg/ml oPRL Conditioned by Granulosa Cells for 2 1/2 days	99	37	38.8	8.8	<0.0001	<0.05
VII	Medium + 10 μg/ml oPRL Conditioned by Granulosa Cells for 2 1/2 days	71	24	61.9	3.2	<0.0001	<0.0001

Porcine granulosa cells were harvested from 2-5 mm porcine follicles and incubated in suspension medium (10% pig serum + 90% TC 199) for 1/2 days with or without ovine prolactin (NIH oPRl S-13). The conditioned medium was centrifuged and filtered and added to cumulus-enclosed oocytes and the oocytes cultred for 2 days. The maturation (Metaphase I to II) was evaluated at the end of the oocyte culture. Data from 4 experiments of granulosa cell cultures and 4 oocyte cultures were pooled to obtain the data shown in the table. Data were analyzed by Chi Square.

Additional studies which we report here demonstrate that ovine prolactin can stimulate granulosa cell secretion of OMI activity (Table 6). For these studies granulosa cells were harvested from 2-5 mm porcine follicles and cultured in 10% pig serum in tissue culture medium 199 A in siliconized glass scintillation vials for $2^1/_2$ days. The conditioned "suspension medium" was subsequently centrifuged and filtered through a 0.22 micron filter and added to cultured cumulus-enclosed porcine oocytes for assay of OMI activity. Two additional experiments in which the conditioned medium was passed through an Amicon PM-10 membrane and charcoal-treated yielded essentially similar results. As shown in Table 6 the addition of 1.0 μg and 10 μg ovine prolactin brough about a significant increase in inhibition, 38.8% and 61.9% inhibition, respectively, of oocyte maturation in the conditioned culture medium. Addition of 0.1 μg prolactin had no effect. Although this study is still in its preliminary stages it does give some indication that prolactin may be involved in control of OMI secretion. The infertility in hyperprolactinemic patients may in part be due to an overproduction of OMI. Prolactin may also lead to alteration in ovarian production of luteinization inhibitor and inhibin which may also contribute to alterations in ovarian function. A stimulation of follicular luteinization inhibitor could indirectly lead to a decrease in progesterone secretion, a situation often found in the presence of elevated prolactin (McNatty, 1975). In the present experiments addition of prolactin alone to granulosa cells had no clear cut effect on progesterone secretion by cultured porcine granulosa cells.

As mentioned above substances cross reacting with prolactin antiserum have been demonstrated in ovarian follicles (Nolin, 1980a). The substance may be prolactin itself and its action may be on secretion of nonsteroidal regulators, akin to its action on mammary tissue-stimulation of lactalbumin and casein biosynthesis. Variability in both basal- and prolactin-stimulated OMI secretion by porcine granulosa cells obtained from slaughterhouse material could be accounted for by differences in in vivo prolactin levels in the pig at the time of slaughter. This is to be expected since prolactin is elevated during stress.

NON-HORMONAL CONTROL OF OOCYTE MATURATION IN VITRO

Examination of meiotic maturation in culture must be carried out under carefully controlled conditions both as an assay system for OMI and as a system for physiological investigations. Consideration of the ionic and nutrient composition of the medium must be made as well as the osmolarity, pH, and nature of buffers and their possible adverse toxic effects.

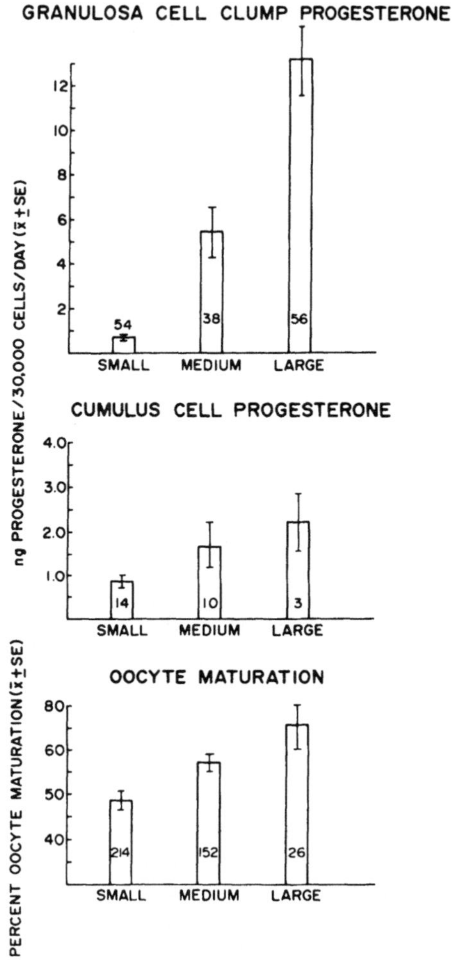

Fig. 5. Effect of porcine follicle size on oocyte maturation and cumulus progesterone secretion in culture. Cumulus-enclosed porcine oocytes and clumps of granulosa cells were aspirated from small (1-2 mm), medium (3-5 mm) and large (6-12 mm) follicles and cultured for 2 days in Medium 199 A plug 15% pig serum. At the end of the culture period the culture medium was frozen and later assayed for progesterone by radioimmunoassay. The oocytes were fixed in acetic acid-methanol 1:3 and stained with aceto-orcein and evaluated for maturation. Cells (nuclei) were counted with crystal violet after treatment with citric acid and progesterone secretion was normalized per 30,000 cells. Taken from Channing et al. (1981) with permission.

Table 7. Effect of Osmolarity of the Culture Medium upon Maturation of Cumulus Enclosed Oocytes

Medium	Total Number	Percent Maturation	% Inhibition Compared to Control 199 A	P vs Control 199 A	P vs #4
1. Control 199 A (280 mOs/l)	161	88.8	---	---	
2. PM 10, AL 35 (1:50)	149	63.8	28.2	<0.001	
3. 260 mOs/l	131	85.5	3.7	NS	NS
4. 280 mOs/l	97	81.4	8.3	NS	
5. 300 mOs/l	134	77.6	12.6	<0.001	NS
6. 320 mOs/l	128	76.6	13.7	<0.001	NS
7. 340 mOs/l	96	71.9	19.0	<0.001	0.03

Oocytes harvested from medium sized porcine follicles were cultured for two days in the presence of varying osmolality media. Medium number 3 was prepared by dilution of control medium 199A (280mOs/l to 260mOs/l). The remaining media numbers 4-7 were prepard by the addition of appropriate amounts of a 100-fold concentrated solution of Medium 199 salts without $NaHCO_3$ to form solutions with osmolarities of 280, 300, 320, and 340 respectively. PM10, AL35 was an OMI standard run as a positive inhibitory control. Statistical comparisons were carried out by use of Chi-square analysis.

Osmolarity

The osmolarity of the culture medium is a critical variable when examining oocyte maturation as documented by studies in the rabbit by Bae and Foote (1980), in the mouse by Brinster (1965), and in the pig by McGaughey (1979). Cultured rabbit ooocytes mature to about the same extent in salt concentrations from 250 mOsm to 310 mOs/kg (Bae and Foote, 1980). We determined the osmolarity required for maturation of cultured cumulus-enclosed porcine oocytes grown in Medium 199 A (Tsafriri and Channing, 1975). This culture medium, consisting of TC 199, 15% pig serum, 1 μg/ml insulin, 2.5 mM Na-lactate, and 0.03 mM Na pyruvate, was chosen since it is routinely used for OMI assays and supports granulosa cumulus cell monolayer formation and progesterone secretion. It is frequently desirable to monitor all of these cell systems along with oocyte maturation throughout purification of OMI activity. In addition the serum may serve to "swamp out" nonspecific adverse effects of small changes in ionic composition as well as aid steroidogeneis by furnishing a carrier protein for secreted progesterone. Osmolarity was varied by the addition of appropriate amounts of a 100-fold concentrated solution of Medium 199 salts without $NaHCO_3$. As seen in Table 7 porcine oocytes mature to nearly the same extent when the osmolarity is varied between 260 and 320 mOs/kg. At osmolarities ≥320 there is a small amount of inhibition of maturation (14-19%). We usually adjust OMI preparations to about 280 mOsm/liter prior to addition to the culture medium although for some purposes we allow osmolarities of up to 310 in the cultures. Cumulus cell outgrowth and progesterone secretion all occurred approximately equally from 260 to 320 mOs. A small decrease in progesterone secretion was observed at 340 and 360 mOsm/liter (data not shown).

Effect of Calcium Ion

It is well documented that calcium ion is required for frog (Kostellow and Morrill, 1980; Moreau et al., 1976; Wasserman et al., 1980) and rat (Batta and Knudson, 1980; Tsafriri and Bar-Ami, 1978) germinal vesicle breakdown. It is possible that Ca^{2+} is required for porcine oocyte maturation but, to our knowledge, no study had been carried out examining the role of Ca^{2+} in porcine oocyte maturation beyond the germinal vesicle breakdown stage. Since it is not feasible to examine the effect of various calcium concentrations upon oocyte maturation using complex serum-containing media it was decided to study the effects of various amounts of calcium upon porcine oocyte maturation using a serum-free Hanks balanced salt solution. The medium was modified to contain balanced Hanks salts plus Na lactate (2.5 mM), Na pyruvate (0.03 mM), glucose (5.55 mM) and bovine serum albumin (0.4%) (Bae and Channing, in preparation) and its osmolarity was adjusted to around 280 mOsm/liter (Bae and Foote, 1980). The porcine oocytes enclosed in their cumuli were harvested from medium-sized follicles, rinsed in Ca^{2+} free balanced modified Hanks salts with 0.1% BSA without pyruvate, lactate and glucose and cultured in medium in the presence of various concentrations of $CaCl_2$ ranging from zero to 17 mM (10 times the normal plasma level). Oocyte maturation was evaluated at the end of a 42 h culture period after fixing the oocytes in acetic acid, ethanol (1:3) and staining with aceto orcein. Changes in Ca^{2+} from 0.053 to 17 mM brought about no significant changes in per cent oocyte maturation. If the $CaCl_2$ was decreased to 0.026 mM (65 oocytes) the oocyte maturation was significantly less (44%) than that observed at 1.7 mM level of $CaCl_2$ (68%). A decrease to 0.0133 mM or to no calcium added led to no maturation at all (70, 72 oocytes respectively) ($p < 0.05$ vs 1.7 mM $CaCl_2$). These data (Bae and Channing, in preparation) demonstrate that some extracellular Ca^{2+} is required for oocyte maturation. This small amount may be supplemented by Ca^{2+} made available from intracellular stores such as endoplasmic reticulum (McGraw et al., 1980). Further studies with calcium ionophore and Ca^{2+} channel blockers are being carried out to more clearly define the role of Ca^{2+} in porcine oocyte maturation in vitro.

Role of Oocyte Maturatin Inhibitor and Calcium

Prior to the preovulatory surge of LH/FSH projections of the cumulus cells have gap junctions with the surface of the oocyte. Under the influence of LH/FSH the cumulus cells withdraw their projections to the oocyte and oocyte maturation progresses (Dekel and

Table 8. Effect of Dibutryl Cyclic AMP and PM-10 Filtrate of Porcine Follicular Fluid upon Porcine Cumulus-Enclosed Oocyte Maturation and Progesterone Secretion

Treatment	Number Oocytes	% Maturation Mean	% Inhibition	P vs Control	Progesterone Secretion ng/ 15 cumuli/day (n=3) Mean	SE	P vs Control
Control	40	75.0	---	---	3.2	0.25	---
PM-10 AL36 (1:50)	41	48.8	34.9	<0.05	0.73	0.1	<0.01
1 mM Dibutryl Cyclic 3′,5′-AMP	40	20	73.3	<0.01	17	1.6	<0.01
10 mM Dibutryl Cyclic 3′,5′-AMP	29	0	100	<0.01	12	1.7	<0.01

Cumulus-enclosed oocytes were harvested from medium sized porcine follicles and cultured for two days. Oocyte maturation and progesterone secretion were estimated at the end of the two day period. (Evans, Conn, and Channing, unpublished observations.)

Beers, 1978). Since the OMI acts to arrest oocyte maturation via the cumulus cells (i.e., addition of partially purified OMI dies not arrest maturation of denuded porcine oocytes, Hillensjö et al., 1979) it is possible that OMI acts to preserve the gap junctions between the projections of the cumulus cells and the oocyte. Since calcium plays an important role in membrane integrity and in the preservation of gap junctions as well as in oocyte maturation, it is possible that the OMI sequesters Ca^{2+} and thereby arrests oocyte maturation. Alternately the gap junctions could act as conduits for OMI so that it can gain access into the oocyte. Further studies with a more purified preparation of OMI should resolve some of these questions on its mechanism of action in bringing about inhibition of oocyte maturation.

Influence of Cyclic AMP

Cyclic AMP has been shown by a number of investigators to exert an inhibitory influence upon oocyte maturation (frog: Morril, 1977; rat: Hillensjö, this volume). Cyclic AMP is a paradoxical compound since when added to rat oocyte cultures it arrests resumption of maturation (Hillensjö, this volume) yet, on the other hand, addition of LH to cultured follicle-enclosed oocytes stimulates cyclic AMP accumulation as well as bringing about oocyte maturation (Lindner et al., 1974). Preliminary studies carried out in our laboratory in cultured cumulus-enclosed porcine oocytes confirmed the paradoxical effect of cyclic AMP on oocyte maturation. Addition of 1 and 10 mM dibutryl cyclic 3',5'-AMP to cultured cumulus-enclosed porcine oocytes significantly inhibited oocyte maturation while at the same time stimulating cumulus cell progesterone secretion (Table 8). In contrast addition of crude OMI (low molecular weight fraction of porcine follicular fluid) inhibited oocyte maturation while also inhibiting progesterone secretion (Table 8 and Hillensjö et al., 1980).

SUMMARY

Oocyte maturation is controlled by a combination of hormonal and local follicular factors. Osmolarity, pH, and perhaps Ca^{2+} concentration of the surrounding medium are also important. Follicular fluid contains a low molecular weight OMI which acts to keep the oocyte from maturing. Luteinizing hormone added to cultured cumulus enclosed porcine oocytes can reverse the inhibitory action of OMI. The level of OMI in the follicular fluid appears to decrease as the follicle matures. Granulosa cells secrete OMI and their potential to secrete OMI in culture decreases as the follicle matures. Addition of FSH and prolactin to cultured granulosa cells stimulates OMI secretion whereas addition of testosterone or dihydrotestosterone brings about a decrease in OMI secretion. Elevated LH in vivo may bring about oocyte maturation before ovulation by (a) an antagonist action on OMI; (b) stimulating the synthesis of testosterone by theca cells and thus inhibiting the synthesis of OMI by granulosa cells; and (c) action on the granulosa cells to promote luteinization which may also cause a decrease in OMI synthesis. The hastened oocyte maturation associated with follicular atresia could be due to a decline in OMI due to granulosa cell death and/or elevated follicular androgens.

ACKNOWLEDGEMENTS

Supported in part by USDA Bard Grant 1-112-79 Ford Foundation Grant 795-0025, NIH Grant HD 12003 (CPC) and NIH Grant HD 14696 (SHP). The able technical assistance of Mr. Timothy Conn and Mr. Jack Vellajo is gratefully acknowledged. We thank the NIH for providing some of the porcine follicular fluid used for these studies.

REFERENCES

Ainsworth, L., Tsang, B. K. , Downey, B. R., Marcus, G. J., Armstrong, D. T., 1980, Interrelationships between follicular fluid steroid levels, gonadotropic stimuli, and oocyte maturation during preovulatory development of porcine follicles. Biol. Reprod., 23:621.

Anderson, L. D., and Stone, S. L., 1980, Hormonal control of oocyte maturation inhibitor (OMI) secretion from porcine granulosa cells. Abstract 358 presented at the 62nd Annual Meeting of the Endocrine Society, Washington, D.C., June 18-20, 1980.

Bae, I.-H., and Foote, R. H., 1975, Carbohydrate and amino acid requirements and ammonia production of rabbit follicular oocytes matured in vitro, Expl. Cell Res., 91:113.

Bae, I.-H., and Foote, R. H., 1980, Maturation of rabbit follicular oocytes in a defined medium of varied osmolality, J. Reprod. Fert., 59:11.

Baker, T. G., and Hunter, R. H. F., 1978, Interrelationships between the oocyte and somatic cells within the Graffian follicle of mammals, Ann. Biol. Anim. Biochem. Biophys., 18:419.

Batta, S. K., and Knudsen, J. F., 1980, Calcium concentration in cumulus enclosed oocytes of rats after treatment with pregnant mares serum, Biol. Reprod., 22:243.

Biggers, J. D., Whittingham, D. G., and Donahue, R. P., 1967, The pattern of energy metabolism in the mouse oocyte and zygote, Proc. Natl. Acad. Sci. (U.S.A.), 58:560.

Brinster, R. L., 1965, Studies on the development of mouse embryos in vitro. I. The effect of osmolarity and hydrogen ion concentration, J. Exp. Zool., 158:49.

Centola, G. M., Anderson, L. D., and Channing, C. P., 1981, Oocyte maturation inhibitor (OMI) activity in porcine granulosa cells, Gamete Research, in press.

Channing, C. P., Batta, S. K., Condon, W., Ganjam, V. K., and Kenney, R. M., 1981a, Levels of inhibin activity and an atretogenic factor(s) in follicular fluid harvested from viable and atretic mare follicles, in: "Dynamics of Ovarian Function," N. B. Schwartz and M. Hunzicker-Dunn, eds., Raven Press, New York, p. 73.

Channing, C. P., Gagliano, P., Hoover, D. J., Tanabe, K., Batta, S. K., Sulewski, S., and Lebech, P., 1981b, Relationship between human follicular fluid inhibin F activity and steroid content, J. Clin. Endocr. Metab., 52:1193.

Channing, C. P., Schaerf, F. W., Anderson, L. D., and Tsafriri, A., 1980, Ovarian follicular and luteal physiology, in: "Reproductive Physiology III. International Review of Physiology, Vol. 22," R. O. Greep, ed., pp. 117-201.

Dekel, N., Kraicer, P. F., Phillips, D. M., Sanchez, R. S., and Segal, S. J., 1978, Cellular associations in the rate oocyte-cumulus cell complex: Morphology and ovulatory changes, Gamete Research, 1:47.

Gwatkin, R. B. L., and Andersen, O. F., 1976, Hamster oocyte maturation in vitro, Life Sciences, 19:527.

Hillensjö, T., Kripner, A. S., Pomerantz, S. H., and Cha-ning, C. P., 1979, Action of porcine follicular fluid oocyte maturation inhibitor in vitro: Possible role of the cumulus cells, in: "Ovarian Follicular and Corpus Luteum Function," C. P. Channing, J. M. Marsh, and A. S. Sadler, eds., Plenum Press, New York, pp. 283-290.

Hillensjö, T., Pomerantz, S. H., Schwartz-Kripner, A., Anderson, L. D., and Channing, C. P., 1980, Inhibition of cumulus cell progesterone secretion by a low molecular weight fraction of porcine follicular fluid which also inhibits oocyte maturation, Endocrinology, 106:584.

Hirshfield, A. N., and Midgley, A. R., Jr., 1978, The role of FSH in the selection of large ovarian follicles in the rat, Biol. Reprod., 19:606-611.

Kent, H., 1978, Inhibition of ovulation by synthetic samples of the tetrapeptide originally isolated from hamster embryos, Biol. Reprod., 19:785.

Kostellow, A. B., and Morrill, G. A., 1980, Calcium dependence of steroid and guanine 3',5'-monophosphate induction of germinal vesicle breakdown in Rana pipiens oocytes, Endocrinol., 106: 1012.

Leibfried, L., and First, N. L., 1980, Effect of bovine and porcine follicular fluid and granulosa cells on maturation of oocytes in vitro, Biol. Reprod., 23:699.

Lindner, N. R., Tsafriri, A., Lieberman, M. E., Zor, U., Koch, Y., Bauminger, S., and Barnea, A., 1974, Gonadotropin action on cultured Graafian follicles: Induction of maturation division of the mammalian oocyte and differentiation of the luteal cell, Recent Progress in Hormone Research, 30:79.

Matson, P. L., Tyler, J. P. P., and Collins, W. P., 1981, Follicular steroid content and oocyte meiotic status after PMSG stimulation of immature hamsters, J. Reprod. Fert., 61:443.

McGaughey, R. W., 1977, The maturation of porcine oocytes in minimal, defined culture media with varied macromolecular supplements and varied osmolarity, Expt. Cell Res., 109:25.

McGraw, C. F., Somlyo, A. V., and Blaustein, M. P., 1980, Localization of calcium in presynaptic nerve terminals: An ultrastructural and electron microprobe analysis, J. Cell Biol., 85:228.

McNatty, K., 1974, A possible role for prolactin in control of steroid secretion by the hyman graafian follicle, Nature, 240:653.

McNatty, K. P., Makris, A., DeGrazia, C., Osathanondh, R., and Ryan, K. J., 1979a, The production of progesterone, androgens, and estrogens by granulosa cells, thecal tissue and stromal tissue from human ovaries in vitro, J. Clin. Endocrinol. Metab., 49:687.

McNatty, K. P., Smith, D. M., Makris, A., Osathanondh, R., and Ryan, K. J., 1979b, The microenvironment of the human follicle: Interrelationships among the steroid levels in antral fluid, the population of granulosa cells, and the status of the oocyte in vivo and in vitro, J. Clin. Endocrinol. Metab., 49:851.

Moreau, M., Doree, M., and Guerria, P., 1976, Electrophoretic introduction of calcium ions into the cortex of xenopus laevis oocytes triggers meiosis reinitiation (1), J. Exp. Zool., 197: 443.

Morrill, G. A., Schatz, F., Kostellow, A. B., and Pouko, J. M., 1977, Changes in cyclic AMP levels in the amphibian ovarian follicle following progesterone induction of meiotic maturation, Differentiation, 8:97.

Nolin, J. M., 1980a, Incorporation of endogenous prolactin by granulosa cells and dictyate oocytes in the postpartum rat: Effects of estrogen, Biol. Reprod., 22:417.

Nolin, J. M., 1980b, The ultimate source of oocyte maturation inhibitor (OMI) may be pituitary prolactin, Abstract presented at the 20th meeting of the American Society for Cell Biology, November 1980.

Nolin, J. M., and Bogdanove, E. M., 1980, Effects of estrogen on prolactin (PRL) incorporation by lutein and milk secretory cells and on pituitary PRL secretion in the postpartum rat: Correlations with target cell responsiveness to PRL, Biol. Reprod., 22:393.

Pomerantz, S. H., Tsafriri, A., and Channing, C. P., 1979, Studies on purification and actions of an oocyte maturation inhibitor from porcine follicular fluid, in: "Peptides: Structure and Biological Function," E. Gross and J. Meinenhofer, eds., Pierce Chemical Co., Rockford, Il., p. 765.

Sato, E., Initani, A., and Nishikawa, Y., 1980, Mechanism of inhibition of germinal vesicle breakdown in pig follicular oocytes with special reference to the role of granulosa cell layer, Jap. J. Fert. and Steril., 25:229.

Sato, E., and Ishibashi, T., 1977, Meiotic arresting action of the substance obtained from cell surface of porcine ovarian granulosa cells, Jap. J. Zootech. Sci., 48:22.

Schuetz, A. W., 1974, Role of hormones in oocyte maturation, Biol. Reprod., 10:150.

Stone, S. L., Pomerantz, S. H., Schwartz-Kripner, A., and Channing, C. P., 1978, Inhibitor of oocyte maturation from porcine follicular fluid: Further purification and evidence for reversible action, Biol. Reprod., 19:585.

Tsafriri, A., 1978, Oocyte maturation in mammals, in: "The Vertebrate Ovary," R. E. Jones, ed., Plenum Press, New York, p. 409.

Tsafriri, A., and Bar-Ami, S., 1978, Role of divalent cations in the resumption of meiosis of rat oocytes, J. Exp. Zool., 205:293.

Tsafriri, A., and Channing, C. P., 1975, An inhibitory influence of granulosa cells and follicular fluid upon porcine oocyte meiosis in vitro, Endocrinology, 96:922.

Tsafriri, A., Channing, C. P., Pomerantz, S. H., and Lindner, H. R., 1977, Inhibition of maturation of isolated rat oocytes by porcine follicular fluid, J. Endocrinol., 75:285.

Tsafriri, A., Pomerantz, S. H., and Channing, C. P., 1976, Inhibition of oocyte maturation by porcine follicular fluid: Partial characterization of the inhibitor, Biol. Reprod., 14:511.

Tsafriri, A., Weinstein, Y., Bar-Ami, S., Channing, C. P., Pomerantz, S. H., and Lindner, H. R., 1979, The control of meiotic maturation of the rat oocyte, in: "Research on Steroids, Vol. 8," Academic Press, London, p. 193.

Wasserman, W. J., Pinto, L. H., O'Connor, C. M., and Smith, L. D., 1980, Progesterone induces a rapid increase in [Ca^{2+}] in xenopus levis oocytes, Proc. Natl. Acad. Sci. U.S.A., 77(3):1534.

INTRAOVARIAN ACTIONS OF GnRH

Fernand Labrie, Carl Séguin,
Fleur-Ange Lefebvre, Jocelyne Massicotte,
Georges Pelletier, Jean-Philippe Borgus,
Paul-A. Kelly, Jerry J. Reeves,
and Alain Bélanger

MRC Group in Molecular Endocrinology
Le Centre Hospitalier de l'Université Laval
Quebec G1V 4G2, Canada

Introduction

Treatment with LHRH agonists can inhibit ovulation and interrupt pregnancy in experimental animals (Banik and Givner, 1975; Corbin et al., 1976; Humphrey et al., 1976; Johnson et al., 1976; Rippel and Johnson, 1976; Kledzik et al., 1978; Rivier et al., 1978).

Following the observation that treatment of male rats with LHRH or its agonists leads to a loss of testicular LH and prolactin receptors (Auclair et al., 1977a, b), it was found that similar treatment of female animals decreased ovarian LH, FSH and prolactin receptors (Kledzik et al., 1978a, b), thus providing the first insight into the mechanisms responsible for the antifertility effects of LHRH agonists in the female. Since the antifertility effects of LHRH agonists are parallel to their LH-releasing activity (Cusan et al., 1979), it was suggested that endogenous LH release leading to a secondary loss of gonadal receptors and blockage of steroidogenesis is responsible for the antifertility effects observed after such treatment (Kledzik et al., 1978a, b). However, the early observation by Rippel and Johnson (1976) that administration of a potent LHRH agonist to hypophysectomized immature rats prevents the hCG-induced increase in ovarian weight already suggested a direct ovarian site of action of the peptide.

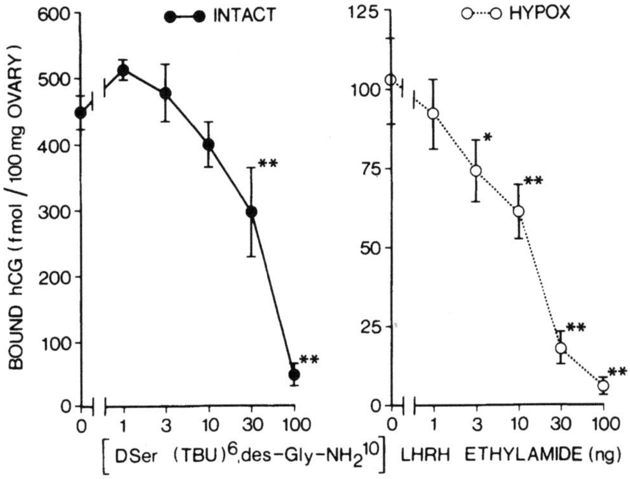

Fig. 1. Effect of daily treatment for 9 days of intact or hypox (3 days earlier) PMSG-treated immature rats with increasing doses of [D-Ser(TBU)^6]LHRH-EA on ovarian LH/hCG receptors. The animals were sacrificed 24 h after the last injection of the peptide (Séguin et al., 1981).

A direct inhibitory action of LHRH agonists at the ovarian level in vivo has also been observed by other groups at the ovarian level in hypophysectomized rats (Mayar et al., 1979; Ying and Guillemin, 1979; Séguin et al., 1980; Jones and Hsueh, 1980). Moreover, LHRH agonists have been shown to inhibit ovarian steroidogenesis in granulosa and luteal cells in culture (Hsueh and Erickson, 1979; Clayton et al., 1979; Massicotte et al., 1980, 1981; Behrman et al., 1980). The direct gonadal site of action of LHRH agonists is supported by the finding of high affinity LHRH binding sites in ovarian homogenate (Reeves et al., 1980; Harwood et al., 1980) as well as luteal (Clayton et al., 1979) and granulosa (Jones et al., 1980) cells.

Valid assessment of the relative importance of the pituitary and gonadal sites of the antifertility action of LHRH agonists can best be achieved under in vivo conditions. Although LHRH and some of its agonists analogues have been shown to have direct gonadal effects in hypophysectomized animals, pharmacological doses of the LH-releasing peptides have been used and no dose-response study has been performed. The present study compares the effects of increasing doses of the LHRH agonist [D-Ser(TBU)^6, des-Gly-NH_2^{10}]LHRH ethylamide (kindly provided by Dr. J. Sandow, Hoechst AG, Frankfurt), on ovarian gonadotropin receptors and steroidogenesis in intact and hypophysectomized female rats. Since hypopysectomy by itself has profound effects on gonadal functions, we have used another approach which minimizes hormonal disturbances in order to assess the importance of the direct action of LHRH agonists, namely by using intact animals treated with an excess of equine anti-bovine LH serum. A study will then be made of the modulation of LHRH receptor levels by pituitary gonadotropins and of the changes of ovarian LHRH receptors which accompany sexual maturation and long-term treatment with LHRH agonists in the rat. This review will next examine the role of the adenylate cyclase in the inhibitory effect of LHRH on ovarian function.

The interest of the above-mentioned studies is supported by the recent data which clearly indicate that treatment with LHRH agonists can also inhibit reproductive functions in the human. In fact, treatment with LHRH or its agonists has been shown to induce luteolysis and inhibit ovulation in normal women (Labrie et al., 1979; Lemay et al., 1978, 1979; Koyama et al., 1978; Nillius et al., 1978).

Relative Importance of the Adenohypophyseal and Ovarian Sites of Action of LHRH Agonists

As illustrated in Fig. 1, the daily injection of 1, 3, or 10 ng of [D-Ser(TBU)^6]LHRH-EA (Buserelin) for 9 days in intact PMSG-treated immature rats has no significant effect on ovarian LH receptor levels while treatment with the 30 and 100 ng doses decreases LH receptor levels by 35 and 95% ($p < 0.01$), respectively. In hypox animals, a significant loss of ovarian LH receptors (30%, $p < 0.05$) is seen in animals treated with as little as 3 ng of the LHRH agonist. The daily 10, 30, 100 ng doses of the peptide decrease ovarian LH binding sites by respectively 40, 82, and 94%.

Previous data suggest that endogenous LH release induced by the LHRH agonists with secondary testicular desensitization is probably the main factor responsible for the loss of gonadotropin receptors, blockage of the steroidogenic pathway as well as inhibition of spermatogenesis observed after treatment of intact adult male animals with LHRH agonists (Séguin et al., 1981). However, in the PMSG-hCG-treated immature female rat, the similar potency of the LHRH agonist to induce loss of ovarian LH receptors in intact and hypox animals suggests that the direct gonadal site of action may play a major role in the ovarian desensitization process.

We next compared the effect of treatment with two 500 ng doses of Buserelin in animals treated with normal horse serum (NHS) or anti-LH serum kindly provided by Dr. W. Hansel, New York. As illustrated in Fig. 2, treatment with the LHRH agonist leads to a 82 and 63% ($p < 0.01$) inhibition of ovarian LH receptor levels in NHS- and A-LH-treated animals, respectively. These data obtained in the

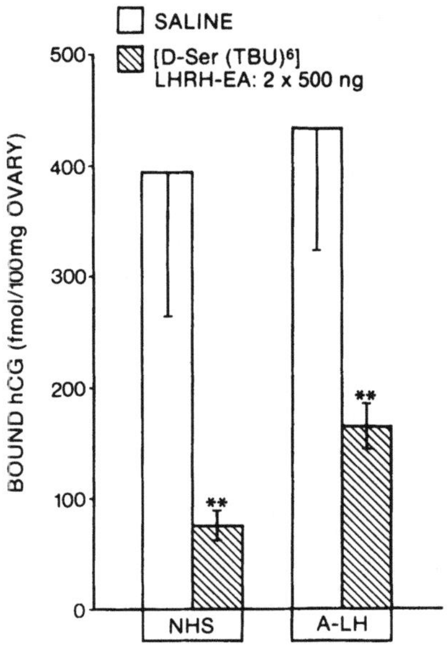

Fig. 2. Effect of a single s.c. injection of anti-LH serum or normal horse serum (NHS) on Buserelin-induced inhibition of ovarian LH receptors in PMSG-primed immature female rats. Buserelin (500 ng) was injected 6 and 48 h after 0.5 ml of the anti-LH serum and the animals were sacrificed 48 h after the second injection of the LHRH agonist.

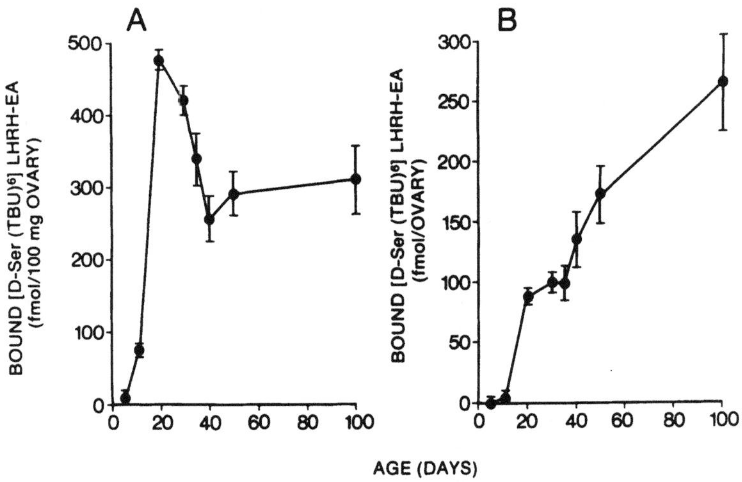

Fig. 3. Age-related changes in ovarian LHRH receptors expressed as concentratin per 100 mg ovary (A) or per two ovaries (B) in the rat.

female rat are again quite different from those observed in the male where the inhibitory effect of treatment with LHRH agonists is markedly reduced in animals treated with the anti-LH serum (data not shown).

Such data clearly suggest that the ovary is more sensitive than the testis to the direct inhibitory action of LHRH agonists. Since the affinity and specificity of the LHRH receptor in ovarian and testicular tissues are similar (Reeves et al., 1980; Lefebvre et al., 1980), the mechanisms responsible for the different sensitivity to LHRH agonists in the two tissues remain to be investigated.

Ovarian LHRH Receptors During Development

Since the first step in the action of LHRH or its agonists is binding to specific LHRH receptors and these receptors are likely to be involved in the physiological control of ovarian function, we next studied possible changes of ovarian LHRH receptors during sexual maturation in the rat. In order to assess the role of the multiple hormonal interactions occurring during ovarian development, changes of LHRH receptor levels were correlated with changes of pituitary and plasma gonadotropin concentration as well as ovarian LH, FSH and prolactin receptor levels.

As illustrated in Fig. 3, ovarian LHRH receptors are almost undetectable at 5 days of age and then increase sharply between days 10 and 20. From day 20, the concentration of ovarian LHRH receptors gradually declines during the prepubertal period to reach adult levels at approximately the time of the first ovulation at about 40 days (Fig. 3A). Thereafter, the concentration of ovarian LHRH receptors remains unchanged up to the last time interval studied (100 days). The concentration of ovarian LHRH receptor remains at the same level in adult cycling animals (Reeves et al., 1980). When LHRH receptor data are expressed in total ovarian content (Fig. 3B), after the rapid increase observed between days 10 and 20, a progressive increase was measured throughout development up to day 100. It was of interest to study the changes of ovarian LH, FSH and prolactin receptors at the same time intervals. Ovarian LH receptor levels increased progressively from day 10 to reach a maximal value on day 40. Thereafter, LH receptors decreased by approximately 40% during the following 10 days and remained constant between days 50 and 100 (data not shown).

This study was complemented by the autoradiographic localization of ovarian LHRH receptors after injection of the potent and stable iodinated LHRH agonist [D-Ser(TBU)^{6}, des-Gly-NH$_2$^{10}]LHRH ethylamide in rats of different ages. In the adult rat, silver grains were observed in corpora lutea, specially on light cells while the few dark cells were devoid of staining. Labeling was also present over mature follicles on both the follicular and thecal cells. Ova and follicular fluid were consistently labelled. Interstitial cells of the steroid-secretory type laying between follicles were as intensively labeled as follicular cells. In growing follicles, similar labeling was observed whereas primordial follicles did not show any significant reaction.

Essentially, the same type of distribution of silver grain was found in the developing ovaries before puberty. Neither mature follicles nor corpora lutea were observed in these ovaries. Silver grains were only observed over granulosa and thecal cells of primary and antral follicles as well as interstitial steroid-secretory cells in the ovaries of rats aged of 11, 15, 20, and 30 days. A few grains were consistently observed over the follicular fluid. No significant labeling could be observed in the ovaries of 2, 6, and 8 day-old rats. As in the adult animals, primordial follicles were never labeled. In all the animals injected with both the radioactive and excess unlabeled LHRH-EA, only grains accounting for background could be detected, thus indicating the specificity of the autoradiographic reaction (data not shown).

The present results clearly show that ovarian LHRH receptor levels markedly decrease in the week preceding the first ovulation while LH and prolactin receptors increase progressively and FSH receptors remain elevated. These findings confirm and extend a recent study which has shown that the number of LHRH receptors in the ovary decreases markedly prior to the time of the first preovulatory surge of gonadotropins (Smith-While and Ojeda, 1981). The ovarian responsiveness to gonadotropins increases before puberty. These changes in ovarian responsiveness could be partly due to the increased ovarian LH and prolactin receptor levels which have been postulated to play a fundamental role in the initiation of puberty in the female rat. It is also quite possible that the decrease of ovarian LHRH receptors which takes place between days 20 and 40 could permit a gradual increase in ovarian responsiveness and steroid output leading to puberty. Moreover, the presence of high levels of ovarian LHRH receptors could provide a mechanism which prevents follicles from progressing through the total maturational process before puberty. This is supported by the autoradiographic finding of LHRH receptors in interstitial and follicular (thecal, granulosa and luteal) cells at all stages of maturation except in the primordial follicles. It is also possible that the inhibitory action of the LHRH receptor could play an important role in the process of atresia. Such a role of the ovarian LHRH receptor remains to be examined in the light of the recent demonstration of LHRH-like activity in ovarian tissue (Godbout and Labrie, unpublished observations).

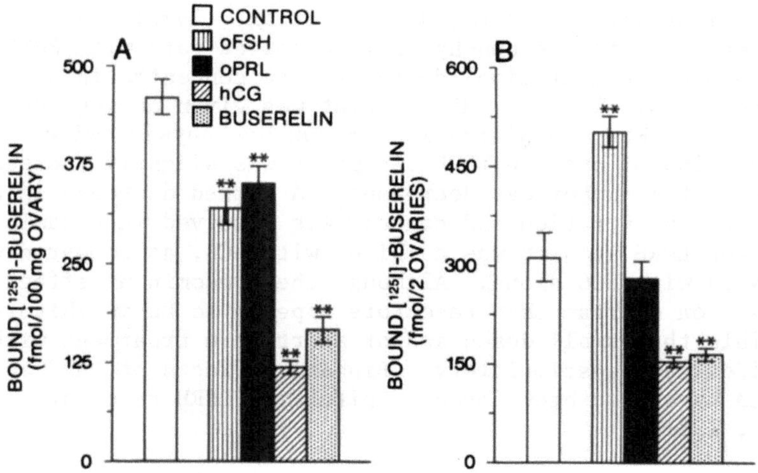

Fig. 4. Effect of treatment with oFSH, oPRL, hCG and Buserelin on LHRH receptor concentration and content in the ovary. Immature female animals treated with PMSG were used in this study. Forty-eight hours after the PMSG injection, groups of treated subcutaneously for a period of five days with: oFSH, 1 μg twice daily; oPFL (NIH-PRL-S14), 100 μg twice daily; hCG, 25 IU/day; or Buserelin, 200 ng/day.

Modulation of Ovarian LHRH Receptors

Since changes of receptor levels are a frequent mechanism of control of cellular activity, particularly by hormones, we have examined the effects of LH, FSH, prolactin, and an LHRH agonist on gonadal LHRH receptor levels in female rats. In order to eliminate interference by endogenous pituitary hormones, hypophysectomized animals were used.

As illustrated in Fig. 4A, treatment of hypophysectomized immature female animals with oFSH, oPRL, hCG or the potent LHRH agonist [D-Ser(TBU)^6, des-Gly-NH$_2$^{10}]LHRH ethylamide (Buserlin), for a period of 5 days, decreased ovarian LHRH receptor concentration by 30, 24, 74, and 64% ($p < 0.01$), respectively. Total ovarian content of LHRH receptors (panel B) was not affected by the administration of oPRL but was increased to 160% of control ($p < 0.01$) following oFSH treatment. Human chorionic gonadotropin and Buserelin inhibited ovarian LHRH receptor content by 51 and 47%, respectively ($p < 0.01$). The observed increase in receptor content after oFSH administration can be explained by the oFSH-induced increase in ovarian weight, from 81 ± 8.6 to 141 ± 8.2 mg ($p < 0.01$). Adminstration of hCG also increased total ovarian weight to 123 ± 8.4 mg ($p < 0.01$).

In agreement with our data, Harwood et al. (1980) have reported that treatment of immature non-hypophysectomized rats with PMSG (an oFSH-like hormonal preparation) decreased the concentratin of ovarian LHRH receptors. When an LHRH agonist was given during the first 3 days following PMSG administration in non-hypophysectomized rats, the concentration of ovarian LHRH receptors was slightly increased, although the total number was decreased. A marked decrease in both LHRH receptor concentration and content was observed when administration of the LHRH agonist was combined with hCG, as compared with groups treated with hCG alone. Although the predominant effect of LHRH agonists on ovarian LHRH receptors appears to be inhibitory, it is possible that small doses and/or short-term treatment with these peptides can be stimulatory. Biphasic effects of LHRH and its agonists have also been found on pituitary LHRH receptors (Ferland et al., 1981).

The present data show that administration of exogenous oPRL, oFSH, or hCG reduces ovarian LHRH receptor concentration in the rat, thus indicating that circulating levels of pituitary hormones known to be involved in the regulation of gonadal functions also play a role in controlling LHRH receptor levels in the ovary. The mechanisms of such an action, as well as the physiological significance of gonadal LHRH receptors, are as yet unknown. In the female, recent data indicate that ovarian LHRH receptor levels decrease immediately before puberty (Fig. 3), at a time when the concentration of receptors for FSH, LH, and prolactin are at peak values, thus suggesting that ovarian functions could become less sensitive to direct inhibition by LHRH at this crucial stage in ovarian development. It will be of great interest to further elucidate how the many possible interactions between pituitary hormones and LHRH, and their respective binding sites in the ovary, affect gonadal development and function.

Inhibition of Cyclic AMP Accumulation by LHRH Agonists

Since the control of ovarian steroidogenesis by gonadotropins is thought to be mediated by cyclic AMP (Kolena and Channing, 1972), we have studied the possible role of the adenylate cyclase system in mediating the inhibitory effects of LHRH agonists on steroidogenesis in rat luteal cells in culture.

As shown in Fig. 5, 100 nm [D-Ser(TBU)^6]LHRH-EA leads to a 40% decrease of the maximal hCG-induced cyclic AMP cell content in luteal cells obtained from immature rats, while an almost (85-90%) complete reversal of the stimulatory effect of the gonadotropin is observed on the concentration of both cyclic AMP (Fig. 5B) and progesterone (Fig. 5C) in the incubation medium.

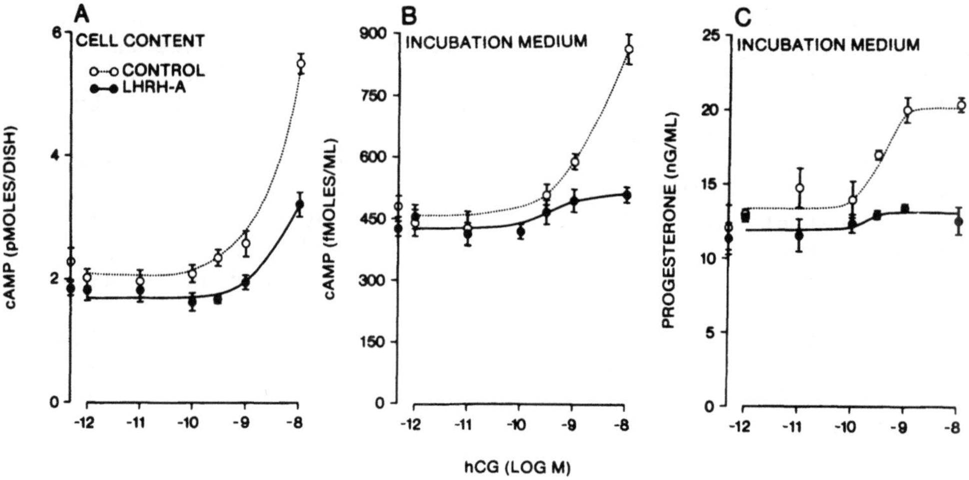

Fig. 5. Effect of increasing concentrations of hCG in the presence (●--●) or absence (o--o) of 100 nM [D-Ser(TBU)[6]]LHRH-EA on cyclic AMP cell content (A), cyclic AMP concentration in the incubation medium (B) and progesterone secretion (C), in immature rat luteal cells in culture. Cyclic AMP in the medium was measured after 2 h of incubation while the cyclic AMP cell content and progesterone secretion were measured after a 3 h incubation period.

In order to investigate the specificity of the inhibitory effect of [D-Ser(TBU)[6]]LHRH-EA on cyclic AMP accumulation, we next studied the effect of an LHRH antagonist, [Dp-Glu[1], D-Phe[2], D-Trp[3], D-Phe[6]]LHRH. While not affecting basal cyclic AMP levels in the incubation medium in the presence or absence of [D-Ser(TBU)[6]]-LHRH-EA or hCG alone, the LHRH antagonist (1 μM) reverses by 50% the inhibitory effect of 100 mM [D-Ser(TBU)[6]]LHRH-EA, and 1 nM hCG-stimulated cyclic AMP levels in the incubation medium after 90 min of incubation. Complementary experiments show that complete reversal of the effect of [D-Ser(TBU)[6]]LHRH-EA can be achieved by the LHRH antagonist when a lower concentration of the agonist is used.

The present findings of a parallel inhibition of the concentration of cyclic AMP in rat luteal cells as well as in the incubation medium and progesterone secretion under the influence of an LHRH agonist strongly suggest that the inhibitory action of the peptide at the ovarian level is exerted through inhibition of cyclic AMP accumulation. A similar inhibitory effect has been previously observed on FSH-induced cyclic AMP accumulation and steroid secretion in porcine granulosa cells in culture (Massicotte et al., 1980).

Specificity of the action of LHRH agonists at the ovarian level is well supported by the reversal by LHRH antagonists of the inhibitory effect of LHRH agonists on both steroid formation (Hsueh and Ling, 1979) and cyclic AMP accumulation (Fig. 5) as well as by the demonstration of similar LHRH receptors in the anterior pituitary gland and ovary (Reeves et al., 1980). The observation of a stimulation of cyclic AMP accumulation by LHRH in adenohypophyseal tissue (Borgeat et al., 1972) and an inhibition of the same parameter in the ovary suggests the involvement of different coupling mechanisms leading to opposite effects of adenylate cyclase activity in the two tissues (Rodbell, 1980). The demonstration of specific effects on the adenylate cyclase system in granulosa and luteal cells suggests that ovarian receptors could play a physiological role in the control of ovarian function by a locally-produced LHRH-like molecule.

References

Auclair, C., Kelly, P. A., Labrie, F., Coy, D. H., and Schally, A. V., 1977a, Inhibition of testicular LH/hCG receptor levels by treatment with a potent LHRH agonist or hCG, Biochem. Biophys. Res. Commun., 76:855.

Auclair, C., Kelly, P. A., Coy, D. H., Schally, A. V., and Labrie, F., 1977b, Potent inhibitory activity of [D-Leu^6, des-Gly-$NH_2{}^{10}$]LHR ethylamide on LH/hCG and PRL testicular receptor levels in the rat, Endocrinology, 101:1890.

Banik, U. K., and Givner, M. L., 1975, Ovulation induction and antifertility effects of an LHRH analog (AY-2505) in cycling rats, J. Reprod. Fertil., 44:87.

Behrman, H. R., Preston, S. L., and Hall, A. K., 1980, Cellular mechanism of the antigonadotropic action of LHRH in the corpous luteum, Endocrinology, 107:656.

Borgeat, P., Chavancy, G., Dupont, A., Labrie, F., Arimura, A., and Schally, A. V., 1972, Stimulation of adenosine 3',5'-cyclic monophosphate accumulation in anterior pituitary gland in vitro by synthetic luteinizing hormone-releasing hormone, Proc. Natl. Acad. Sci. U.S.A., 69:2677.

Clayton, R. N., Harwood, J. P., and Catt, K. J., 1979, Gonadotropin-releasing hormone analogue binds to luteal cells and inhibits progesterone production, Nature, 282:90.

Corbin, A., Beattie, C. W., Yardley, J., and Toell, T. J., 1976, Postcoital contraceptive effects of an agonistic analogue of luteinizing hormone-releasing hormone, Endocr. Res. Commun., 3:359.

Cusan, L., Auclair, C., Bélanger, A., Ferland, L., Kelly, P. A., Séguin, C., and Labrie, F., 1979, Inhibitory effects of long-term treatment with a luteinizing hormone-releasing hormone agonist on the pituitary-gonadal axis in male and female rats, Endocrinology, 104:1369.

Ferland, L., Marchetti, B., Séguin, C., Reeves, R., and Labrie, F., 1981, Dissociated changes of pituitary LHRH receptors and responsiveness to the neurohormone induced by 17β-estradiol and LHRH in vivo in the rat, Endocrinology, in press.

Harwood, J. P., Clayton, R. N., Chen, T. T., Knox, G., and Catt, K. J., 1980, Ovarian gonadotropin-releasing hormone receptors. II. Regulation and effects on ovarian development, Endocrinology, 107:414.

Hsueh, A. J. W., and Erickson, G. F., 1979, Extrapituitary action of gonadotropin-releasing hormone: direct inhibition of ovarian steroidogenesis, Science, 204:854.

Hsueh, A. J. W., and Ling, N. C., 1979, Effect of an antagonistic analog of gonadotropin-releasing hormone upon ovarian granulosa cell function, Life Sci., 25:1223.

Humphrey, R. R., Windsor, B. L., Bausley, F. G., and Edgren, R. A., 1976, Antifertility effects of an analog of luteinizing hormone-releasing hormone (LHRH) in rats, Contraception, 14:625.

Johnson, E. S., Gendrich, R. L., and White, W. F., 1976, Delay of puberty and inhibition of reproductive processes in the rat by a gonadotropin-releasing hormone agonist analog, Fertil. Steril., 27:853.

Jones, P. B. C., and Hsueh, A. J. W., 1980, Direct inhibitory effect of gonadotropin-releasing hormone upon luteal luteinizing hormone receptor and steroidogenesis in hypophysectomized rats, Endocrinology, 107:1930.

Jones, P. B. C., Conn, P. M., Marian, J., and Hsueh, A. J. W., 1980, Binding of gonadotropin-releasing hormone agonist to rat ovarian granulosa cells, Life Sci., 27:2125.

Kledzik, G. S., Cusan, L., Auclair, C., Kelly, P. A., and Labrie, F., 1978a, Inhibition of ovarian LH and FSH receptor levels by treatment with [D-Ala^6, des-Gly-$NH_2{}^{10}$]LHRH ethylamide during the estrous cycle in the rat, Fertil. Steril., 30:348.

Kledzik, G. S., Cusan, L., Auclair, C., Kelly, P. A., and Labrie, F., 1978b, Luteinizing hormone (LH)-releasing hormone agonist on rat ovarian LH and follicle-stimulating hormone receptor levels during pregnancy, Fertil. Steril., 29:560-567.

Kolena, J., and Channing, C. P., 1972, Stimulatory effects of LH, FSH, and prostaglandins upon cyclic 3',5'-AMP levels in porcine granulosa cells, Endocrinology, 90:1543.

Koyama, T., Okhura, T., Humasaka, T., and Saito, M., 1978, Effect of post-ovulatory treatment with a luteinizing hormone-releasing hormone analog on the plasma level of progesterone in women, Fertil. Steril., 30:549.

Labrie, F., Auclair, C., Cusan, L., Lemay, A., Bélanger, A., Kelly, P. A., Ferland, L., Kledzik, G. S., Azadian-Boulanger, G., and Raynaud, J. P., 1979, Inhibitory effects of treatment with LHRH or its agonists on ovarian receptor levels and function, in: "Ovarian follicular and corpus luteum function," C. P. Channing, J. Marsh, and W. Sadler, eds., Advances in Experimental Biology and Medicine, Vol. 112, p. 687, Plenum Press, N.Y.

Lefebvre, F. A., Reeves, J. J., Séguin, C., Massicotte, J., and Labrie, F., 1980, Specific binding of a potent LHRH agonist in rat testis, Mol. Cell. Endocrinol., 20:127-134.

Lemay, A., Labrie, F., Raynaud, J. P., and Azadian-Boulanger, G., 1978, Action lutéolytique de la LHRH chez la femme, C. R. Acad. Sci. [D] Paris, 286:527.

Lemay, A., Labrie, F., Ferland, L., and Raynaud, J. P., 1979, Possible luteolytic effects of luteinizing hormone (LH)-releasing hormone in normal women, Fertil. Steril., 31:29.

Massicotte, J., Veilleux, R., Lavoie, M., and Labrie, F., 1980, An LHRH agonist inhibits FSH-induced cyclic AMP accumulation and steroidogenesis in porcine granulosa cells in culture, Biochem. Biophys. Res. Commun., 94:1362.

Massicotte, J., Borgus, J. P., Lachance, R., and Labrie, F., 1981, Inhibition of hCG-induced cyclic AMP accumulation and steroidogenesis in rat luteal cells by an LHRH agonist, J. Steroid Biochem., 14:239.

Mayar, M. Q., Tarnavsky, G. K., and Reeves, J. J., 1979, Ovarian growth and uptake of iodinated [D-Leu^6]LHRH-EA in hCG-treated rat, Proc. Soc. Exp. Biol. Med., 161:216.

Nillius, S. J., Bergguist, C., and Wide, L., 1978, Inhibition of ovulation in women by chronic treatment with a stimulatory LHRH analogue — A new approach to birth control, Contraception, 17:537.

Reeves, J. J., Séguin, C., Lefebvre, F. A., Kelly, P. A., and Labrie, F., 1980, Similar luteinizing hormone-releasing hormone binding sites in rat anterior pituitary and ovary, Proc. Natl. Acad. Sci. U.S.A., 77:5567.

Rippel, R. H., and Johnson, E. S., 1976, Inhibition of hCG-induced ovarian and uterine weight augmentation in the immature rat by analogs of BnRH, Proc. Soc. Exp. Biol. Med., 152:432.

Rivier, C., Rivier, J., and Vale, W., 1978, Chronic effects of [D-Trp^6, NEt]LHRH on reproductive processes in the female rat, Endocrinology, 103:2299.

Rodbell, M., 1980, The role of hormone receptors and GTP-regulatory proteins in membrane transduction, Nature, 284:17.

Séguin, C., Reeves, J. J., Lefebvre, F. A., and Labrie, F., 1980, Direct effects of an LHRH agonist on the rat ovary and testis, in: "Proc. 62nd Annual Meeting of the Endocrine Society," Washington, D.C. (Abstr.).

Séguin, C., Bélanger, A., Cusan, L., Pelletier, G., Reeves, J. J., Lefebvre, F. A., Kelly, P. A., and Labrie, F., 1981, Relative importance of the adenophypophyseal and gonadal sites of inhibitory action of LHRH agonists, Biol. Reprod., in press.

Smith-White, S., and Ojeda, S. R., 1981, Changes in ovarian LHRH receptor content during the onset of puberty in the female rat, Endocrinology, 108:347.

Ying, S. Y., and Guillemin, R., 1979, [D-Trp^6-Pro^9-NEt]-luteinizing hormone-releasing hormone factor inhibits follicular development in hypophysectomized rats, Nature, 280:593.

REGULATION OF OVARIAN GRANULOSA AND LUTEAL CELL FUNCTIONS BY GONADOTROPIN RELEASING HORMONE AND ITS ANTAGONIST

Aaron J. W. Hsueh and Phillip B. C. Jones

Department of Reproductive Medicine, M-025
University of California, San Diego
La Jolla, California 92093

Introduction

Since the purification and sequencing of gonadotropin releasing hormone (GnRH) and the subsequent synthesis of potent GnRH agonists and antagonists, many laboratories have been involved in investigating the effects of these peptides on reproductive functions. In women, administration of GnRH at a restricted time in the cycle results in preovulatory-like surges of gonadotropin (Yen et al., 1975; Grimes et al., 1975). Paradoxically, long-term administration of potent GnRH agonists is always associated with the inhibition of a variety of female reproductive functions such as ovarian steroidogenesis, ovulation, ovum transport, ovum implantation, pregnancy, uterine growth and ovarian-dependent mammary tumorigenesis.

In early studies employing GnRH as an ovulation-inducing agent in women undergoing insemination therapy or in ewes during anestrus, GnRH was found to induce significant LH surges accompanied by a paradoxical decrease in luteal function as reflected by a reduction of progesterone production (Haresign et al., 1975; Grimes et al., 1975). Likewise, Corbin and Beattie (1975) reported the abortion-inducing effect of large doses of GnRH in pregnant female rats. A 100% inhibition of pregnancy was observed in rats treated with 1 mg GnRH/day from day 1-7 of pregnancy, whereas an 80% inhibition was found in rabbits treated during days 1-7 of pregnancy. These results suggested the potential of using GnRH as a post-coital and post-implantation contraceptive. Lin and Yoshinaga (1976) also demonstrated a similar effect of GnRH in blocking pregnancy and delaying implantation. Hilliard and co-workers (1976) further reported the ability of GnRH to serve as a luteolytic agent and to decrease fetal survival in rabbits.

Subsequent studies using various potent GnRH agonists have extended these observations. Chronic administration of [D-Leu^6,des-Gly^{10}-Pro^9-NHEt]-GnRH inhibited normal ovarian development as evidenced by arrested uterine growth, delay of vaginal opening and absence of normal cycling (Johnson et al., 1976a). Rippel and Johnson (1976) also reported the inhibitory effect of this GnRH agonist on hCG-induced ovarian and uterine weight augmentation in immature rats. Their demonstration of the inhibitory effect of [D-Leu^6,des-Gly^{10}-Pro^9-NHEt]-GnRH on hCG-induced ovarian weight increases in hypophysectomized rats was the first suggestion of a possible extra-pituitary effect of these GnRH agonists. Prolonged administration of the same GnRH agonist was also shown to induce regression of ovarian-dependent mammary tumors in rats (Johnson, 1976b). Similarly, pharmacological doses of GnRH agonists and native GnRH were shown to delay puberty, disrupt normal ovarian cycles and to possess pre- and post-coital contraceptive properties (Humphrey et al., 1976; Beattie and Corbin, 1977; Beattie et al., 1977; Arimura et al., 1977; Rivier et al., 1978). Furthermore, the serum estradiol surge necessary for normal implantation in rats was delayed by GnRH treatment (Humphrey et al., 1978) and the inhibitory effect of GnRH during pregnancy was attributed to a luteolytic effect of GnRH (Humphrey et al., 1977). These studies were extended by Kledzik et al. (1978) who reported the potent inhibitory effect of [D-Ala^6-Pro^9-NHEt]-GnRH in decreasing ovarian LH and FSH receptor levels, uterine weight and plasma progesterone during pregnancy.

The paradoxical inhibitory effect of GnRH and GnRH agonists on female reproductive functions was also demonstrated in women. A potent GnRH agonist [D-Ser(tBu)^6-des-Gly^{10}-NHEt]-GnRH, was shown to inhibit ovulation in normal cycling women (Nillius et al., 1978). Furthermore, injection of 100 μg [des-Gly^{10},Pro^9-NHEt]-GnRH daily for 5 days in women during the post-ovulatory period suppressed serum progesterone levels (Koyama et al., 1978). Likewise, the administration of five 250 μg doses of native GnRH at 4 h intervals during the luteal phase also decreased serum progesterone levels and shortened the length of the luteal phase by 1 to 4 days (Lemay et al., 1979). A similar luteolytic effect of a GnRH agonist, [D-Trp^6,Pro^9-NHEt]-GnRH, was also demonstrated in women during luteal phase of the cycle (Casper and Yen, 1979). These studies suggest a luteolytic effect of GnRH and indicate the promising potential of GnRH and GnRH agonists as contraceptive agents in women.

To explain the action of pharmacological doses of GnRH and its agonists in inhibiting female reproductive functions, one can offer at least three possible mechanisms.

A. Chronic stimulation of pituitary cells by high doses of GnRH may result in "desensitization" of the pituitary gonadotropes followed by a subsequent decrease in circulating gonadotrophins (Belchetz et al., 1978; de Koning et al., 1978).

B. GnRH may induce an elevation of LH secretion which, in turn, results in LH-induced "desensitization" of the gonadal cells (Conti et al., 1976; Hsueh et al., 1977, Rao et al., 1977).

C. An extrapituitary, direct inhibitory effect of GnRH upon gonadal cells.

The present report describes our studies demonstrating an extrapituitary, direct inhibitory action of GnRH upon ovarian granulosa and luteal cells.

Direct Effect of GnRH and Its Antagonist on Ovarian Granulosa Cell Functions in vitro

In order to demonstrate the direct inhibitory effect of GnRH upon ovarian functions, we employed an in vitro primary culture of ovarian granulosa cells. Granulosa cells obtained from preantral follicles of immature hypophysectomized rats have been shown by us and other investigators to respond to FSH in the production of estrogen and progestins (Dorrington et al., 1975; Armstrong and Dorrington, 1976; Erickson and Hsueh, 1978; Nimrod and Lindner, 1976; Hillier et al., 1977). As we reported earlier (Wang et al., 1979), treatment with purified ovine FSH stimulated increases in estrogen, progesterone and 20α-hydroxy-pregn-4-en-3-one (20α-OH-P) production by granulosa cells cultured in the presence of 10^{-7} M Δ^4-androstenedione. In contrast, high doses of LH and PRL were without effect in these cultured cells. FSH treatment also stimulated increases in LH/hCG and PRL receptors (Erickson et al., 1979; Wang et al., 1979). The increases in LH and PRL receptor contents are accompanied by an enhancement in the ability of LH and PRL to stimulate granulosa cell steroidogenesis.

Employing this in vitro experimental model, we tested the direct inhibitory effect of GnRH and its antagonist on various FSH-induced functional parameters in cultured granulosa cells (Hsueh and Erickson, 1979; Hsueh and Ling, 1979; Hsueh et al., 1980). As shown in Figs. 1 and 2, purified ovine FSH (100 ng/ml), in the presence of Δ^4-androstenedione, stimulated substantial increases in estrogen and progesterone production by cultured granulosa cells. Concomitant treatment with increasing concentrations of GnRH (10^{-11} M to 10^{-7} M) inhibited granulosa cell steroidogenesis in a dose-dependent manner with ED_{50} values of 5×10^{-10} and 1.5×10^{-9} M for estrogen and progesterone production, respectively (Figs. 1 and 2). In contrast, increasing concentrations (3×10^{-9} to 10^{-6} M) of a potent GnRH antagonist, [D-pGlu^1,D-Phe^2,D-Trp^{3,6}]-GnRH, did not affect the FSH stimulation of estrogen and progesterone production. In cultures containing FSH and an inhibiting dose (10^{-8} M) of GnRH, concomitant addition of increasing concentrations (10^{-8} to 10^{-6} M) of the antagonist caused a dose-dependent blockage of GnRH action by pre-

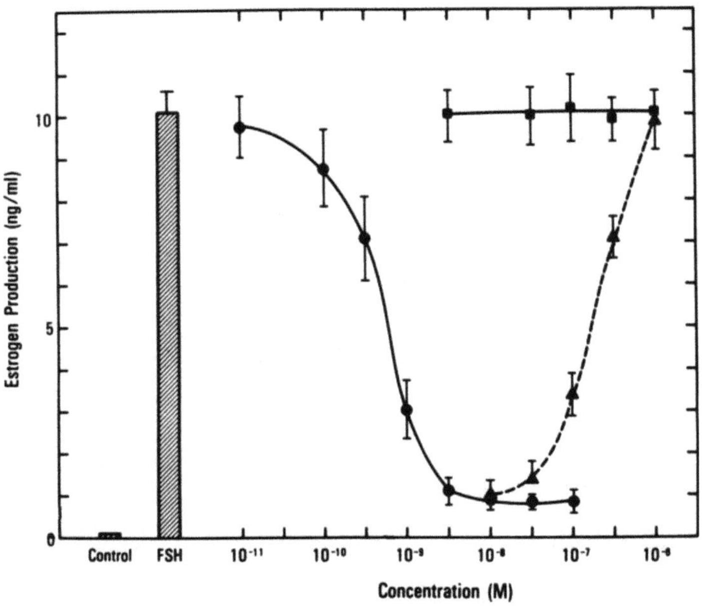

Fig. 1. Effects of GnRH and the GnRH antagonist upon FSH-induced increase in estrogen production by rat granulosa cells in vitro. Granulosa cells were cultured with purified ovine FSH (100 ng/ml). Increasing concentrations of GnRH (10^{-11}-10^{-7} M; o--o) or the GnRH antagonist (3 × 10^{-9}-10^{-6} M; □--□) were added with FSH to study their effects on FSH-induced increase in estrogen production. To test the interaction between the GnRH antagonist and GnRH, granulosa cells were treated with increasing concentrations of the antagonist (10^{-8}-10^{-6} M) concomitantly with FSH and 10^{-8} M of GnRH (Δ--Δ). Media concentrations of estrogen were determined by radioimmunoassay. Each data point represents mean ± SE of 4 cultures. (Reprinted with permission from A. J. W. Hsueh and N. C. Ling, Life Science, 25:1223, 1979, Pergamon Press, Ltd.)

venting the GnRH inhibition of estrogen and progesterone production. The antagonist reduced the GnRH effect by 50% at a molar ratio (IDR_{50}) of 20/1 and 12/1 ([antagonist]/[GnRH]) for estrogen and progesterone production, respectively. At 10^{-6} M, the antagonist completely prevented the GnRH (10^{-8} M) inhibition of FSH-stimulated estrogen and progesterone production by granulosa cells. These results indicated the direct inhibitory effect of GnRH on ovarian steroidogenesis and the potent antagonistic effect of the GnRH antagonist on GnRH action in ovarian granulosa cells.

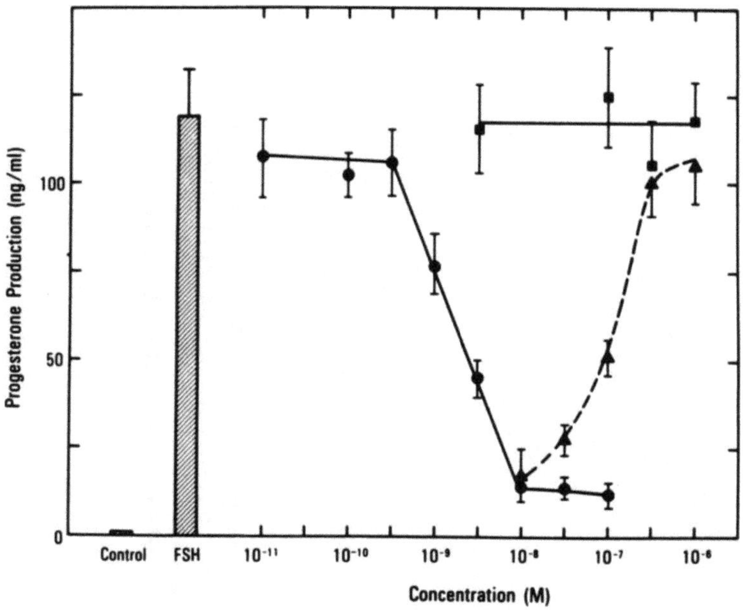

Fig. 2. Effect of GnRH and the GnRH antagonist upon FSH-induced increase in progesterone production by rat granulosa cells in vitro. The experimental procedures were the same as in Fig. 1. Media progesterone concentrations were determined by radioimmunoassay at the end of the 2 day culture. Each data point represents mean ± SE of 4 cultures. (Reprinted with permission from A. J. W. Hsueh and N. C. Ling: Life Science, 25:1223, 1979, Pergamon Press, Ltd.)

We have also demonstrated that the effect of GnRH on steroidogenesis is not due to interference with the estrogen or progesterone radioimmunoassa by GnRH. As shown in Fig. 3, GnRH at $\sim 10^{-7}$ M did not interfere with the binding of estradiol and progesterone to their respective antisera.

As shown in Fig. 4, FSH induced substantial increases in LH/hCG receptor content in granulosa cells cultured in the serum-free medium. Concomitant treatment with 10^{-8} M GnRH completely inhibited the FSH effect, decreasing the $[^{125}I]$-hCG binding ability of granulosa cells to control levels (Fig. 4). In contrast, concomitant treatment with 10^{-6} M of the antagonist did not affect the FSH induction of LH/hCG receptor. Moreover, when added together with FSH and 10^{-8} M GnRH, the antagonist (10^{-6} M) completely abolished the inhibitory effect of GnRH; FSH treatment induced a full complement

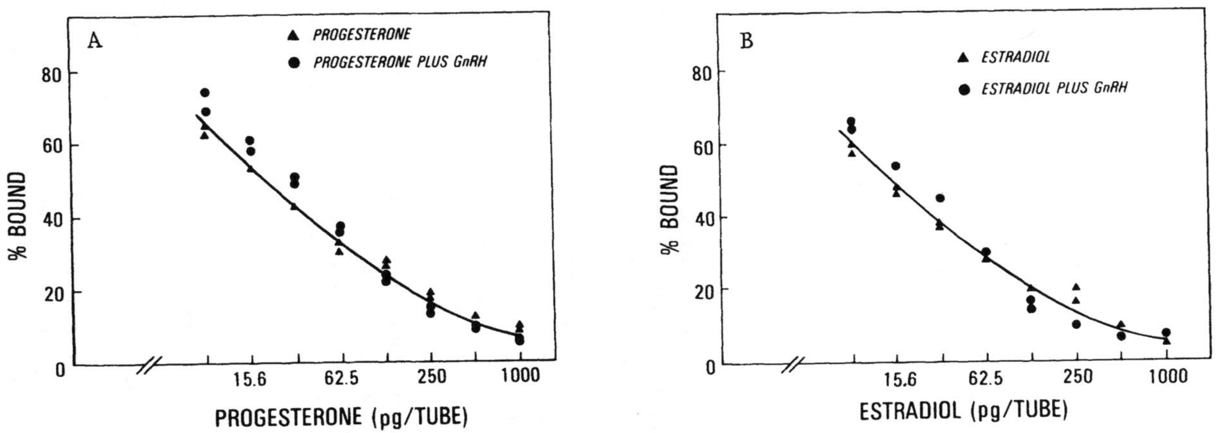

Fig. 3. Lack of interference of GnRH upon estrogen and progesterone radioimmunoassay. GnRH (10^{-7} M) was added to samples containing increasing concentrations of estradiol or progesterone (▲--▲) and compared with assay tubes containing the same concentrations of steroids without the addition of GnRH (●--●). Individual determinations were plotted. (A) Progesterone. (B) Estrogen.

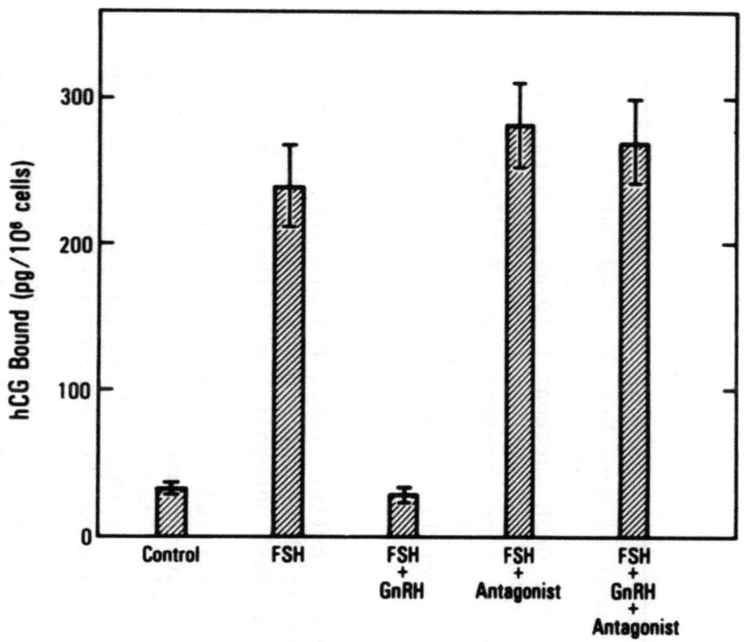

Fig. 4. Effect of GnRH and the GnRH antagonist upon FSH-induced increase in LH/hCG receptor in rat granulosa cells in vitro. Granulosa cells were cultured with purified ovine FSH (100 ng/ml), GnRH (10^{-8} M) and/or the antagonist (10^{-6} M). At the end of the 2 day culture period, the concentration of LH/hCG receptors in the granulosa cells was determined by [^{125}I]-hCG radioreceptor assay and expressed as pg hCG bound/10^6 granulosa cells. Each data point represents mean ± SE of at least 4 determinations. (Reprinted with permission from A. J. W. Hsueh and N. C. Ling: <u>Life Science</u>, 25:1223, 1979, Pergamon Press, Ltd.)

Table 1. Effect of GnRH and Its Antagonist upon FSH Induction of PRL Receptor in Cultured Granulosa Cells in vitro

Treatment	^{125}I-PRL Bound (pg/10^6 cells)
Control	86.1 ± 7.7
FSH	341.4 ± 3.4
FSH plus GnRH (10^{-8}M)	121.1 ± 30.7
FSH + GnRH (10^{-8}M) + GnRH Antagonist (10^{-6}M)	341.4 ± 1.7

Granulosa cells from immature hypophysectomized rats were cultured for 2 days with FSH (100 ng/ml) in the presence or absence of GnRH and/or the GnRH antagonist (D-pGlu^1-D-Phe^2,D-Trp^{3,6}-GnRH). PRL receptor content was measured by incubating cells with a saturating concentration of ^{125}I-PRL in the presence or absence of 100-fold excess of nonlabelled PRL for 16 h at 23°C. Specific binding is expressed as pg/10^6 cells.

of LH/hCG receptors in these cells. These results demonstrate the inhibitory effect of GnRH on LH receptor formation and the potent antagonistic effect of the GnRH antagonist upon the GnRH action.

FSH treatment also stimulated substantial (6.7-fold) increases in PRL receptor in the cultured granulosa cells (Table 1). Concomitant treatment with 10^{-8} M GnRH inhibited the PRL effect by 63% whereas treatment with 10^{-6} M of the GnRH antagonist blocked the GnRH inhibition of FSH-stimulated PRL receptor content.

Chloera toxin, an agent known to induce cAMP production in granulosa cells, was shown to stimulate progesterone and estrogen production in cultured rat granulosa cells. To examine the effect of GnRH and its antagonist on cholera toxin-stimulated steroidogenesis, granulosa cells were cultured for 2 days with cholera toxin (10 µg/ml) and the appropriate hormones. As shown in Fig. 5 (left panel), cholera toxin treatment increased progesterone production 172-fold from 0.28 to 48.5 ng/ml. Concomitant treatment with GnRH decreased progesterone production by 46%. When the antagonist was added together with cholera toxin and GnRH, the GnRH inhibition was completely blocked. Treatment with the antagonist and cholera toxin did not affect the cholera toxin-induced increase in progesterone production.

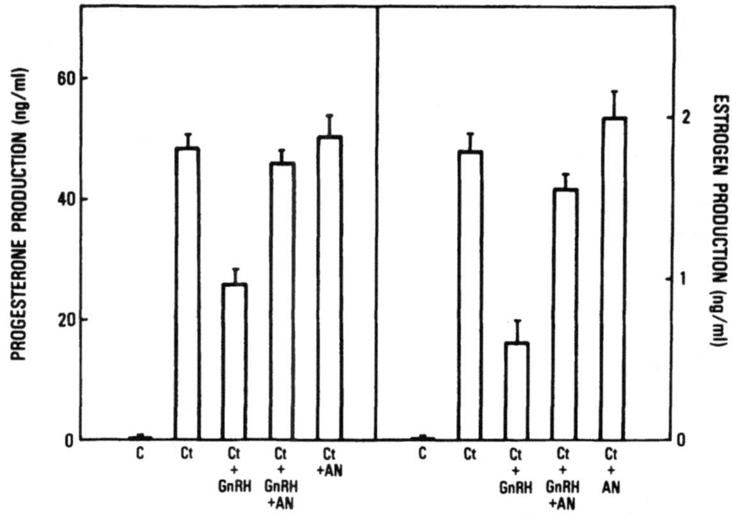

Fig. 5. Effect of treatment with GnRH and the antagonist upon cholera toxin-stimulated steroidogenesis by rat granulosa cells in vitro. Granulosa cells (~10^5 viable cells/culture) were cultured in 1 ml of McCoy's 5a medium containing 100 nM androstenedione in the absence (control, C) or presence of cholera toxin (Ct; 10 μg/ml) or cholera toxin plus 10^{-8} M GnRH, 10^{-6} M antagonist (AN), or both. Media concentrations of progesterone and estrogen were measured by radioimmunoassay. Data points represent mean ± SEM of four determinations. (Reprinted with permission from P. B. C. Jones and A. J. W. Hsueh: Biol. Reprod., 24:747, 1981, Society for the Study of Reproduction.)

As shown in Fig. 5 (right panel), cholera toxin treatment increased estrogen production from undetectable levels (<70 pg/ml) to 1.80 ng/ml. Concomitant treatment with 10^{-8} M GnRH decreased estrogen production by 66%. Treatment with 10^{-6} M antagonist, together with cholera toxin and GnRH, blocked the GnRH inhibition. In contrast, treatment of cells with cholera toxin and the antagonist did not affect cholera toxin stimulation of estrogen production.

To examine the possibility that GnRH treatment may be detrimental to the granulosa cells, we determined cell viability at the beginning and the end of the 2 day culture period by the trypan blue exclusion method. The percentage of viable cells varied from 50-70% at the beginning of the experiment. In a representative experiment in which 55% of the viable cells was initially cultured, the cell viability was studied at the end of the 2 day culture period by counting 1500 cells/dish and 6 individual dishes/treatment group.

Table 2. Effect of Treatment with GnRH, Its Antagonist and FSH upon the Protein Content of Rat Granulosa Cells in vitro

Treatment	Protein Content (μg/Culture)
Control	18.1 ± 0.6
GnRH (10^{-8} M)	25.0 ± 1.4
AN (10^{-6} M)	20.5 ± 0.6
GnRH (10^{-8} M) + AN (10^{-6} M)	20.5 ± 0.8
FSH	22.9 ± 1.3
FSH + GnRH (10^{-8} M)	25.2 ± 0.3

Granulosa cells (∿200,000 viable cells/culture) were incubated for 2 days in medium alone (control), with GnRH, a GnRH antagonist (AN; [D-pGlu^1, D-Phe^2, D-Trp^{3,6}-GnRH]), FSH (10 ng/ml), or combinations thereof. After the 2 day culture, media were removed and the cells were scraped from the dishes with a rubber policeman. The Bio-Rad protein assay was used to determine the protein content of each culture and bovine gamma globulin was used as the standard. Data is presented as mean ± SEM of four determinations.

The percentages of viable cells were determined to be 52.9 ± 2.8% and 54.1 ± 2.0% for cultures treated with FSH and FSH plus 10^{-6} M GnRH, respectively. Also, no substantial decrease in total cell number was detected in either treatment group. These results indicate that GnRH treatment does not preferentially affect cell viability. Moreover, the effect of GnRH treatment on granulosa cell protein content was studied (Table 2). Treatment with a high concentration of GnRH stimulated protein synthesis in cultured granulosa cells. At 10^{-8} M, GnRH treatment resulted in a 38% increase in protein content. The stimulatory effect of GnRH was also shown to be blocked by concomitant treatment with the GnRH antagonist. FSH also stimulated a slight increase in granulosa cell protein content (26% increase). Concomitant GnRH treatment did not inhibit the FSH-stimulated increase in protein synthesis.

These studies demonstrated the direct inhibitory effect of GnRH in regulating various granulosa cell functions in vitro and the ability of the GnRH antagonist to block the inhibitory effect of GnRH.

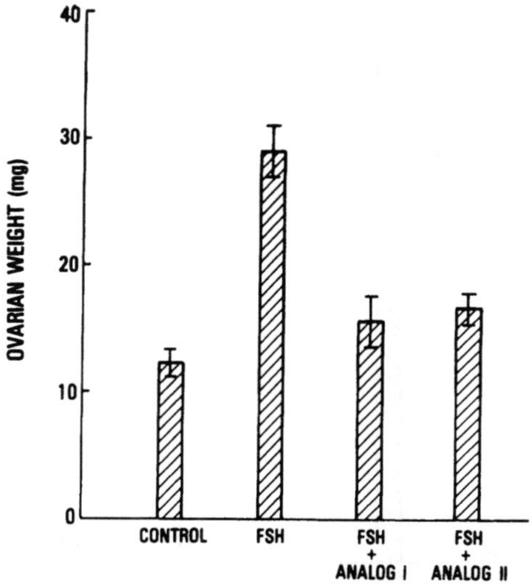

Fig. 6. Effect of treatment with GnRH and its agonists on FSH stimulation of ovarian weight in immature hypophysectomized rats. n = 8. Mean ± SE.

Furthermore, GnRH treatment appears to inhibit these functions specifically without decreasing cell viability or overall cellular protein synthesis.

Extrapituitary Action of GnRH and Its Antagonist on Ovarian Functions in Hypophysectomized Female Rats in vivo

We further extended these in vitro studies to in vivo experiments employing immature hypophysectomized female rats (Hsueh et al., 1980).

Four days after hypophysectomy, immature, estrogen-treated female rats were injected subcutaneously, twice daily, for 2 days with 100 µg of ovine FSH (NIH-FSH-S12) or the same amount of FSH plus a GnRH agonist (50 µg of analog 1 ([D-Leu^6(N$^\alpha$Me)Leu^7]GnRH) or 10 µg of analog 2 ([des-Gly^{10},D-Leu^6(N$^\alpha$Me)Leu^7,Pro^9-NHEt]GnRH). The ovarian weight, as well as the LH receptor content and steroidogenic potential of granulosa cells were determined 12 hours after the last injection. FSH treatment induced a significant increase in ovarian weight (Fig. 6). In contrast, concomitant treatment with the GnRH agonists inhibited the FSH-induced increase in ovarian weight.

FSH (100 µg/injection) increased LH/hCG receptor content in granulosa cells by 66-fold, whereas concomitant treatment with the

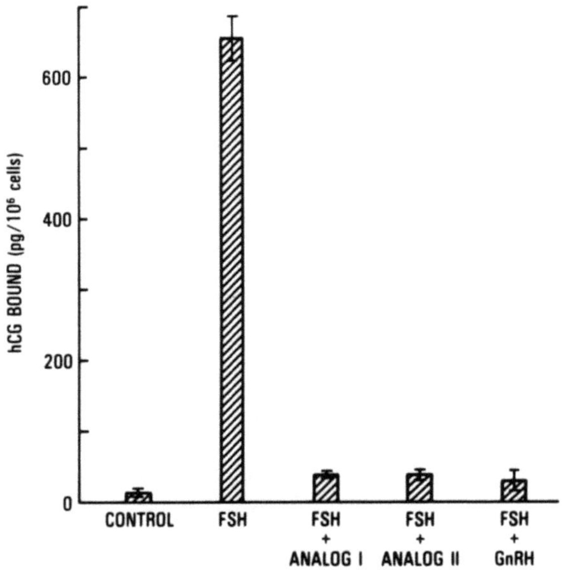

Fig. 7. Inhibitory effect of GnRH and its agonist upon FSH induction of LH/hCG receptor in vivo. Immature, hypophysectomized, diethylstilbestrol-treated female rats were treated with 100 μg of FSH with or without GnRH or the GnRH agonist (100 μg GnRH, 50 μg analog 1, or 10 μg analog 2). At the end of the 2-day treatment period, granulosa cells were collected and used for determination of LH/hCG receptor content. Eight to 10 ovaries were pooled for each treatment group. Granulosa cells (0.5-1 × 10^6 cells/tube) were incubated for 16 h at 23°C with saturating concentrations of [^{125}I]hCG (∿10 ng) in the presence or absence of an excess of unlabelled hCG (100 IU Pregnyl/tube). Specific binding was calculated. Each data point represents the mean ± SE of 3-6 determinations. (Reprinted with permission from A. J. W. Hsueh et al.: Endocrinology, 106:1697, 1980, The Endocrine Society.)

GnRH decapeptide (100 μg/injection) or GnRH agonist (50 μg of analog 1 or 10 μg of analog 2) inhibited the FSH-induced increase in LH/hCG receptor content by 95% (Fig. 7). Incubation of granulosa cells obtained from FSH-treated animals with graded doses of aromatase substrate (Δ^4-androstenedione) resulted in a dose-dependent increase in estrogen production (maximal production, 1.65 ng/10^5 cells; ED_{50} = 3 × 10^{-8} M) (Fig. 8). In contrast, concomitant treatment with GnRH agonists depressed estrogen production. Histological examination of the ovaries from animals treated with FSH plus the GnRH agonists indicated the presence of antrum-containing follicles.

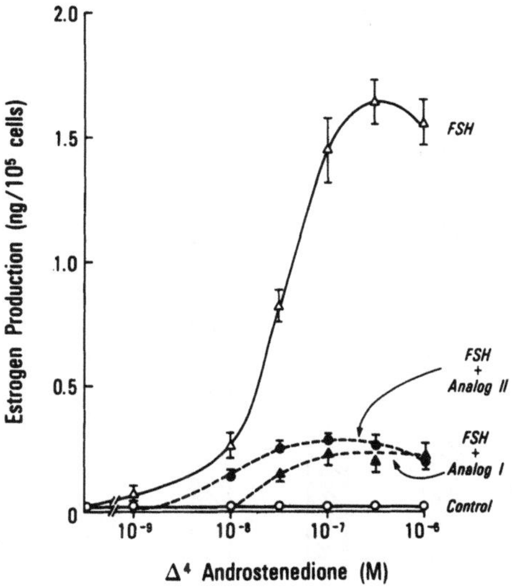

Fig. 8. Inhibitory effect of GnRH agonists upon FSH induction of aromatase activity in vivo. Immature, hypophysectomized, diethylstilbestrol-treated rats were treated with 100 μg oFSH (twice daily) for 2 days with or without concomitant treatment with GnRH analog 1 (50 μg) or GnRH analog 2 (10 μg). At the end of the 2-day treatment, granulosa cells were pooled from six to eight ovaries from each treatment group and incubated for 3 h at 37°C in Hepes-0.1% BSA buffer containing increasing concentrations of androstenedione as substrate for aromatase. At the end of the incubation, estrogen concentration in the medium was measured by RIA. Each data point represents the mean ± SE of three or four samples. (Reprinted with permission from A. J. W. Hsueh et al.: Endocrinology, 106:1697, 1980, The Endocrine Society.)

Some ova in these follicles had resumed meiotic maturation and appeared to be undergoing fragmentation. The stimulatory effect of GnRH on meiotic maturation in rat ovaries was further studied in vivo and in vitro (Erickson and Hsueh, 1980; Hillensjö and LeMaire, 1980). Administration of GnRH or its potent agonist, in the presence or absence of FSH, induced aberrant patterns of meiosis and parthenogenetic-like cleavage in a large population of the fully-grown oocytes in ovaries of hypophysectomized and intact immature rats.

These in vivo studies in hypophysectomized rats confirm the in vitro studies in cultured granulosa cells in demonstrating the extrapituitary inhibitory effect of GnRH and its agonist on ovarian estrogen production and LH receptor formation.

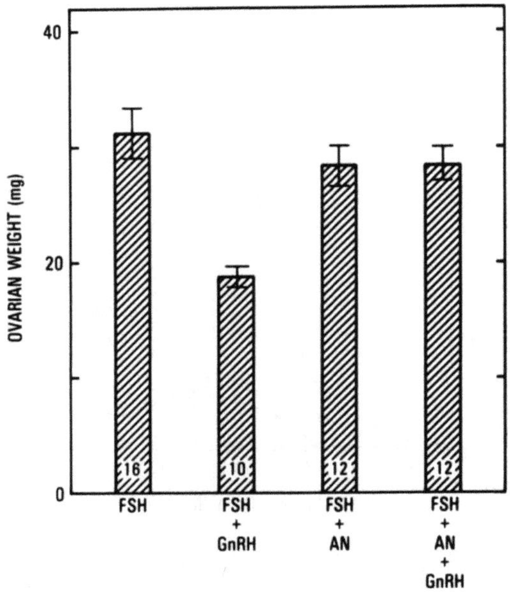

Fig. 9. Effect of treatment with GnRH and its antagonist upon the FSH-induced increase in ovarian weight in hypophysectomized female rats. Immature, hypophysectomized rats were treated twice daily for 2 days with 7.0 μg of oFSH along with 30 μg of GnRH and/or 500 μg of the GnRH antagonist (AN). Ovarian weights were determined at the end of the 2 day treatment period. The number of ovaries used per group is indicated within each column. Data points represent mean ± SEM. (Reprinted with permission from P. B. C. Jones and A. J. W. Hsueh: Biol. Reprod., 24:747, 1981, Society for the Study of Reproduction.)

Furthermore, we studied the effect of a GnRH antagonist on GnRH inhibition of ovarian functions in vivo (Jones and Hsueh, 1981a). As shown in Figs. 9-11, GnRH treatment in vivo significantly decreased ovarian weight (40%) as well as granulosa cell hCG binding capacity (85%) and estrogen production (78%) in FSH-treated hypophysectomized rats. In contrast, concomitant treatment with the GnRH antagonist did not affect ovarian weight or hCG binding capacity in the FSH-treated animals. The in vitro estrogen production by granulosa cells of these animals was decreased by 35% as compared with FSH-treated animals. When the GnRH antagonist was administered concomitantly with GnRH and FSH to hypophysectomized rats, the antagonist completely abolished the inhibitory effect of GnRH upon ovarian weight (Fig. 9). Likewise, the antagonist treatment signi-

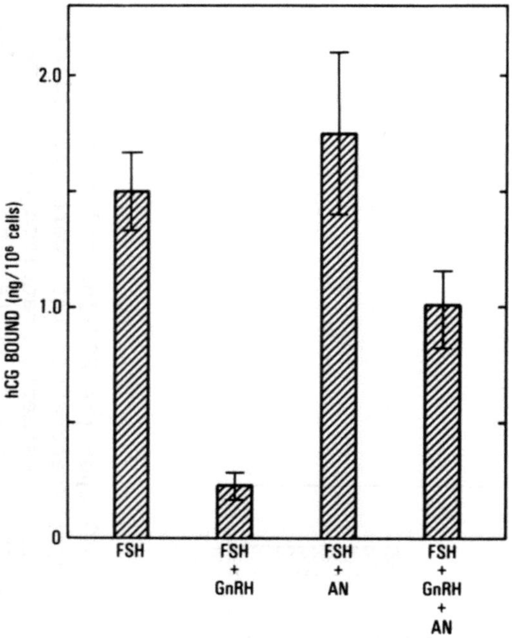

Fig. 10. Effect of treatment with GnRH and its antagonist (AN) upon the FSH-stimulated increase in [^{125}I]hCG binding capacity of ovarian granulosa cells. Hypophysectomized rats were treated as described in the legend of Fig. 9. Granulosa cells were collected and incubated with [^{125}I]hCG (total binding) or [^{125}I]hCG plus excess unlabelled hCG (non-specific binding) for 16 h at 23°C. Specific [^{125}I]hCG binding to granulosa cells was determined. Each bar represents the mean ± SEM of four determinations. (Reprinted with permission from P. B. C. Jones and A. J. W. Hsueh: Biol. Reprod., 24:747, 1981, Society for the Study of Reproduction.)

ficantly inhibited the GnRH-induced decreases in hCG binding capacity (Fig. 10) and estrogen production (Fig. 11) by rat granulosa cells.

Direct Effect of GnRH and Its Antagonist on Granulosa-Luteal Cell Functions in vitro

We have demonstrated that FSH treatment of ovarian granulosa cells in vitro induced an increase in LH and PRL receptor content as well as in the responsiveness of granulosa cells to LH and PRL. Subsequent treatment of FSH-primed cells with LH resulted in in-

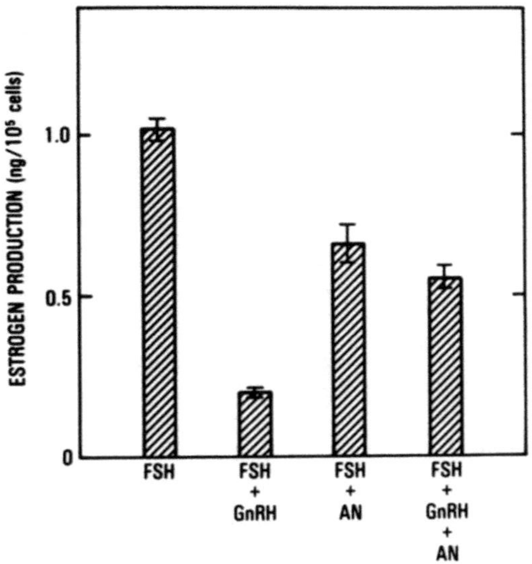

Fig. 11. Effect of treatment with GnRH and its antagonist (AN) upon the FSH-induced increase in aromatase activity in ovarian granulosa cells. Hypophysectomized rats were treated as described in the legend of Fig. 9. At the end of the treatment, granulosa cells were obtained and incubated for 3 h at 37°C in HEPES-0.1% BSA buffer containing 10^{-7} M androstenedione. The concentration of estrogen in the media was measured by radioimmunoassay. Each bar represents the mean ± SEM for four determinations. (Reprinted with permission from P. B. C. Jones and A. J. W. Hsueh: Biol. Reprod., 24:747, 1981, Society for the Study of Reproduction.)

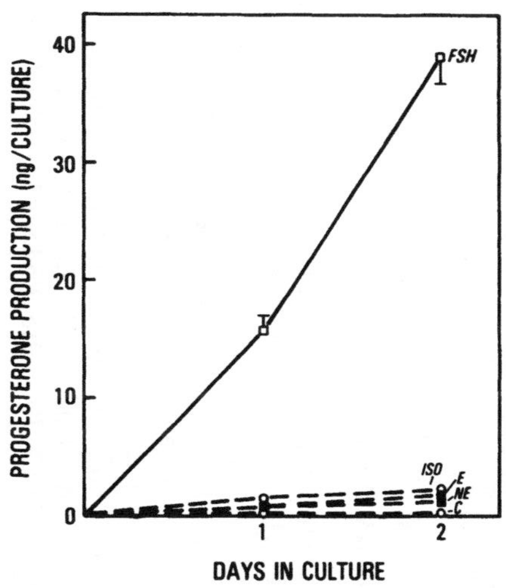

Fig. 12. Effect of treatment with adrenergic agents on progesterone production in granulosa cells obtained from immature hypophysectomized diethylstilbestrol-treated rats. Granulosa cells (1×10^5 viable cells/culture) were cultured for 2 days in the presence or absence of FSH (10 ng/ml), (-)isoproterenol (ISO; 10^{-5} M), (-)epinephrine (E; 10^{-5} M) or (-)-norepinephrine (NE; 10^{-5} M). Medium was collected at 24 h intervals at which time fresh media and hormones were added to appropriate dishes. Androstenedione (10^{-7} M) was present throughout the experimental period. Progesterone production was assayed by RIA. The results represent the mean ± SEM of 4 determinations. C = control cultures. (Reprinted with permission from E. Y. Adashi and A. J. W. Hsueh: Endocrinology, 108:2170, The Endocrine Society.)

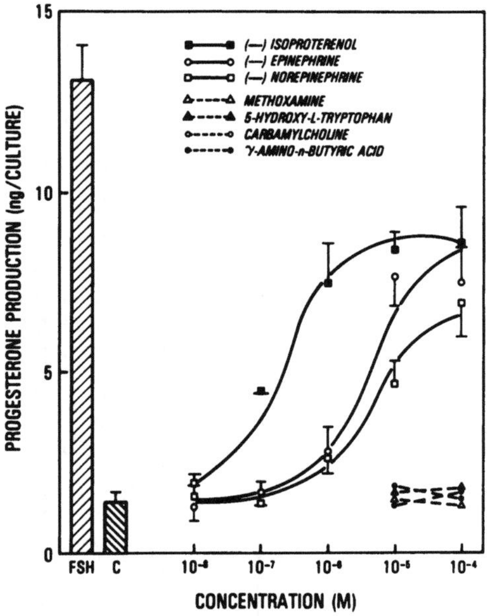

Fig. 13. Effect of treatment with adrenergic agents on progesterone production by granulosa cells pretreated with FSH in vitro. Granulosa cells (1×10^5 viable cells/culture) were cultured for 2 days in the presence of oFSH (10 ng/ml), washed and then reincubated for an additional 24 h in the presence or absence of oFSH (10 ng/ml) or increasing concentrations (10^{-8}–10^{-4} M) of the specified agent. (Reprinted with permission from E. Y. Adashi and A. J. W. Hsueh: Endocrinology, 108:2170, The Endocrine Society.)

increases of estrogen and progestin production (Erickson et al., 1979). In contrast, treatment with PRL stimulated progesterone production without affecting aromatase activity (Wang et al., 1979). Furthermore, we have also demonstrated an increase in β_2-adrenergic responsiveness in ovarian granulosa cells following FSH treatment in vitro (Adashi and Hsueh, 1981). As shown in Fig. 12, progesterone production by granulosa cells obtained from immature hypophysectomized diethylstilbestrol-treated rats was stimulated by FSH but was only minimally affected by various adrenergic agents. In contrast, FSH treatment for 2 days in vitro resulted in an increase in the adrenergic responsiveness of granulosa cells. Subsequent treatment with (-)epinephrine, (-)norepinephrine or (-)isoproterenol (a potent β-adrenergic agonist) resulted in dose-dependent increases in progesterone production (Fig. 13). In contrast, high doses of methoxamine (an α-adrenergic agonist), 5-hydroxy-L-tryptophan (the precursor of serotonin), carbamylcholine (a cholinergic agonist) and

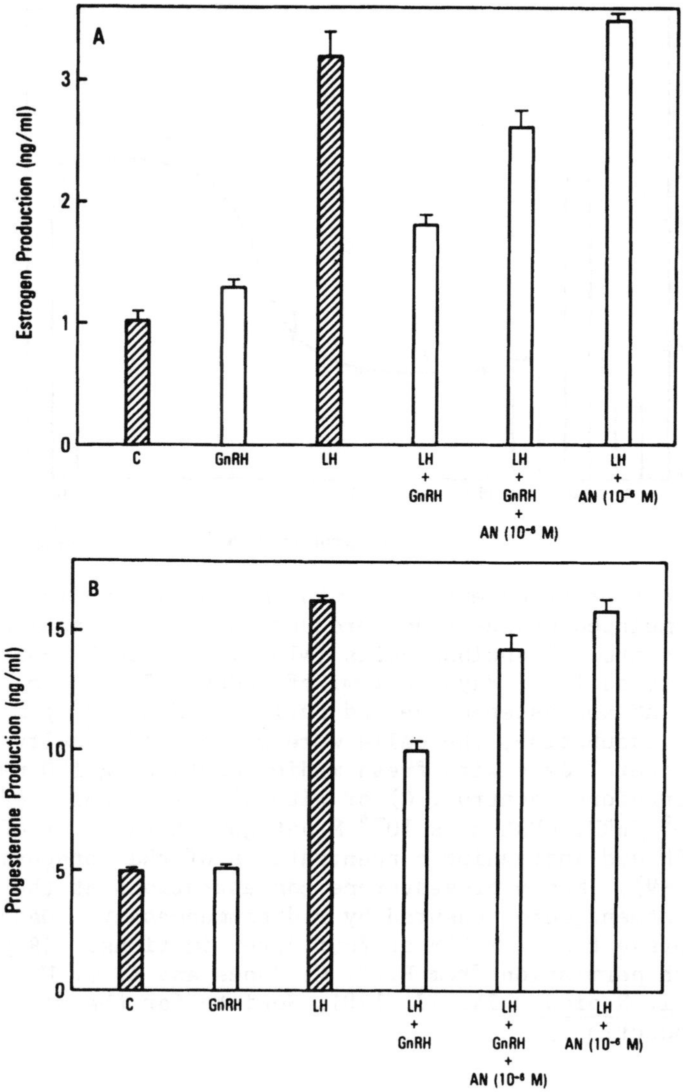

Fig. 14. Effect of treatment with GnRH and the antagonist upon LH-stimulated steroidogenesis by rat granulosa cells in vitro. Granulosa cells were primed for 2 days with oFSH. The cells were washed and recultured for another 2 days with fresh medium containing 100 nM androstenedione (control, C) and 10^{-8} M GnRH or oLH (50 ng/ml) alone or with GnRH, 10^{-6} M antagonist (AN), or both. Estrogen (panel A) and progesterone (panel B) in the media were measured by radioimmunoassay. Data points represent mean ± SEM of four determinations. (Reprinted with permission from P. B. C. Jones and A. J. W. Hsueh: Biol. Reprod., 24:747, 1981, Society for the Study of Reproduction.)

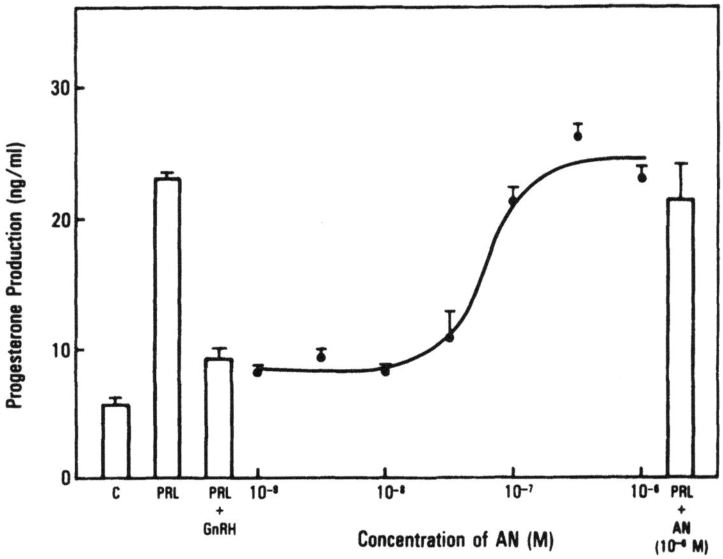

Fig. 15. Effect of treatment with GnRH and the antagonist upon PRL-stimulated progesterone production by rat granulosa cells in vitro. Granulosa cells (∿10^5 viable cells/culture) were cultured for 2 days in 1 ml of McCoy's 5a medium containing 100 nM androstenedione and purified oFSH (10 ng/ml). After the incubation, the cells were washed and recultured for another 2 days with fresh medium containing 100 nM androstenedione (control, C) or with oPRL (1 μg/ml), oPRL plus 10^{-8} GnRH, oPRL plus 10^{-6} M antagonist (AN), or oPRL plus GnRH and increasing concentrations of the antagonist (●--●). Media progesterone concentrations at the end of treatment were measured by radioimmunoassay. Data points represent mean ± SEM of four determinations. (Reprinted with permission from P. B. C. Jones and A. J. W. Hsueh: Biol. Reprod., 24:747, 1981, Society for the Study of Reproduction.)

GABA did not affect progesterone production. Furthermore, a selective β_2-adrenergic agonist (terbutaline) stimulated progesterone production in FSH-treated granulosa cells whereas a β_1-adrenergic agonist (dobutamine) was 10,000 fold less effective (data not shown).

Although the functional properties of FSH-primed granulosa cells may differ from those of luteal cells in vivo (because the latter cells are derived from both granulosa and theca cells), these granulosa-luteal cells possess functional LH, PRL, and adrenergic receptors similar to those of the luteal cells. We therefore employed these granulosa-luteal cells to study the direct GnRH regulation of luteal cell functions in vitro.

To examine the effect of GnRH and its antagonist on LH-stimulated steroidogenesis, granulosa cells were treated with oFSH for 2 days to induce functional LH receptors. Cells were washed and incubated for another 2 days in the absence or presence of 10^{-8} M GnRH, oLH (50 ng/ml), the GnRH antagonist (10^{-6} M), or appropriate combinations thereof. As shown in Fig. 14A, LH treatment increased estrogen production threefold from the control level whereas concomitant treatment with GnRH decreased estrogen production by 43%. Treatment with 10^{-6} M antagonist, together with LH and GnRH, significantly blocked the GnRH inhibition resulting in a 44% increase in estrogen production compared with cells treated with LH and GnRH. In contrast, treatment of cells with LH and the antagonist did not result in a significant change in estrogen production compared with LH-treated cells.

As shown in Fig. 14B, LH treatment of these FSH-primed cells also increased progesterone production by 3.3-fold whereas concomitant treatment with GnRH decreased the LH-induced increases in progesterone production by 39%. When the antagonist was added together with LH and GnRH, the GnRH inhibition of progesterone production was significantly blocked. In contrast, treatment of cells with antagonist and LH did not affect LH stimulation of progesterone production.

The effect of GnRH and its antagonist on PRL stimulation of progesterone production was also studied. Granulosa cells were treated with oFSH (10 ng/ml) for 2 days. Then, the cultures were washed and the cells were reincubated for 2 days with oPRL (1 μg/ml) with or without the appropriate hormones. As shown in Fig. 15, the media concentrations of progesterone in control cultures (or in the presence of 10^{-6} M antagonist) was ∿5 ng/ml. PRL treatment for 2 days increased progesterone production by about four-fold whereas concomitant treatment with 10^{-8} M GnRH decreased progesterone production by 60%. In contrast, increasing concentrations of the antagonist, when added together with PRL and GnRH, blocked the inhibitory effect of GnRH in a dose-dependent manner with an ED_{50} of 5.5×10^{-8} M. Thus, the antagonist blocked the GnRH effect by 50% at a molar ratio (IDR_{50}) of 5.5/1 ([antagonist]/[GnRH]). There was no significant difference in progesterone production among cultures treated with PRL alone, PRL plus 10^{-8} M GnRH and 10^{-6} M antagonist, or PRL plus 10^{-6} M antagonist.

PRL treatment in vitro increased LH receptor content in rat granulosa cells. We examined the effect of treatment with GnRH and its antagonist on the PRL-induced increase in LH/hCG receptor content. Granulosa cells were treated with oFSH (10 ng/ml) for 2 days, washed and reincubated for 2 days with oPRL (1 μg/ml) with or without the appropriate hormones. PRL treatment induced a 5.4-fold increase in the hCG binding capacity compared with the control levels (Fig. 16). Concomitant treatment with 10^{-8} M GnRH resulted

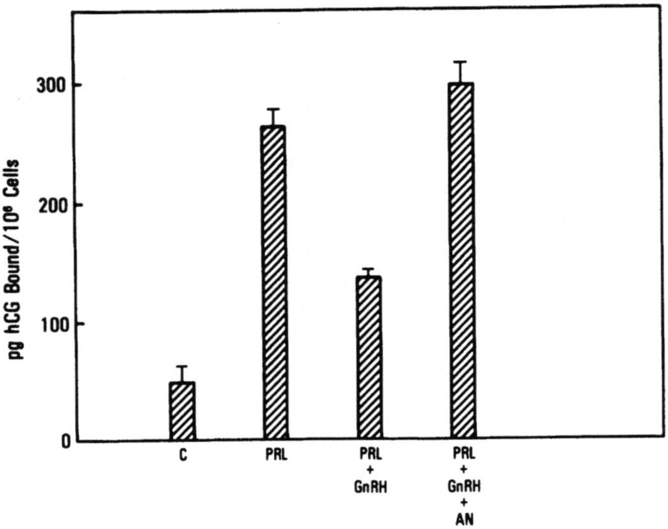

Fig. 16. Effect of treatment with GnRH and the antagonist upon PRL maintenance of LH/hCG receptor in rat ovarian granulosa cells in vitro. Granulosa cells (3×10^5 viable cells/culture) from hypophysectomized rats were cultured for 2 days in McCoy's 5a medium containing 100 nM androstenedione and purified oFSH (10 ng/ml). After the incubation, the cells were washed and recultured for 2 days in fresh medium containing 100 nM androstenedione (control, C) or with oPRL (1 μg/ml), oPRL plus 10^{-8} M GnRH or oPRL plus GnRH and the antagonist (AN; 10^{-6} M). At the end of incubation, the concentration of LH/hCG receptors in the granulosa cells was determined by [^{125}I]hCG radioreceptor assay and expressed as pg hCG bound/10^6 granulosa cells. Data points represent mean ± SEM of four determinations. (Reprinted with permission from P. B. C. Jones and A. J. W. Hsueh: Biol. Reprod., 24:747, 1981, Society for the Study of Reproduction.)

in a 48% decrease in hCG binding compared with PRL-treated cells. In contrast, treatment with 10^{-6} M antagonist, together with PRL and GnRH, completely blocked the GnRH inhibition. The hCG binding capacity in cells treated with the antagonist, PRL, and GnRH was not significantly different from that of the PRL-treated group. Thus, the antagonist blocked the GnRH inhibition of hCG binding capacity maintained by PRL.

Adrenergic compounds stimulated progesterone production in granulosa-luteal cells in vitro (Adashi and Hsueh, 1980). We further

Table 3. Effect of GnRH and Its Antagonist upon Terbutaline Stimulation of Progesterone Production in Cultured Granulosa Cells

Treatment	Progesterone Production (ng/Culture)
Control	1.17 ± 0.14
Terbutaline	2.65 ± 0.28
Terbutaline + GnRH (10^{-7} M)	1.51 ± 0.16
Terbutaline + GnRH (10^{-8} M)	1.94 ± 0.17
Terbutaline + AN (10^{-6} M)	2.47 ± 0.12
Terbutaline + GnRH (10^{-8} M) + AN (10^{-6} M)	2.58 ± 0.13

Granulosa cells (~ 100,000 viable cells/culture) were incubated for 2 days with FSH (10 ng/ml). After the 2 day incubation, the cultures were washed and reincubated for another 2 days with fresh medium alone (Control), 10^{-7} M terbutaline, or terbutaline with GnRH and/or the GnRH antagonist (AN). The concentrations of progesterone in the media were measured by radioimmunoassay. Data is presented as mean ± SEM of four determinations.

tested the GnRH inhibition of progesterone production induced by adrenergic compounds in FSH-primed granulosa cells (Table 3). Treatment with 10^{-7} M of terbutaline, a β_2-adrenergic agonist, for 2 days stimulated a 2.3-fold increase in progesterone production whereas concomitant treatment with GnRH inhibited the action of the β_2-adrenergic agent in a dose-dependent manner. In addition, the inhibitory effect of GnRH was prevented by concomitant treatment with the GnRH antagonist (10^{-6} M).

Inhibitory Effect of GnRH upon Luteal LH Receptor and Progesterone Production in Hypophysectomized Rats in vivo

To extend the observed direct effect of GnRH on granulosa-luteal cells in vitro, we further tested the inhibitory effect of GnRH on luteal functions in vivo (Jones and Hsueh, 1980). Hypophysectomized, diethylstilbestrol-treated female rats were primed with FSH for 2 days, followed by an ovulating dose of LH or hCG. Two days later, ovarian luteal functions were maintained by daily injections of 250 μg PRL for 3 days (Fig. 17). PRL treatment increased ovarian weight, serum progesterone level (from 13.0 ± 0.5 to 298 ± 24 ng/ml) and ovarian hCG-binding capacity (from 5.8 ± 1.3 to 584 ± 86 ng

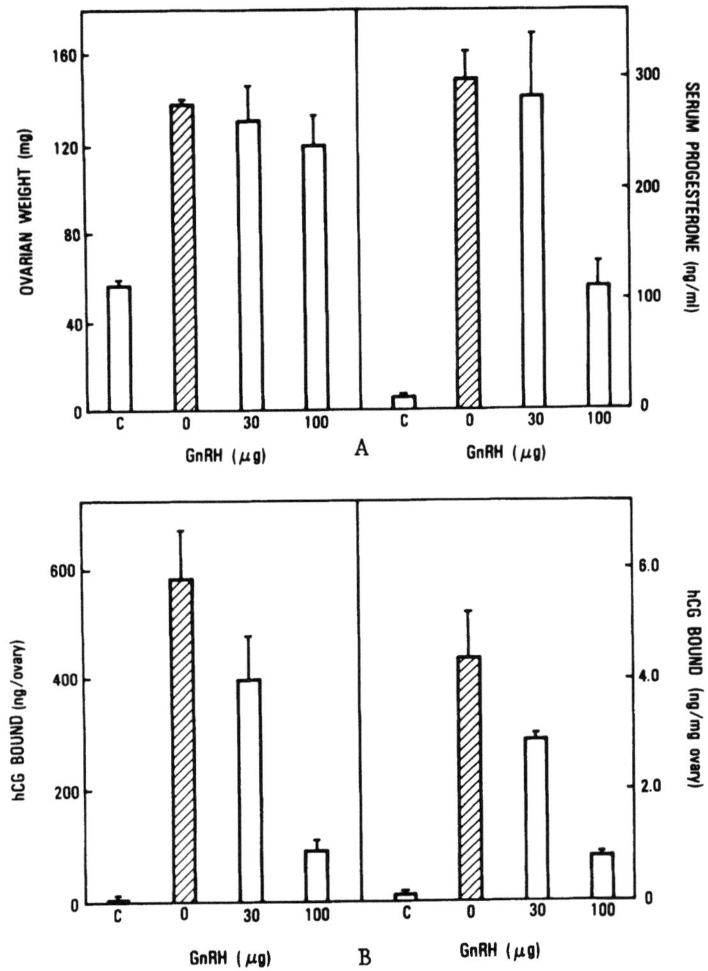

Fig. 17. Inhibitory effect of GnRH upon PRL-stimulated increases in ovarian weight, serum progesterone, and LH receptor levels in hypophysectomized female rats. Immature hypophysectomized, diethylstilbestrol-treated rats received injections of 6.7 μg oFSH every 12 h for 2 days, followed by an ovulating dose (48 μg) of oLH. Forty-eight hours later, animals received either saline [control (C)], 250 μg/day oPRL alone (0), or oPRL plus GnRH (30 or 100 μg/day) for 3 days. Ovarian weight, serum progesterone, and LH receptor levels were determined. Each bar represents the mean ± SEM of four to eight determinations. (Reprinted with permission from P. B. C. Jones and A. J. W. Hsueh: Endocrinology, 107:1930, 1980, The Endocrine Society.)

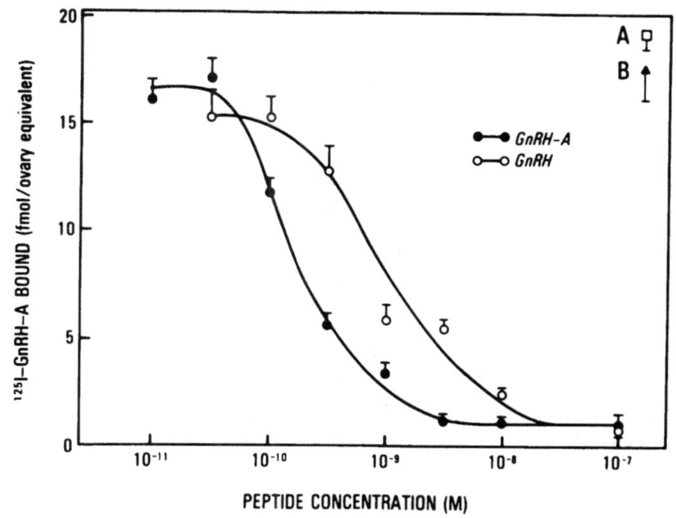

Fig. 18. Competitive effect of GnRH and the GnRH agonist (des-Gly^{10},D-Ser(tBu)^{6},Pro^{9}-NHEt-GnRH) upon the binding of ^{125}I-GnRH agonist to ovarian membranes. Increasing concentrations of GnRH (o--o) or the GnRH agonist (o--o) were added to tubes containing ^{125}I-GnRH agonist and ovarian membrane preparations. Each data point represents the mean ± SEM of four determinations. ^{125}I-GnRH agonist binding in tubes containing 10^{-6} M of GnRH partial peptides is also shown (A:GnRH$^{1-5}$NH$_2$; B:GnRH$^{5-10}$). (Reprinted with permission from P. B. C. Jones et al.: Life Science, 27:2125, 1980, Pergamon Press.)

bound hCG/ovary). In contrast, concomitant treatment with GnRH resulted in dose-dependent decreases in the PRL-induced increase of serum progesterone and ovarian LH/hCG receptor content. GnRH at 100 μg/day caused a 60% decrease in serum progesterone and an 80% decrease in ovarian LH/hCG receptor content. GnRH did not affect the ovarian weight increase maintained by PRL or the binding affinity (K_d) of ovarian LH receptors. Similar results were obtained when animals were treated with a GnRH agonist.

The observed direct effect of GnRH and its agonist on luteal functions maintained by PRL in hypophysectomized rats serves as the basis to explain the postcoital contraceptive effect of these peptides in pregnant rats and may explain, at least partially, the potent luteolytic effect of GnRH and its agonists in women during the luteal phase of the cycle.

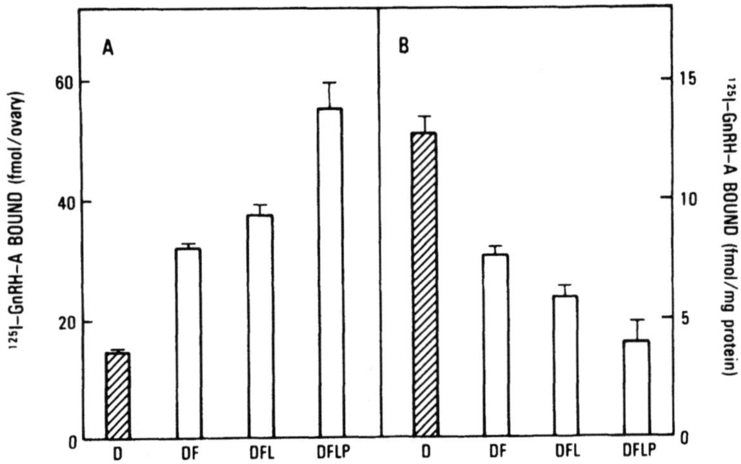

Fig. 19. Effect of in vivo hormone treatments upon ovarian GnRH binding capacity. Immature hypophysectomized rats were treated sequentially with diethylstilbestrol, FSH, LH, and PRL. Groups of animals were sacrificed during the course of treatment following diethylstilbestrol priming (D), 2 days of FSH treatment (DF), 1 day of LH treatment (DFL), and 2 days of PRL treatment (DLFP). ^{125}I-GnRH agonist binding capacity was expressed as fmol GnRH agonist bound per ovary (panel A) or as fmol GnRH agonist bound per mg protein (panel B). Each data point represents the mean ± SEM of four determinations. (Reprinted with permission from P. B. C. Jones et al.: Life Science, 27:2125, 1980, Pergamon Press.)

Binding of Radiolabeled GnRH Agonist to Rat Ovarian Cells

The demonstration of the direct effect of GnRH and its agonists on ovarian cell functions, coupled with the observed antagonistic effect of the GnRH antagonist, suggest the presence of stringent stereospecific GnRH recognition sites in the rat ovarian cells. A GnRH agonist, [des-Gly^{10},D-Ser(tBu)^{6},Pro^{9}-NHEt]GnRH, was labeled with ^{125}I to examine the possibility that the inhibitory actions of GnRH were mediated through specific ovarian receptors which recognize GnRH (Jones et al., 1980). The GnRH agonist was shown to be 215 times more potent than GnRH in inhibiting granulosa cell estrogen production in vitro. The use of the potent GnRH agonist presumably minimizes possible degradation by ovarian peptidase and eliminates binding to low affinity binding sites. Ovarian membrane fractions obtained from immature, hypophysectomized diethylstilbestrol-treated rats were incubated with 20,000-25,000 cpm of the ^{125}I-GnRH agonist in the presence or absence of increasing concentrations of GnRH or

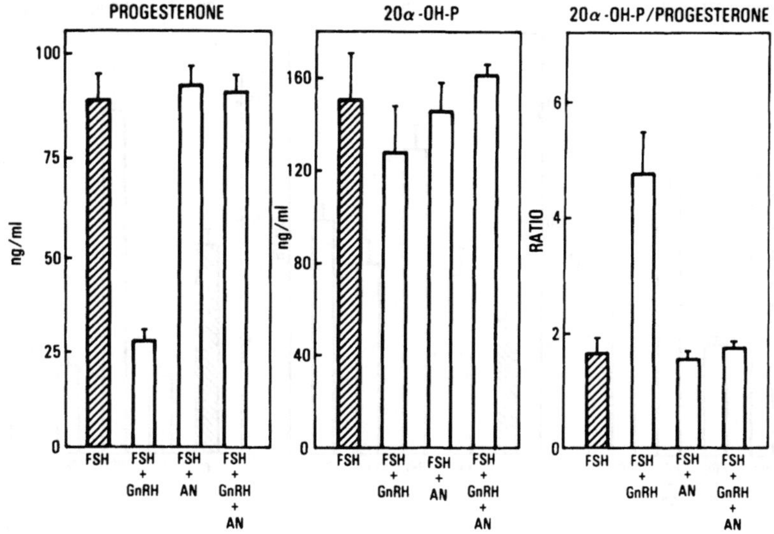

Fig. 20. Effect of treatment with GnRH upon the FSH stimulation of progesterone and 20α-OH-P production in granulosa cells in vitro. Granulosa cells (1×10^5 viable cells/dish) were cultured for 2 days in 1 ml of culture medium containing 100 nM androstenedione and ovine FSH (10 ng/ml) alone or with 10^{-8} M GnRH, 10^{-6} M antagonist (AN), or both. Progesterone (left panel) and 20α-OH-P (middle panel) in the media were measured by radioimmunoassay. These values were used to calculate the 20α-OH-P/progesterone ratio (right panel). Data points represent mean ± SEM of four determinations.

the GnRH agonist. Specific binding was determined by a filtration assay. Stereospecific, high affinity binding was detected in the ovarian membranes (Fig. 18); the dissociation constant for the labeled GnRH agonist was determined to be $0.84 \pm 0.33 \times 10^{-10}$ M and the binding capacity was calculated to be 12.9 fmol/mg ovarian protein, or 0.142 fmol/μg ovarian DNA. The binding affinity for the native GnRH decapeptide was 3.3 times lower than that of the GnRH agonist whereas two GnRH partial peptides (A: $GnRH^{1-5}NH_2$; B: $GnRH^{5-10}$) did not compete for the ^{125}I-agonist binding.

After sequential treatment with FSH, LH, and PRL to the hypophysectomized female rats, the ovarian GnRH binding capacity increased per ovary, but decreased per mg ovarian protein (Fig. 19). Furthermore, ovarian granulosa cells were isolated and their binding capacity was determined to be 25.2 fmol/mg protein, or 0.133 fmol/μg DNA, suggesting that the granulosa cells contain ovarian

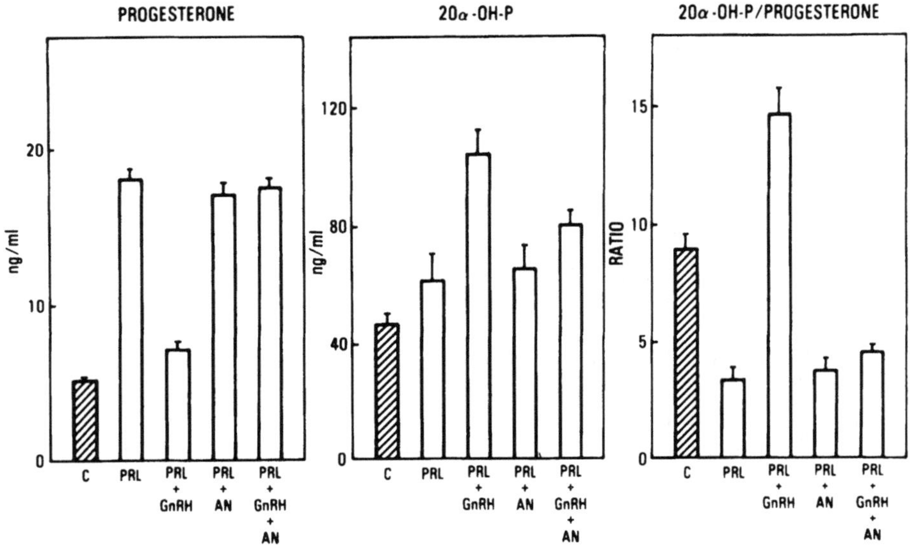

Fig. 21. Effect of treatment with GnRH upon the PRL stimulation of progesterone and 20α-OH-P production by FSH-primed granulosa cells in vitro. Granulosa cells (1×10^5 viable cells/dish) were cultured for 2 days in 1 ml of culture medium with ovine FSH (10 ng/ml). After the incubation, the cells were washed and recultured for another 2 days with fresh medium containing 100 nM androstenedione (control (C)). Some cells were also treated with ovine PRL (1 μg/ml) alone or PRL plus 10^{-8} M GnRH, 10^{-6} M antagonist (AN), or both. Progesterone (left panel) and 20α-OH-P (middle panel) in the media were measured by radioimmunoassay. These values were used to calculate the 20α-OH-P/progesterone ratio (right panel). Data points represent mean ± SEM of four determinations.

GnRH binding sites. Thus, the presence of stereospecific, high affinity GnRH binding sites in the rat ovarian granulosa cells was demonstrated.

Mechanism of the GnRH Inhibition of Gonadotropin Stimulation of Progesterone Production

The GnRH-induced decrease in progesterone production by ovarian cells treated with gonadotropins may be related to alterations in the activity of enzymes in the steroidogenic pathway. We examined, in the cultured rat granulosa cells, the effect of GnRH treatment on the activity of 20α-hydroxysteroid dehydrogenase (EC 1.1.1.149; 20α-

OH-SDH) which converts progesterone to the less active metabolite 20α-hydroxypregn-4-en-3-one (20α-OH-P) (Jones and Hsueh, 1981b).

We first measured the production of progesterone and 20α-OH-P by granulosa cells treated with FSH and/or GnRH or its antagonist. FSH treatment stimulated progesterone and 20α-OH-P production (Fig. 20). Concomitant treatment with 10^{-8} M GnRH resulted in a 70% decrease in progesterone production without significantly affecting 20α-OH-P production. The antagonist (10^{-6} M), when added together with FSH and 10^{-8} M GnRH, blocked the GnRH inhibition while treatment with the antagonist and FSH did not affect FSH stimulation of progestin production. In cultures treated with FSH plus GnRH, the 20α-OH-P/progesterone ratio was increased by 2.8-fold. Concomitant treatment with the antagonist blocked the GnRH-induced increase in 20α-OH-P/progesterone ratio while treatment with the antagonist and FSH did not affect this ratio.

To examine the effects of GnRH and its antagonist upon PRL stimulation of progesterone and 20α-OH-P production, granulosa cells were first primed for 2 days with oFSH (10 ng/ml) to induce functional PRL receptors. Then, the cells were washed and recultured for another 2 days with ovine PRL (1 μg/ml) and the appropriate hormones. PRL treatment increased progesterone production 3.5-fold as compared with control cultures (Fig. 21). Concomitant treatment with 10^{-8} M GnRH resulted in a 61% decrease in progesterone production. PRL treatment also caused a slight increase in 20α-OH-P production. In contrast, concomitant GnRH treatment further enhanced 20α-OH-P production. Treatment with 10^{-6} M antagonist, along with PRL and GnRH, blocked the GnRH inhibition of progesterone production and the GnRH stimulation of 20α-OH-P production. Furthermore, treatment with the antagonist and PRL did not affect the PRL-stimulated progesterone production while, an apparent, although statistically insignificant, increase in 20α-OH-P production was observed. PRL treatment decreased the 20α-OH-P/progesterone ratio by 62%, as compared to the untreated controls. Concomitant treatment with GnRH resulted in a 4.3-fold increase in the 20α-OH-P/progesterone ratio as compared to the PRL-treated group and, in fact, the ratio was greater than that observed with the untreated control group. This increase in the 20α-OH-P/progesterone ratio was blocked by concomitant treatment with the antagonist while treatment of cells with the antagonist and PRL did not affect the ratio.

The observed increase in 20α-OH-P/progesterone ratio in media of granulosa cells treated with GnRH suggested a possible stimulatory effect of GnRH upon 20α-OH-SDH enzyme activity. We directly measured the activity of 20α-OH-SDH in cultures treated with FSH and the appropriate hormones for 2 days. As shown in Fig. 22, FSH treatment increased 20α-OH-SDH activity 1.3-fold as compared with control

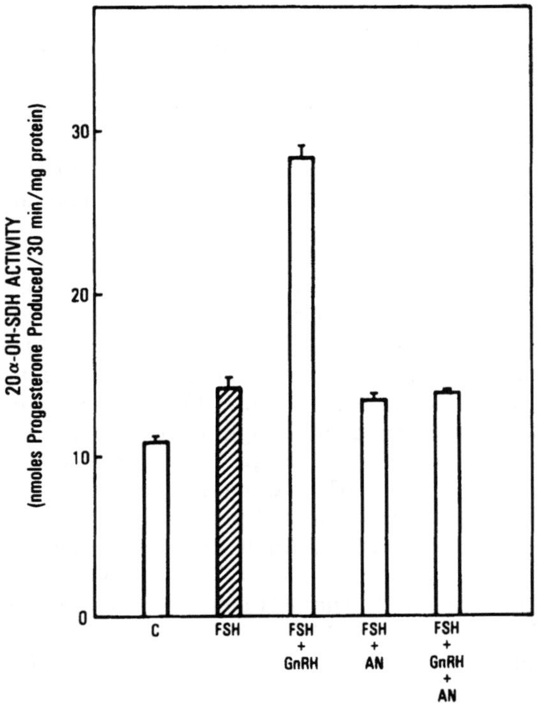

Fig. 22. Effect of treatment with GnRH and its antagonist upon FSH stimulation of 20α-hydroxysteroid dehydrogenase activity. Granulosa cells (3×10^5 viable cells/dish) were cultured for 2 days in 1 ml of culture medium containing 100 nM androstenedione (control (C)), with ovine FSH (10 ng/ml) alone or FSH plus 10^{-8} M GnRH, 10^{-6} M antagonist (AN), or both. Enzyme activities were determined after the 2 day incubation. Data points represent mean ± SEM of three determinations.

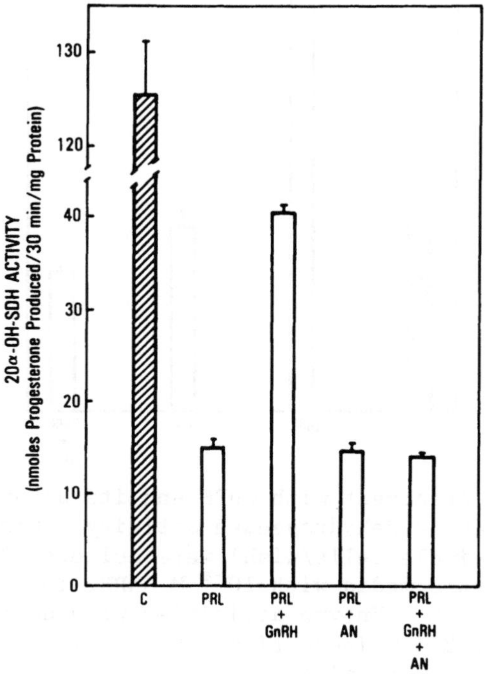

Fig. 23. Effect of treatment with GnRH and its antagonist on the PRL inhibition of 20α-hydroxysteroid dehydrogenase activity. Granulosa cells (3×10^5 viable cells/dish) were cultured for 2 days in 1 ml of culture medium with ovine FSH (10 ng/ml). After the incubation, the cells were washed and recultured for another 2 days with fresh medium containing 100 nM androstenedione in the absence (control (C)) or presence of ovine PRL (1 μg/ml) or PRL plus 10^{-8} M GnRH, 10^{-6} M antagonist (AN), or both. Enzyme activities were determined at the end of the incubation. Data points represent mean ± SEM of six determinations.

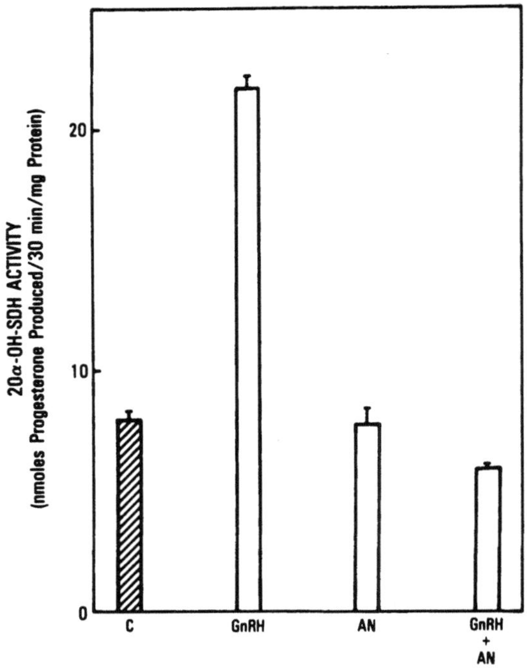

Fig. 24. Effect of treatment with GnRH and its antagonist on 20α-hydroxysteroid dehydrogenase activity. Granulosa cells (3×10^5 viable cells/dish) were cultured for 2 days in 1 ml of culture media with 10^{-8} M GnRH, 10^{-6} M antagonist (AN), or both. Enzyme activities were determined at the end of the 2 day incubation. Data points represent mean ± SEM of three determinations.

untreated cells. Concomitant treatment with 10^{-8} M GnRH further increased enzyme activity to 2.6-fold of the control value whereas concomitant treatment with 10^{-6} M of the antagonist did not affect the FSH stimulation in enzyme activity. The antagonist, when added together with FSH and GnRH, completely prevented the GnRH stimulation of 20α-OH-SDH activity. Further studies indicated that FSH and GnRH affect the enzyme activity without altering the affinity (Km) of this enzyme to its substrate.

To examine the effects of treatment with GnRH, its antagonist and PRL on 20α-OH-SDH activity, granulosa cells were first primed for 2 days with FSH in vitro and then reincubated for 2 days with various hormones. As shown in Fig. 23, 20α-OH-SDH activity was significantly increased in control cells at the end of the 4-day culture as compared to cultures treated for 2 days with FSH (Fig. 22). PRL treatment inhibited this increase, resulting in an 88% decrease in enzyme activity. In contrast, concomitant treatment with GnRH partially blocked the inhibitory effect of PRL resulting in a 2.7-fold increase in enzyme activity as compared to cells treated with PRL alone. When cultures were treated with PRL, GnRH and the antagonist, the GnRH-induced increase in enzyme activity was blocked by the antagonist; by itself, the antagonist did not affect PRL action.

Direct Modulation of Progesterone Metabolizing Enzyme by GnRH and Its Antagonist

To determine whether GnRH itself stimulated the enzyme activity in the absence of FSH or PRL, granulosa cells were treated for 2 days with medium alone, 10^{-8} M GnRH, 10^{-6} M antagonist, or GnRH plus the antagonist. As shown in Fig. 24, GnRH treatment resulted in a 2.7-fold increase in enzyme activity as compared to control untreated cells, whereas the GnRH-induced increase in enzyme activity was blocked by concomitant treatment with the antagonist. Treatment with the antagonist (10^{-6} M) alone did not affect enzyme activity. Thus, it appears that GnRH treatment increases 20α-OH-SDH activity and furthermore, this increase appears to be mediated through stereospecific receptors.

These studies demonstrated, for the first time, the direct stimulation of an ovarian functional parameter by the hypothalamic deeapeptide. This phenomenon serves as the basis to explain, at least partially, the inhibitory effect of GnRH on progesterone production induced by FSH and PRL. The increase in 20α-OH-SDH activity results in the conversion of biologically active progesterone to the 20α-OH-derivative which is completely devoid of progestational activity.

The observation of a stimulatory effect of GnRH on ovarian function is in direct contrast to earlier reports consistently demonstrating the direct inhibitory effects of GnRH and its agonists on various ovarian and testicular functions. Therefore, our study indicates that pharmacological doses of GnRH and its agonists do not exert an overall inhibition of ovarian function but rather cause specific stimulation of the progesterone-metabolizing enzyme to divert the normal steroidogenic pathway. This provides one mechanism for the GnRH inhibition of progesterone production but does not explain the GnRH inhibition of estrogen production and LH receptor content in granulosa cells. Thus, the multiple inhibitory effects of GnRH on gonadal functions may be mediated through diverse mechanisms. Furthermore, the present study does not rule out the possibility that GnRH and its agonist may affect additional steroidogenic enzymes resulting in decreases in progesterone production.

Summary

The present report summarizes our recent studied on the mechanism of the paradoxical inhibitory effect of GnRH and its agonists on ovarian granulosa and luteal cell functions. GnRH and its synthetic analogs were shown to directly regulate the functions of cultured ovarian granulosa and granulosa-luteal cells in vitro as well as to modulate granulosa and luteal functions in hypophysectomized rats in vivo. The direct effects of these peptides were probably mediated through specific GnRH recognition sites in the ovary.

Table 4. GnRH Modulation of Ovarian Granulosa Cell Functions Stimulated by FSH and Cholera Toxin

Trophic Agent	Functional Parameter	Modulation by GnRH
FSH	↑ aromatase	↓
	↑ progesterone production	↓
	↑ 20α-OH-P production	↓
	↑ 20α-OH-SDH	↑
	↑ LH receptor	↓
	↑ PRL receptor	↓
Cholera Toxin	↑ aromatase	↓
	↑ progesterone production	↓

As shown in Table 4, FSH stimulates multiple functional parameters in cultured granulosa cells. Concomitant treatment with GnRH inhibited FSH stimulation of aromatase activity, progesterone production and the number of LH and PRL receptors. In contrast, GnRH treatment enhanced the FSH stimulation of 20α-OH-SDH activity. The observed GnRH stimulation of progesterone metabolizing enzyme provided a possible mechanism for explaining the GnRH inhibition of progesterone production.

FSH priming resulted in increases in the responsiveness of granulosa cells to LH, PRL, and β_2-adrenergic agents (Table 5). In FSH primed granulosa-luteal cells, LH treatment stimulated aromatase activity whereas treatment with LH, PRL, and β_2-adrenergic agents

Table 5. GnRH Modulation of Ovarian Granulosa-Luteal Cell Functions Stimulated by LH, PRL, and β-Adrenergic Agents

Trophic Agent	Functional Parameter	Modulation by GnRH
LH	↑ aromatase	↓
	↑ progesterone production	↓
PRL	↑ progesterone production	↓
	↓ 20α-OH-SDH	↑
	↑ LH receptor	↓
β-adrenergic agents	↑ progesterone production	↓

stimulated progesterone production. In addition, PRL treatment maintained LH receptor content in these cells. These stimulatory effects of LH, PRL, and β_2-adrenergic agents on the functional parameters of the granulosa-luteal cells were also modulated by GnRH. GnRH inhibited LH stimulation of aromatase and progesterone production as well as progesterone production stimulated by PRL and β_2-adrenergic agents. In addition, PRL stimulated progesterone production, at least partially, by decreasing 20α-OH-SDH activity whereas GnRH inhibited the PRL stimulation of progesterone production by antagonizing the inhibitory effect of PRL on the 20α-OH-SDH activity. PRL stimulation of LH receptor content was also blocked by concomitant treatment with GnRH.

The observed direct actions of GnRH serve as the basis for explaining the paradoxical inhibitory effect of GnRH and its agonist upon ovarian functions and other related female reproductive functions in intact animals. In addition to its ability to modulate the actions of gonadotropins, GnRH was also shown to stimulate 20α-OH-SDH in cultured granulosa cells without gonadotropin treatment. Thus, the observed direct GnRH regulation of ovarian steroidogenic enzyme, coupled with the identification of GnRH binding sites in the granulosa cells, suggest that GnRH or a GnRH-like peptide may be a potential bonafide "hormone" at the granulosa cells. Since peripheral concentrations of GnRH are believed to be lower than 10^{-11} M (Eskay et al., 1977) and the minimal effective dose of GnRH in cultured granulosa cells is at least 10-fold higher, a hypothalamic control of ovarian function through the general circulation is not possible. However, a possible local production of GnRH-like peptide by ovarian tissues has been suggested (Ying et al., 1981). This may be analogous to recent reports of the paraendocrine control of the functions of human placenta and frog sympathetic ganglia by GnRH or GnRH-like peptides (Siler-Khodr and Khodr, 1979; Jan et al., 1979). Thus, endogenous local production of a GnRH-like peptide or its antagonist at the ovarian level may play an important physiological role in the regulation of the maturation of ovarian follicles.

Acknowledgements

This work was supported by NIH Research Grant HD-14084 and program project grant HD-12303. We thank Mrs. Linda Tucker and Cheryl Fabics for excellent technical assistance and Ms. Carol Yoza for typing the manuscript. A. J. W. Hsueh is the recipient of Research Career Development Award HD-00375. This work is done in partial fulfillment of the Ph.D. degree of P. B. C. Jones.

REFERENCES

Adashi, E. Y., and Hsueh, A. J. W., 1981, Stimulation of β_2-adrenergic responsiveness by follicle-stimulating hormone in rat granulosa cells in vitro and in vivo, Endocrinology, 108:2170.

Arimura, A., Pedroza, E., Vilchez-Martinez, J. A., and Schally, A. V., 1977, Preventation of implantation by [D-Trp[6]]-LH-RH in the rat: Comparative study with the effects of large doses of hCG on pregnancy, Endocrine Res. Commun., 4:357.

Armstrong, D. T., and Dorrington, J. H., 1976, Androgens augment FSH-induced progesterone secretion by cultured rat granulosa cells, Endocrinology, 99:1411.

Beattie, C. W., and Corbin, A., 1977, Pre- and postcoital activity of LH-RH in the rat, Biol. Reprod., 16:333.

Beattie, C. W., Corbin, A., Cole, G., Corry, S., Jones, R. C., Koch, K., and Tracy, J., 1977, Mechanism of the postcoital contraceptive effect of LH-RH in the rat. 1. Serum hormone levels during chronic LH-RH administration, Biol. Reprod., 16:322.

Belchetz, P. E., Plant, T. M., Nakai, T., Keogh, E. J., and Knobil, E., 1978, Hypophysial responses to continuous and intermittent delivery of hypothalamic gonadotropin-releasing hormone, Science, 202:631.

Casper, R. F., and Yen, S. S. C., 1979, Induction of luteolysis in the human with a long-acting analog of luteinizing hormone-releasing factor, Science, 205:408.

Conti, M., Harwood, J. P., Hsueh, A. J. W., Dufau, M. L., and Catt, K. J., 1976, Gonadotropin-induced loss of hormone receptors and desensitization of adenylate cyclase in the ovary, J. Biol. Chem., 251:7729.

Corbin, A., and Beattie, C. W., 1975, Post-coital contraceptive and uterotrophic effects of luteinizing hormone releasing hormone, Endocrine Res. Commun., 2:445.

de Koning, J., van Dieten, J. A. M. J., and van Rees, G. P., 1978, Refractoriness of the pituitary gland after continuous exposure to luteinizing hormone releasing hormones. J. Endocrinol., 79:311.

Dorrington, J. H., Moon, Y. S., and Armstrong, D. T., 1975, Estradiol-17β biosynthesis in cultured granulosa cells from hypophysectomized immature rats: Stimulation by follicle stimulating hormone, Endocrinology, 97:1328.

Erickson, G. F., and Hsueh, A. J. W., 1978, Stimulation of aromatase activity by follicle stimulating hormone in rat granulosa cells in vivo and in vitro, Endocrinology, 102:1275.

Erickson, G. F., and Hsueh, A. J. W., 1980, GnRH induces abnormal oocyte maturation in hypophysectomized rats, Soc. Gynecol. Invest., 27th Meeting, Abstract 371.

Erickson, G. F. Wang, C., and Hsueh, A. J. W., 1979, FSH induction of functional LH receptor in rat granulosa cells cultured in a chemically-defined medium, Nature, 279:336.

Eskay, R. L., Mical, R. S., and Porter, J. C., 1977, Relationship between luteinizing hormone releasing hormone concentration in hypophyseal portal blood and luteinizing hormone release in intact, castrated, and electrochemically stimulated rats, Endocrinology, 100:263.

Grimes, E. M., Taymor, M. L., and Thompson, I. E., 1975, Induction of timed ovulation with synthetic luteinizing hormone-releasing hormone in women undergoing insemination therapy. 1. Effect of a single parenteral administration at midcycle, Fert. Steril., 26:277.

Haresign, W., Foster, J. P., Haynes, N. B., Crighton, D. B., and Lamming, G. E., 1975, Progesterone levels following treatment of seasonally anoestrous ewes with synthetic LH-releasing hormone, J. Reprod. Fert., 43:269.

Hillensjö, T., and LeMaire, W. J., 1980, Gonadotropin releasing hormone agonists stimulate meiotic maturation of follicle-enclosed rat oocytes in vitro, Nature, 287:145.

Hilliard, J., Pang, C.-N., and Sawyer, C. H., 1976, Effects of luteinizing hormone-releasing hormone on fetal survival in pregnant rabbits, Fert. Steril., 27:421.

Hillier, S. G., Knazek, R. A., and Ross, G. T., 1977, Androgenic stimulation of progesterone production by granulosa cells from preantral ovarian follicles: Further in vitro studies using replicate cell cultures, Endocrinology, 100:1539.

Hsueh, A. J. W., Dufau, M. L., and Catt, K. J., 1977, Desensitization of testicular responses-negative control of LH/hCG receptors, cAMP production and steroidogenesis by exogenous hCG in vivo, Proc. Natl. Acad. Sci., 74:592.

Hsueh, A. J. W., and Erickson, G. F., 1979, Extrapituitary action of gonadotropin releasing hormone: Direct inhibition of ovarian steroidogenesis, Science, 204:854.

Hsueh, A. J. W., and Ling, N. C., 1979, Effect of an antagonistic analog of gonadotropin releasing hormone upon ovarian granulosa cell function, Life Sci., 25:1223.

Hsueh, A. J. W., Wang, C., and Erickson, G. F., 1980, Direct inhibitory effect of gonadotropin-releasing hormone upon follicle-stimulating hormone induction of luteinizing hormone receptors and aromatase activity in rat granulosa cells, Endocrinology, 106:1697.

Humphrey, R. R., Windsor, B. L., Bousley, F. G., and Edgren, R. A., 1976, Antifertility effects of an analog of luteinizing hormone releasing hormone in rats, Contraception, 14:625.

Humphrey, R. R., Windsor, B. L., Reel, J. R., and Edgren, R. A., 1977, The effects of luteinizing hormone releasing hormone (LH-RH) in pregnant rats. 1. Postnidatory effects, Biol. Reprod., 16:614.

Humphrey, R. R., Windsor, B. L., Jones, D. C., Reel, J. R., and Edgren, R. A., 1978, The effects of luteinizing hormone releasing hormone (LH-RH) in pregnant rats. 2. Prenidatory effects and delayed parturition, Biol. Reprod., 19:84.

Humphries, J., Wan, Y.-P., Folkers, K., Bowers, C. Y., Presence of proline in position 3 for potent inhibition of the activity of the luteinizing hormone releasing hormone and of ovulation, Biochem. Biophys. Res. Commun., 72:939.

Jan, Y. N., Jan, C. Y., and Kuffler, S. W., 1979, A peptide as a possible transmitter in sympathetic ganglia of the frog, Proc. Natl. Acad. Sci. U.S.A., 76:1501.
Johnson, E. S., Gendrich, R. L., and White, W. F., 1976a, Delay of puberty and inhibition of reproductive processes in the rat by a gonadotropin-releasing hormone agonist analog, Fert. Steril., 27:853.
Johnson, E. S., Seely, J. H., White, W. F., and DeSombre, E. R., 1976b, Endocrine-dependent rat mammary tumor regression: Use of a gonadotropin releasing hormone analog, Science, 194:329.
Jones, P. B. C., Conn, P. M., Marian, J., and Hsueh, A. J. W., 1980, Binding of gonadotropin-releasing hormone agonist to rat ovarian granulosa cells, Life Sci., 27:2125.
Jones, P. B. C., and Hsueh, A. J. W., 1980, Direct inhibitory effect of gonadotropin-releasing hormone upon luteal LH receptor and steroidogenesis in hypophysectomized rats, Endocrinology, 107:1930.
Jones, P. B. C., and Hsueh, A. J. W., 1981a, Direct effects of GnRH and its antagonist upon ovarian functions stimulated by FSH, prolactin and LH. Biology of Reproduction, 24:747.
Jones, P. B. C., and Hsueh, A. J. W., 1981b, Direct stimulation of ovarian progesterone-metabolizing enzyme by gonadotropin-releasing hormone in cultured granulosa cells, J. Biol. Chem., 256:1248.
Kledzik, G. S., Cusan, L., Auclair, C., Kelly, P. A., and Labrie, F., 1978, Inhibition of ovarian luteinizing hormone (LH) and follicle-stimulating hormone receptor levels with an LH-releasing hormone agonist during the estrous cycle in the rat, Fert. Steril., 30:348.
Kledzik, G. S., Cusan, L., Auclair, C., Kelly, P. A., and Labrie, F., 1978, Inhibitory effect of a luteinizing hormone releasing hormone agonist on rat ovarian LH and FSH receptor levels during pregnancy, Fertil. Steril., 29:560.
Koyama, T., Ohkura, T., Kumasaka, T., and Saito, M., 1978, Effect of postovulatory treatment with a luteinizing hormone-releasing hormone analog on the plasma level of progesterone in women, Fertil. Steril., 30:549.
Kuhl, H., and Taubert, H.-D., 1976, LH-RH-inactivating enzyme system in the rat hypothalamus and pituitary: Activity changes during the oestrus cycle with and without treatment with sex hormones, Abstract #311, p. 128.
Lemay, A., Labrie, F., Ferland, L., and Raynaud, J.-P., 1979, Possible luteolytic effects of luteinizing hormone-releasing hormone in normal women, Fertil. Steril., 31:29.
Lin, Y. C., and Yoshinaga, K., 1976, Inhibitory effect of GnRH on ovum implantation in the rat, Proc. 58th Annual Meeting, Endocrine Society, Abstract 174, p. 143.
Nillius, S. J., Bergquist, C., and Wide, L., 1978, Inhibition of ovulation in women by chronic treatment with a stimulatory LRH analogue — a new approach to birth control? Contraception, 17:537.

Nimrod, A., and Lindner, H. R., 1976, A synergistic effect of androgen on the stimulation of progesterone secretion by FSH in cultured rat granulosa cells, Mol. Cell. Endocrinol., 5:315.

Rao, M. C., Richards, J. S., Midgley, A. R., and Reichert, L. E., 1977, Regulation of gonadotropin receptors by luteinizing hormone in granulosa cells, Endocrinology, 101:512.

Rippel, R. H., and Johnson, E. S., 1976, Regression of corpora lutea in the rabbit after injection of a gonadotropin-releasing peptide, Proc. Soc. Exp. Biol. Med., 152:432.

Rivier, C., Rivier, J., and Vale, W., 1978, Chronic effects of [D-Trp^6-Pro^9-NEt]luteinizing hormone-releasing factor on reproductive processes in the female rat, Endocrinology, 103:2299.

Siler-Khodr, R. M., and Khodr, G. S., 1979, Extrahypothalamic luteinizing hormone-releasing factor: Release of immunoreactive LRF in vitro, Fertil Steril., 32:294.

Wang, C., Hsueh, A. J. W., and Erickson, G. F., 1979, Induction of functional prolactin receptors by follicle-stimulating hormone in rat granulosa cells in vivo and in vitro, J. Biol. Chem., 254:11330.

Yen, S. S. C., Lasley, B. L., Wang, C. F., Leblanc, H., and Siler, T. M., 1975, The operating characteristics of the hypothalamic-pituitary system during the menstrual cycle, Recent Prog. Horm. Res., 31:321.

Ying, S.-Y., Ling, N., Bohlen, P., and Guillemin, R., 1981, Gonadocrinins: peptides in ovarian follicular fluid stimulating the secretion of pituitary gonadotropins, Endocrinology, 108:1206.

LH-BINDING INHIBITORS FROM THE CORPUS LUTEUM

Darrell N. Ward, Wan-Kyng Liu,
Stephan D. Glenn,
and Cornelia P. Channing*

Division of Biochemistry
The University of Texas System Cancer Center
M. D. Anderson Hospital and Tumor Institute
6723 Bertner Avenue
Houston, Texas 77030

The initial observations of LH receptor binding inhibitor activity were quite serendipitous. A graduate student (Dr. Kuo-Pao Paul Yang) attempted to use some ovaries from psuedopregnant rats as a source of receptors for luteinizing hormone. These ovaries had been stored for quite some time at -20°. When Paul attempted to measure LH-binding in a crude homogenate of these ovaries no significant binding was observed. However, in the residuum after 1,2 or 3 washings there was significant specific binding for LH. Our first studies (Yang et al., 1976a) demonstrated that the binding inhibition resided in both a low molecular weight (dialyzable) and a high molecular weight (non-dializable) fraction. We were aware of the earlier studies of Bhalla and Reichert (1974) concerning an aqueous ethanol solubilized fraction from testicular tissue which interacted directly with gonadotropins as shown by physical properties. Reichert and Bhalla had suggested the material behaved as though it were a solubilized portion of the hormone receptor. We demonstrated, at least for the low molecular weight form of the LH.RBI, that such a direct interaction of our material with LH was not obtained. This was demonstrated by molecular exclusion processes and inability to observe an association complex for labeled LH with the inhibitor (Yang et al., 1976a).

---

*Department of Physiology, University of Maryland School of Medicine, 660 West Redwood Street, Baltimore, Maryland 21201.

To preserve the receptors all of our initial studies had been conducted under extraction conditions that utilized reduced temperature, i.e., 4°C. It was subsequently demonstrated with these conditions of extraction that the freezing of the ovaries and storage was an essential condition (Yang et al., 1976a, 1979) in order to extract the LH.RBI and measure its presence. The importance of freezing to the detection of LH.RBI has been observed also by at least two other laboratories (Sakai et al., 1977; Kumari et al., 1980). Subsequently it was shown that with more drastic extraction conditions, i.e., room temperature or 37°, the LH.RBI activity was extractable without a prior prolonged period of storage frozen (Yang et al., 1979).

Our initial studies were with the LH.RBI activity from ovaries from psuedopregnant rats. Such ovaries are heavily luteinized. The studies of Sakai et al. (1977) established the existence of an LH.RBI activity detectable in porcine corpora lutea. This study first pinpointed the corpora lutea as the portion of the ovary to contain significant quantities of LH.RBI activity. Recently, however, Ranganathan and Sairam (1979) have reported that LH.RBI activity is also detectable in bovine or porcine follicular fluid. LH.RBI in the corpora lutea of species such as the cat, dog, sheep, and goat (Yang, 1978) was subsequently demonstrated. In these cases the corpora lutea were collected from pregnant animals rather than animals made pseudopregnant by hormone manipulation. Recently Kumari et al. (1980) have shown that corpora lutea isolated during the normal menstrual cycle or corpora lutea of pregnancy in the human both contain an LH.RBI activity. In their study LH.RBI activity from normal menstrual cycle corpora lutea appeared to inhibit binding of hCG to a greater extent than was observed with extracts from the corpora lutea of pregnancy. Yang et al. (1976a) had shown that the ovaries of mature, non-pregnant rats had very little LH.RBI activity. However, this was determined with the extract from the total ovary and not from the individual corpora lutea. In studies by Kumari et al. (1979) and Tucker et al. (1979) it was shown that LH.RBI activity was greater in aqueous extracts from old porcine corpora lutea than from extracts of young porcine corpora lutea.

We demonstrated that LH.RBI not only inhibited binding of LH but also inhibited the LH-stimulated progesterone biosynthesis by ovarian slices from pseudopregnant rats (Yang et al., 1978, 1979). Kumari et al. (1979) demonstrated that LH.RBI would also inhibit the in vitro production of progesterone following LH stimulation of cultures of porcine granulosa cells.

A recent report by Channing et al. (1981) has provided evidence that LH.RBI as a partially purified fraction from porcine corpora lutea was capable of inhibiting pregnancy in the rabbit. In order to observe this inhibition it was necessary to administer the LH.RBI preparation either immediately prior to mating or on day 1 or 2 following mating.

The observations currently available now indicate that LH.RBI preparations can, in fact, affect normal physiological processes. It has not been established, however, whether LH.RBI per se plays an important physiological role. Indeed, a critical evaluation of the possible physiological role of LH.RBI would be facilitated by the isolation and characterization of the chemical nature of LH.RBI.

The nature of the transition that occurs during freezing to improve the extractability of the LH.RBI is not understood. We expect the transition somehow involves a lipid phase since it has been shown (Yang et al., 1978) that Triton X-100 as 0.1% solutions facilitates the extractability of LH.RBI. Moreover, recent evidence obtained by Dr. Glenn in our laboratory indicates that the storage at -20° was again a fortuitous occurrence, since preliminary findings suggest that storage at -70° allows the transition to occur either not at all or very slowly. We have also noted (Ward et al., 1981) that cold butanol extraction removes a significant portion of LH.RBI activity.

In our studies of the LH.RBI obtained from the ovaries of pseudopregnant rats there remains a curious dichotomy between the responses observed in receptor preparations from ovaries as compared with crude receptor preparations from testes. The initial findings were reported in 1976 (Yang et al., 1976b) and were further extended in two subsequent reports (Yang et al., 1978, 1979). Basically the observations were that inhibition was observable in a crude homogenate of ovaries from psuedopregnant rats as the LH receptor source, but not in a crude homogenate from testes of the male rat. It was further shown that while the LH.RBI preparation inhibited the LH-stimulated production of progesterone by ovarian slices or granulosa cells, the LH.RBI preparations did not inhibit LH stimulation of testosterone production in a washed testicular mince preparation. Not only did an extract containing LH.RBI augment LH stimulation of production of testosterone in this system, but by itself the LH.RBI preparation, obtained from ovary, stimulated the production of testosterone (Yang, 1979). As a result of these observations we have been led to speculate on a possible difference in male and female LH receptors and their allosteric control. A suggested model to account for this behavior has been described (Yang, 1976b and Yang et al., 1978). This dichotomy of behavior for the rat LH.RBI preparations is being further investigated in Dr. Yang's laboratory in the Department of Medicine at our Institution.

We have observed that this dualistic behavior is not obtained with the LH.RBI preparations from porcine corpora lutea with which we have been working (Ward et al., 1981). With the porcine material it has been possible to demonstrate a binding inhibition in crude testicular homogenate preparations. We will extend these observations in the present paper.

FRACTIONATION SCHEME I

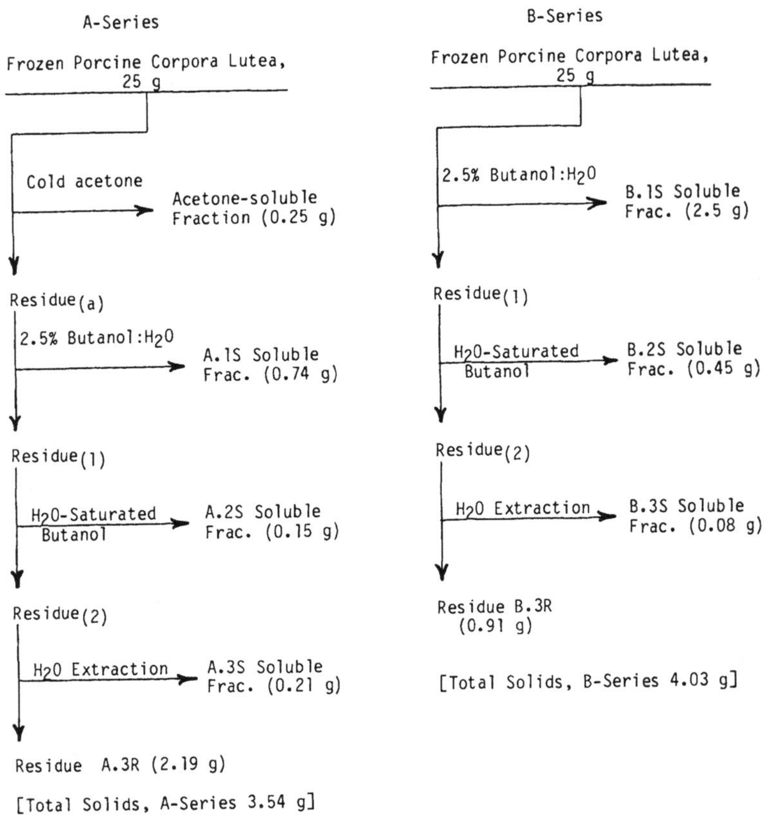

Much of the work which has been summarized relating to the LH.RBI has dealt with relatively crude preparations. Moreover, when one attempts fractionation one encounters what apparently represents difference forms of the LH.RBI. One distinction was observed in the initial report from our laboratory, i.e., there is a high and low molecular weight form of LH.RBI (Yang et al., 1976a). In our studies with the extract of the LH.RBI from porcine corpora lutea the amount of the low molecular weight form is vanishingly small, although under certain conditions it may still be detected. To clarify some of these questions as to form and function of LH.RBI we have been attempting to purify the LH.RBI, or as is obviously more precise, the LH.RBIs. We summarized the status of these purification attempts in a recent conference (Ward et al., 1981).

Ranganathan and Sairam, in their abstract of 1978, described fractionation studies and were able to make some conclusions as to the nature of the individual fractions they encountered. (For further discussion of that report see below.) Unfortunately that study is only available in abstract form at this time.

We have recently been able to differentiate some of the types of LH.RBI by procedures which were a further development of the butanol solubility studies initiated by Dr. Glenn. It will be the purpose of our presentation to describe some of the new approaches this has allowed us to pursue with respect to our characterization of the LH.RBI.

Fractionation Scheme I summarizes the procedures we have applied to fractionate LH.RBI in our latest endeavors. These extractions were undertaken to answer two questions: (1) Could we prepare an acetone-dried preparation from the corpora lutea as a convenience to shipment from the point of collection to our laboratory (it is much more convenient to ship an acetone-dried powder than corpora lutea packed in dry ice); and (2) to determine whether we could improve upon the separation of the aqueous soluble and butanol-soluble fractions we had observed previously (Ward et al., 1981). In this scheme, Procedure A represents the fractionation of the acetone dried corpora lutea, Procedure B represents those corpora lutea which were homogenized directly in the 2.5% butanol solution. Twenty-five grams of corpora lutea were used in both series. In each instance the corpora lutea had been stored frozen at -20° for well beyond 10 weeks, the period our lab earlier had shown necessary for maximum extractability of LH.RBI (Yang et al., 1976a, 1979). After allowing for removal of about 84% water content, the acetone removed approximately 7% of the remaining tissue solids. However, as we soon learned, this treatment materially affected the LH.RBI activity recovery and the type of fractionation obtained with the residual material. The approximate dry weight obtained in each fraction is given on the fractionation scheme summary. These are approximate weights since the acetone soluble and the butanol soluble fractions are sticky, waxy materials that are difficult to handle for direct weighing.

Before we discuss the bioassay of these fractions we should introduce some unpublished observations made by Dr. Hiromu Sugino during an extensive study of LH.RBI fractionation (Ward et al., 1981). Dr. Sugino had observed, particularly with some of the fractions obtained in his DEAE-chromatography studies, that certain fractions would stimulate the binding of LH rather than inhibit it. In some of his less purified fractions he would observe initially an inhibition followed by stimulation or sometimes an initial stimulation followed by inhibition as the dose of the fraction was increased. We began to refer to this operationally as the "stimulator fraction." Mechanistically it was difficult to decide what might cause such a stimulation of LH-binding. Our best hypothesis, and it was only an hypothesis, was that we were encountering some

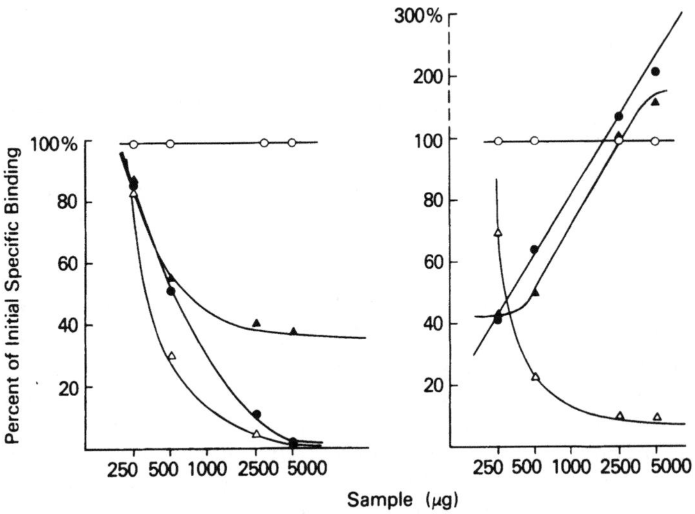

Fig. 1. Competitive inhibition curves of the corpus luteum extracts (Fractionation Scheme I) in a rat ovary homogenate preparation. The assay employed 1 ng of ^{125}I-oLH and 10 mgs of ovary tissue homogenate in a total volume of 0.5 ml. Incubation was for 2 h at 37°C. A-series, left panel; B-series, right panel: -●-●-, A.1S or B.1S; -▲-▲-, A.2S or B.2S; -Δ-Δ-, A.3S or B.3S; -o-o-, A.3R or B.3R.

of the receptor-like solubilized material which had been described earlier by Bhalla and Reichert (1974), under conditions that would allow insertion of additional receptor material into the preparations we were using. Alternatively, the fractions might be deblocking receptor sites. Be that as it may, the fractionation scheme we have shown in which we introduced acetone extraction gave us the first clear-cut separation of this stimulator effect.

In Fig. 1 is presented the LH-binding inhibition data for the two extraction procedures shown in Scheme I. On the left-hand side are shown the data for the acetone-treated material. The A.1S, A.2S, and A.3S fractions all showed inhibition. The A.2S (the butanol soluble fraction) plateaued at approximately 60% inhibition, but at the highest doses none of the fractions showed stimulatory activity. In this series the acetone-soluble fraction was not assayed. In contrast, again testing with the ovarian homogenate receptor preparation, the corresponding fractions from the corpora lutea that had not been treated with acetone we found only the B.3S fraction contained inhibitor activity without stimulator activity. The B.1S and B.2S fractions, the 2.5% butanol-water extracted fraction and the water-saturated butanol extracted fraction, both showed an initial binding

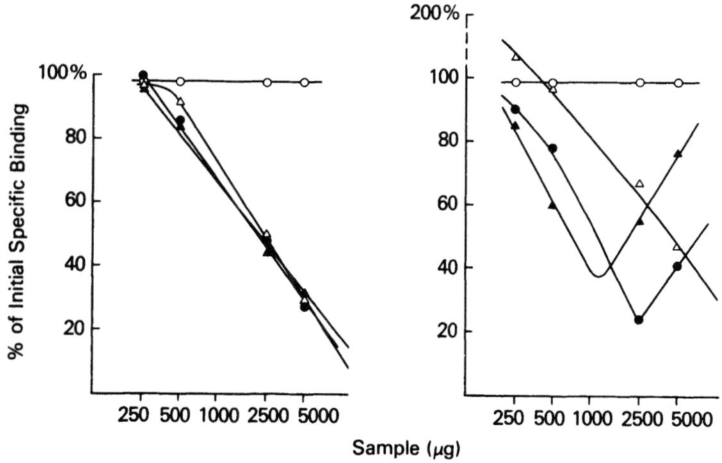

Fig. 2. Competitive inhibition curves of the corpus luteum extracts (Fractionation Scheme I) in a rat testicular homogenate preparation. The assay employed 2.5 ng ^{125}I-oLH and 25 mgs of testicular tissue homogenate in a total volume of 0.5 ml. Symbols are the same as in Fig. 1.

inhibition of approximately 60%, followed by an increase in the amount of binding as the quantity of the fraction tested was increased.

When we tested these same fractions in a testicular homogenate preparation we found essentially similar results with the exception that the testicular receptors seemed somewhat less sensitive to the binding inhibition, i.e., a greater quantity of inhibitor was required. On the other hand, the testicular preparation responded in a more linear fashion if one compares the left-hand side of Figs. 1 and 2. As in the ovarian system fractions B.1S and B.2S showed an initial binding inhibition over a somewhat more extended range than observed in the ovarian system, but eventually at about a 2.5 mg dose some of the stimulator effect was apparent. In this case, however, it was still less than the control level and the stimulator simply reversed inhibition somewhat.

Since the acetone desiccation had substantially shifted the amount of material in the initial 2.5% butanol extract (i.e., A.1S and B.1S), vis-a-vis the amount of material in the subsequent aqueous extraction, i.e., the A.3S or B.3S fraction, we decided to combine the 1.S and 3.S fractions in each series in order to simplify the comparison of the water-extractable fractions under the two sets of conditions.

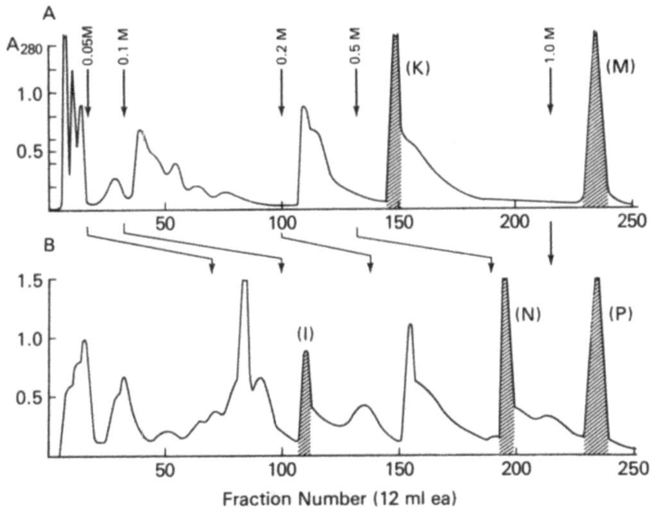

Fig. 3. Chromatography of the aqueous extracts on a DEAE-cellulose (Whatman DE-32), column (2.5 × 15 cm) developed with stepwise increase of ammonium bicarbonate concentration, as indicated by the arrows. Column chromatography was carried out at 4°C with a flow rate of 60 mls/hr. Activity was located in the cross-hatched fractions. A, upper panel: 750 mg of combined fraction A.1S + A.3S. B, lower panel: 1000 mg of combined fraction B.1S + B.3S. Samples were dissolved in the initial 0.01 M-ammonium bicarbonate buffer, pH 8, and clarified by centrifugation to remove any insoluble material before application to the column. At points indicated by the arrows the ionic strength of the developing buffer was increased. As dictated by the emerging pattern the buffer applications were delayed in B, compared to A, except the final buffer.

In Fig. 3 we present a comparison of the chromatogram on DEAE-cellulose of the aqueous soluble fractions from the acetone desiccated material as compared to the aqueous soluble fractions from the corpora lutea which had not been submitted to acetone desiccation. For these chromatograms the sample was dissolved in 0.01 M ammonium bicarbonate. The column was equilibrated in this same buffer. Material which was insoluble in this low ionic strength buffer was removed prior to application to the chromatogram. In the case of the B-series material this insoluble fraction contained the majority of the stimulatory fraction. Development of the chromatograms was done with increasing molarity of ammonium bicarbonate applied in a stepwise fashion. The most significant difference in the patterns for the aqueous-soluble fractions from Scheme A and B was in the nature

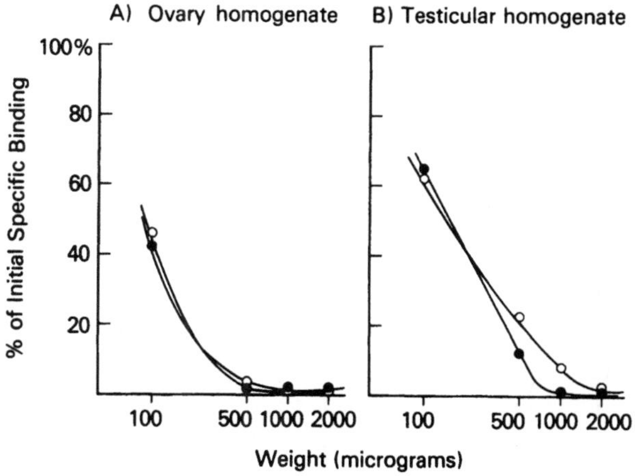

Fig. 4. Competitive inhibition curves of fraction "K" and "M" in: A) ovary homogenate (1 ng ^{125}I-oLH + 2.5 mgs of tissue) and B) testicular homogenate (2.5 ng ^{125}I-oLH + 25 mgs of tissue). Fraction "K," -o-o-, fraction "M," -●-●-.

of the pattern during the earlier (lower ionic strength) elution development. The distribution of peaks was quite different (Fig. 3). There was also some stimulator activity detected in the initial fraction emerging from the column of the Procedure B material (Fig. 3B).

The fractions which contained the majority of the LH.RBI activity have been indicated by cross-hatches in Fig. 3. In the A-series this represents the material in peaks K and M as designated for the series and in the B-series the peaks I, N, and P. (These are simply sequential letter designations of the peaks emerging from each chromatogram.) There was also some additional inhibitor activity in the trailing areas following these cross-hatched peaks, but we concentrated our further evaluation on the peaks which had the highest specific activity for the isolated LH.RBI fractions.

In Fig. 4 we present the activity comparison of the K and M fractions from the A-series as tested in both the ovarian and testicular homogenate receptor systems. It can be seen that in the ovarian system between 100 and 500 micrograms of material was sufficient to produce complete inhibition of binding, while in the testicular homogenate between 1 and 2 mg of these same preparations were required for complete inhibition of binding.

In Fig. 5 we show a similar comparison with the three inhibitor peaks obtained in the B-series submitted to DEAE-chromatography. The 0.1 M bicarbonate eluted a peak designated "I" which we have now termed the "solvent-sensitive inhibitor fraction." This frac-

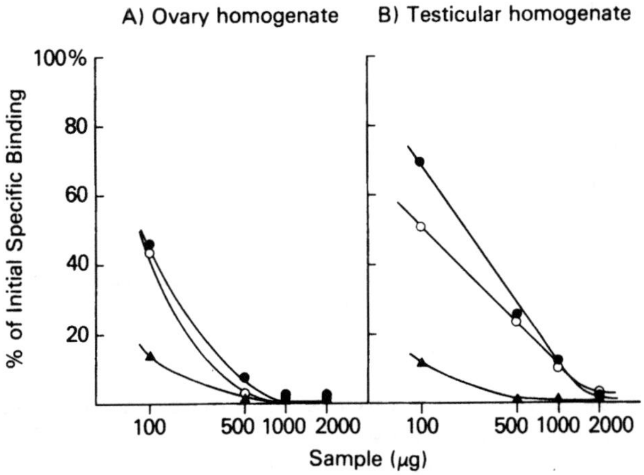

Fig. 5. Competitive inhibition curves of fraction "I," "N,", and "P," in: A) ovary homogenate and B) testicular homogenate. Assay conditions same as in Fig. 4. -▲-▲-▲-, fraction "I"; -o-o-, fraction "N"; -●-●-, fraction "P."

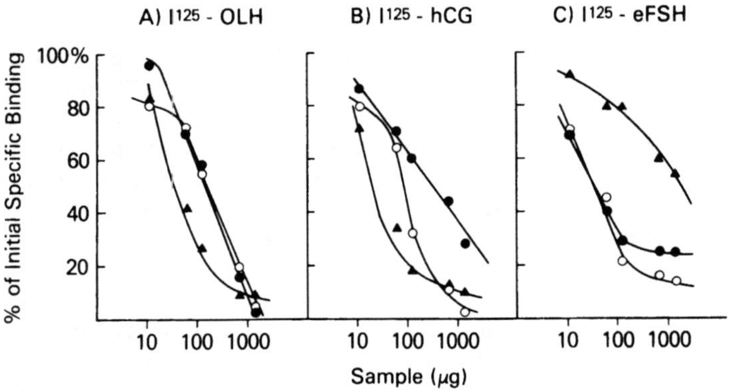

Fig. 6. Competitive inhibition curves of fraction "I," "N," and "P" in ovary homogenate. A) 1 ng ^{125}I-oLH + 2.5 mgs of tissue; B) 1 ng ^{125}I-hCG + 2.5 mgs of tissue; C) 1 ng of ^{125}I-eFSH + 2.5 mgs of tissue; -▲-▲-, fraction "I"; -o-o-, fraction "N"; -●-●-, fraction "P.".

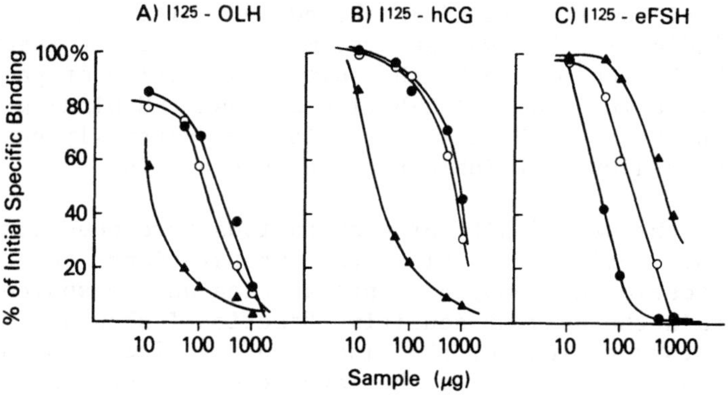

Fig. 7. Competitive inhibition curves of fraction "I," "N," and "P" in rat testicular homogenate. A) 2.5 ng ^{125}I-oLH + 25 mgs of tissue; B) 2.5 ng ^{125}I-hCG + 25 mgs of tissue; C) 2.6 ng of ^{125} I-eFSH + 25 mgs of tissue; -▲-▲-, fraction "I," -o-o-, fraction "N"; -●-●-, fraction "P."

tion had unique properties in both the ovarian and testicular homogenate system. It can be seen in Fig. 5 that the I-fraction with as little as 100 micrograms produced from 85 to 90% inhibition of labeled LH-binding to either the ovarian or the testicular homogenate system. The N and P fractions, eluted with 0.5 M and 1 M bicarbonate respectively, showed essentially the same inhibition potency as the K and M fractions from the A-series as shown in Fig. 4. The curves presented in Figs. 4 and 5 represent the assays as they were obtained on screening the initial fractions. Doses had been chosen to anticipate the upper end of the inhibition curve in order that activity could be assessed for even the low-potency fractions.

In Fig. 6 we present assay results for the same three fractions as shown in Fig. 5 but under conditions which evaluated the binding over the full range of inhibition responses. Again the K and M peaks from the A-series showed binding inhibition quite comparable to that for the N and P peaks from the B-series, therefore further data from the A-series has not been presented in Fig. 6. However, in this series we also compared the binding inhibition for labelled hCG and for labeled equine FSH. (The labeled equine FSH preparation was very kindly furnished by Dr. George Bousfield.) In this series of assays with the ovarian homogenates, fraction-I, the solvent sensitive inhibitor, was the strongest inhibitor for both iodinated LH as tracer or iodinated hCG as tracer, i.e., the left-hand side of each panel. The N and P peak fractions were essentially superimposable as inhibitors for the labeled LH. In the hCG panel on the right-hand side it can be seen that fraction P was a lower potency inhibitor of hCG binding.

Although all of the fractions showed an inhibition against equine FSH binding in the ovarian system, a significant inversion occurred with the inhibition by the solvent-sensitive or peak I material. Here a much lower inhibition against FSH binding was observed. Again the N and P peak materials were virtually superimposable in their degree of inhibition for FSH binding.

In Fig. 7 the same inhibitor preparations have been tested against the same series of labeled trace preparations, but this time using a testicular homogenate preparation as the source of the receptor. Again the solvent-sensitive fraction I showed a much greater inhibitory action for the binding of labeled LH or hCG. Moreover, in the case of hCG as the label the binding inhibition by the N and P peak material is much less pronounced, i.e., the curves are shifted further to the right.

In the testicular homogenate system the response to the solvent-sensitive Peak I material is far less effective in the case of inhibition of labeled equine FSH than the other LH.RBI fractions. Thus, the insensitivity of FSH binding to inhibition by the solvent-sensitive inhibitor fraction was observed in both ovarian and testicular systems.

Fraction P in the testicular system seemed more effective as an FSH inhibitor than was Fraction N. Recall, it was Peak P material that was less inhibitory against hCG in the ovarian homogenate system (re., Fig. 6, center panel). At this point in our studies further comment relative to these different responses in the different test systems is unwarranted.

In order to define a provisional unit and to enable us to make some estimate of our activity recoveries we have arbitrarily designated one unit of LH.RBI activity as that quantity of inhibitor which is equivalent to the binding inhibition of 1 nanogram of purified ovine LH. As was shown in an earlier publication (Ward et al., 1981) the inhibition curve with the LH.RBI preparations is somewhat steeper than the ovine LH material, thus at best this is only an approximate unit. However, some idea of relative potency can be gained by comparing the $I.D._{50}$ on the respective curves. If we make this type of calculation we can estimate a specific activity and a total activity recovery for the A- and B-series of the fractionation scheme presented earlier. In Table 1 we summarize these calculations. It is immediately apparent that in the A-series the acetone treatment eliminated over 90% of the aqueous soluble activity. Moreover, almost 75% of the activity observed in the B-series could be attributed to the very potent fraction I, the solvent-sensitive peak. This material had a specific activity of 615 RBI units/mg compared to 30 to 80 RBI units/mg in the K, M, N, and P fractions.

Table 1A. Recovery of LH.RBI Activity from A-Series Fractionation Study

Fraction	ID_{50}[a] (μg)	Specific Activity (u/mg)	Total Units[b]	NOTES
Acetone-Soluble	N.D.	------	-----	
A.1S	2000	4.0	3000	
A.3S	2000	4.0	800	
DEAE-Fr.K	150	53.3	2132	71% of Total/column
DEAE-Fr.M	150	53.3	427	14% " "

[a] Measured in the rat testicular homogenate system. The ID_{50} of ovine LH was 8 ng under conditions of the assay.

[b] The aqueous-soluble LH.RBI only. The total activity applied to the DEAE cellulose column was 3800 units.

Table 1B. Recovery of LH.RBI Activity from B-Series Fractionation Study

Fraction	ID_{50}[a] (μg)	Specific Activity (u/mg)	Total Units[b]	NOTES
B.1S	(1200[c])	(> 7[c])	>17,500	Indirect estimate from DEAE fraction indicates at least 38,000 units in the aqueous fractions.
B.3S	(? [c])	(?)	(?)	
DEAE-Fr.I	13	615	18,450	67% of Total/column
DEAE-Fr.N	100	80	4,160	15% " "
DEAE-Fr.P	275	30	690	3% " "

[a] See Table 1A

[b] See Table 1A

[c] An estimate of specific activity cannot be made due to interference from the stimulator activity (see text).

It was noted earlier that Dr. Glenn had shown the butanol soluble fraction from corpora lutea which had not been treated with acetone, accounted for a very significant portion of the LH.RBI activity. At this point a more quantitative statement is difficult to make due to the severe solubility problems in both dissolving, handling and inserting this fraction back into the aqueous test systems we used with out receptor preparations. What data we have been able to collect on a butanol-soluble preparation is summarized in

Table 2. Estimated Recovery of LH.RBI Activity in the Butanol-Extractable Fraction from 1 kg of Porcine Corpora Lutea

Fraction	Weight (mg)	ID_{50}[a] (μg)	Specific[b] Activity (u/mg)	Total Activity[b] (Units)
Bu-I (Butanol-only extract)	12,600	750	16	200,000
Bu-I.R (Aqueous extractable activity in the residue)	N.D.[c]	960	12.5	N.D.[c]
Bu-I Ethanol-insoluble	4,620	740	16.3	N.D.[c]
Bu-I Ethanol-soluble	7,980 [1,900 applied to Sephadex LH-60 Column, Fig. 8]		N.D.[c]	N.D.[c]
Fr. A, Fig. 8	800	660	18.2	14,600
Fr. B, Fig. 8	620	410	29.1	18,000
Fr. C, Fig. 8	390	---	Inactive	-0-

[a]Estimated with the homogenate preparation from ovaries from pseudopregnant rats. The ID_{50} for ovine LH in this system is 12 ng.

[b]Activity estimates are complicated by the relative water-insolubility of the fractions studied. The values are probably all under-estimates for this reason.

[c]Not determined.

Table 2. From 1 kg of frozen corpora lutea we obtained 12.6 g of butanol-soluble material (Bu-I), with a relatively low specific activity. However, it must be emphasized these are minimal activity estimates due to the intractible nature of these fractions. We often put them on a shaker or stirrer for hours to achieve only partial solution — and relatively dilute solutions, evenso.

An aliquot of the ethanol-soluble portion of the BU-I fraction was chromatographed on Sephadex LH-60 column to obtain the pattern shown in Fig. 8. From the UV absorption pattern only three fractions were identified, designated A, B, and C. Fraction C, the apparent low-molecular weight material, was inactive. Across-the-curve assays of individual fractions were not attempted. Both fractions A and B contained LH.RBI activity. Fraction B showed the highest specific activity, suggesting perhaps the peak of LH.RBI activity was centered toward the front of fraction B.

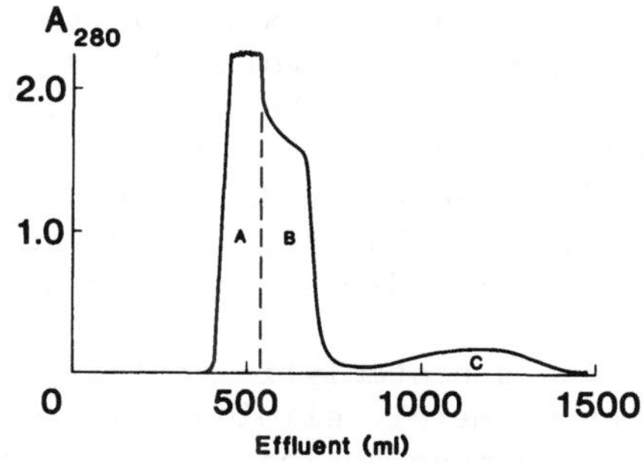

Fig. 8. Chromatography of the ethanol-soluble portion of fraction Bu-I (Table 2); 1,900 mg were applied to a Sephadex LH-60 column (5 × 50 cm) developed with ethanol at a flow rate of approximately 15 ml/hr. Flow rates were difficult to maintain.

At this point in our studies our emphasis centers on the characterization of the solvent-sensitive LH.RBI and the fractions obtained from DEAE-cellulose chromatography by 0.5 M and 1.0 M ammonium bicarbonate elution.

We have been content to work with an operational definition of LH.RBIs to this point in our studies. However, with these recent results it is probably timely to consider our results in terms of the reports from other laboratories.

The studies of Bhalla and Reichert (1974) have been noted. Their receptor-like material solubilized by 30 percent ethanol treatment of testes was detectable in similar extracts of liver, kidney, and blood. This preparation was studied further by Bhalla et al. (1976) using a dextran-coated charcoal procedure to separate factor-bound LH, hCG, or FSH from free hormone.

We cannot eliminate receptor-related components as being involved in at least some of the LH.RBI fractions we are studying. There are, however, certain distinctions in the activity we are pursuing. We do not detect significant LH.RBI activity in rat testes extracted under our conditions, including storage for greater than 10 weeks frozen. An LH.RBI-like activity was not observed in testes, non-luteal ovarian tissue, heart, or lung extracts. Liver extracts showed some inhibition of LH binding.

The preparations described by Bhalla and colleagues (1976) showed greater inhibition of FSH binding than LH binding. In our preparations the converse was observed, as seen in some of the studies presented above. In neither instance, however, has an absolute specificity for LH or FSH been demonstrated.

Reichert and colleagues have an extensive series of studies of FSH.RBI (see FSH-BI chapter, this volume). They have characterized preparations from serum (Reichert et al., 1979), from testicular extracts (Reichert and Abou-Issa, 1977), and follicular fluid (Darga and Reichert, 1978). These studies are of considerable interest in terms of our own work on LH.RBI, but further discussion will be deferred to the chapter by Reichert, in this volume.

In the simplest sense a proteolytic activity could provide an apparent LH.RBI activity by merely eliminating the labeled hormone. The role of proteases in hormone action may even be an integral part of the basic mechanisms as has become evident from the studies of Ryan and colleagues (Reichert et al., see also the chapter by Ryan, this volume). The possibility that a protease activity is involved in some of the fractions we are examining would seem to be unlikely in that pre-incubation of the labeled LH with the LH.RBI fraction did not lead to enhanced binding inhibition. However, direct examination for protease activity in all the LH.RBI fractions discussed above has not yet been done.

In 1976, Mullin and colleagues reported that gangliosides were implicated in the mechanism of thyrotropin binding to the thyroid plasma membrane. Bhalla (1977) subsequently demonstrated that gangliosides could bind FSH and LH as measured by his charcoal-coated dextran procedure. The gangliosides contain a hydrophilic carbohydrate portion and a lipophilic portion. Their presence in some of the fractions we are studying has not been excluded, and, indeed, would not be surprising.

It has been reported that asialo-hCG acts as a competitive antagonist of thyrotropin action at the thyroid membrane (Carayon et al., 1980). The possibility is thus suggested that specific degradation products of the glycoprotein hormones (e.g., FSH, LH) could provide an LH.RBI activity. Of course, unaltered LH could also inhibit labeled LH binding but this would not be consistent with inhibition by LH.RBI of LH stimulation of progesterone production which has been noted for several in vitro systems (Yang et al., 1978, Kumari et al., 1979, 1980).

In this connection, the report of Ranganathan and Sairam (1979) suggested that the LH.RBI and FSH.RBI in follicular fluid with molecular weight >10,000 behaved like FSH and LH, using the criteria of Con A adsorption-elution and radio-immunoassay. These criteria could also apply to a slightly altered form of both hormones. Two classes of lower molecular weight RBIs were also detected in their study.

Rao (1975) has observed that nucleotides in the concentration range 10^{-3} to 10^{-4} would inhibit hCG binding to corpora lutea membranes. This provides yet another possible LH.RBI. The expected chromatographic behavior of several nucleotides would not preclude their presence in the DEAE fractions where we have detected acitivity. However, it is not likely the activity in our porcine corpora lutea extracts represents an activity attributable to the presence of nucleotides, since these would be dialyzable. Virtually all the activity in the porcine corpora lutea extracts is non-dialyzable. The dialyzable LH.RBI activity in rat corpora lutea extracts (Yang et al., 1976) could be related to this observation, however. In the case of guanyl nucleotides Abou-Issa and Reichert (1979) observed essentially the opposite effect on FSH stimulation of rat testicular adenylate cyclase, i.e., an augmentation rather than an inhibition.

Behrman and Hichens (1976) reported prostaglandin F_2 alpha in an in vivo system inhibited hCG binding to corpora lutea receptors within 30 minutes. The inhibition persisted over 24 hours. Although this is undoubtedly a physiologically important LH receptor binding inhibition it is probably not directly related to the RBI effects we are observing, unless we are dealing with a non-dialyzable, prostaglandin binding fraction, which is effective in our test systems. In this sense we must consider both the LH.RBI effects and the inhibition of LH-stimulated progesterone production.

In summary, we hope these recent refinements in the fractionation of LH.RBI provide us with the means to evaluate the true nature of LH.RBI, and to assess its possible relationship to the abovementioned instances of demonstrated binding inhibition. To do this we must pursue our purification studies still further, since none of the fractions described above has reached a stage of homogeneity.

Acknowledgements

This work was supported in part by Grants AM-09801 and HD-08338 (Project 14) from the National Institute of Arthritis, Metabolism and Digestive Diseases, and National Institute for Child Health and Human Development, National Institutes of Health, respectively; the Ford Foundation, Grant No. 790-0656, and from The Robert A. Welch Foundation, Grant G-147 (DNW), plus National Institute for Child Health and Human Development, National Institutes of Health HD-98834 (CPC).

References

Abous-Issa, H., and Reichert, Jr., L. E., 1979, Modulation of Follicle-Stimulating Hormone-Sensitive Rat Testicular Adenylate Cyclase Activity by Guanyl Nucleotides, Endocrinology, 104:189.

Behrman, H. R., and Hichens, M., 1976, Rapid Block of Gonadotropin Uptake by Corpora Lutea In Vivo Induced by Prostaglandin F_2, Prostaglandins, 12:83.

Bhalla, V. K., 1977, The Physiology of Growth Hormone, Prolactin, Follicle-Stimulating Hormone and Luteinizing Hormone: Some Thoughts on the Role of Soluble Factors in the Mechanism of Gonadotropin Action, in: "The Pituitary — A Current Review," Allen, M. B., Jr., and Mahesh, V., eds., Chapter 6, pp. 97-119, Academic Press, New York.

Bhalla, V. K., and Reichert, Jr., L. E., 1974, Gonadotropin Receptors in Rat Testes: Interaction of an Ethanol-Soluble Testicular Factor with Human Follicle-Stimulating Hormone and Luteinizing Hormone, J. Biol. Chem., 249:7996.

Bhalla, V. K., Haskell, J., Grier, H., and Mahesh, V. B., 1976, Gonadotropin Binding Factor(s): Extraction of High Affinity Gonadotropin Binding Sites from Rat Testis and Partial Characterization of Their Interaction with Human Follitropin, Lutropin, and Choriogonadotropin, J. Biol. Chem., 251:4947.

Carayon, P., Amr, S., and Nisula, B., 1980, A Competitive Antagonist of Thyrotropin: Asialo-Choriogonadotropin, Biochem. Biophys. Res. Commun., 97:69.

Channing, C. P., Batta, S. K., and Bae, I.-H., 1981, Inhibitory Effect of Charcoal-Treated Aqueous Porcine Corpus Luteum Extract Upon Ovulation in the Rabbit, Proceedings of the Society for Experimental Biology and Medicine, in press.

Darga, N. C., and Reichert, Jr., L. E., 1978, Some Properties of the Interaction of Follicle Stimulating Hormone with Bovine Granulosa Cells and Its Inhibition by Follicular Fluid, Biol. Reprod., 19:235.

Kumari, G. L., Tucker, S., and Channing, C. P., 1979, Changes in Levels of a LH Binding Inhibitor in Aqueous Extracts of Porcine Corpora Lutea as a Function of Aging of the Corpus Luteum, Biol. Reprod., 21:1043.

Kumari, G. L., Vohra, S., Joshi, L., and Roy, S., 1980, Demonstration of Luteinizing Hormone/Human Chorionic Gonadotropin Receptor-Binding Inhibitor in Aqueous Extracts of Frozen Human Corpus Luteum, Hormone Res., 13:57.

Mullin, B. T., Fishman, P. H., Lee, G., Aloj, S. M., Ledley, F. D., Winard, R. J., Kohn, L. D., and Brady, R. O., 1976, Thyrotropin-ganglioside Interactions and Their Relationship to the Structure and Function of Thyrotropin Receptors, Proc. Natl. Acad. Sci., 73:842.

Ranganathan, M. R., and Sairam, M. R., 1979, Studies on the Characterization of the Gonadotropin Receptor Binding Inhibitors in Follicular Fluid, Biol. Reprod., 20[Supple. 1], #194, p. 108A.

Rao, C.-F., 1975, Mechanism of Nucleotide Inhibition of Gonadotropin Binding to Cell Membranes of Bovine Corpus Luteum, Molec. Cell. Endocrinol., 3:255.

Reichert, Jr., L. E., and About-Issa, H., 1977, Studies on a Low Molecular Weight Testicular Factor Which Inhibits Binding of FSH to Receptor, Biol. Reprod., 17:614.

Reichert, Jr., L. E., Sanzo, M. A., and Darga, N. S., 1979, Studies on a Low Molecular Weight Follicle-Stimulating Hormone Binding Inhibitor from Human Serum, J. Clin. Endocrinol. Metab., 49:866.

Richert, N. D., Bramley, T. A., and Ryan, R. J., 1978, Hormone Binding Proteases and the Regulation of Adenylate Cyclase Activity, in: "Novel Aspects of Reproductive Phyisology," C. H. Spilman and J. W. Wilks, eds., pp. 81-106, Spectrum Publiations, New York.

Sakai, C. N., Engel, B., and Channing, C. P., 1977, Ability of an Extract of Pig Corpus Luteum to Inhibit Binding of ^{125}I-Labeled Human Chorionic Gonadotropin to Porcine Granulosa Cells, Proc. Soc. Exper. Biol. Med., 155:373.

Tucker, S., Kumari, L., and Channing, C. P., 1979, Evidence of a Greater Activity of LH/hCG-Binding Inhibitor Present in Aqueous Extracts from Old Compared to Young Porcine Corpus Luteum, in: "Ovarian Follicular and Corpus Luteum Function," C. P. Channing, J. M. Marsh, and W. A. Sadler, eds., Plenum Press, New York.

Ward, D. N., Nahm, H. S., Shalek, R. J., Schwartz, N. B., Lorenzen, J., Moore, R. B., Grady, R. G., Channing, C. P., Glenn, S. D., Sugino, H., and Yang, K.-P., 1981, In Pursuit of Physiological Inhibitors of and from the Ovary, in: "Bioregulators of Reproduction," G. Jagiello and H. J. Vogel, eds., pp. 371-387, Academic Press, New York.

Yang, K.-P., Samaan, N. A., and Ward, D. N., 1976a, Characterization of an Inhibitor for Luteinizing Hormone Receptor Site Binding, Endocrinology, 98:233.

Yang, K.-P., Samaan, N. A., and Ward, D. N., 1976b, Lutropin Receptors from Male and Female Tissues: Different Responses to a Lutropin Receptor Binding Inhibitor, Proc. Soc. Exper. Biol. Med., 152:606.

Yang, K.-P., Samaan, N. A., and Ward, D. N., 1979, Effects of Luteinizing Hormone Receptor-Binding Inhibitor on the In Vitro Steroidogenesis by Rat Ovary and Testis, Endocrinology, 104:552.

Yang, K.-P., Gray, K. N., Jardine, J. H., Yen, H.-L. N., Samaan, N. A., and Ward, D. N., 1978, LH-RBI — An Inhibitor of In Vitro Luteinizing Hormone Binding to Ovarian Receptors and LH-Stimulated Progesterone Synthesis by Ovary, in: "Novel Aspects of Reproductive Physiology," C. H. Spilman and J. W. Wilks, eds., pp. 61-80, Spectrum Publications, New York.

CHARACTERIZATION OF LH/hCG RECEPTOR BINDING INHIBITOR IN CORPORA LUTEA OF HUMAN AND SHEEP OVARIES

G. L. Kumari,* Narender Kumar,†
S. Duraiswami,† J. K. Datta,*
I. C. Vidyasagar,* S. Vohra,‡ and S. Roy*

New Delhi, India

INTRODUCTION

Yang, Samaan, and Ward (1976) were the first to report the presence of a luteinizing hormone receptor binding inhibitor (LHRBI) in heavily luteinized rat ovaries. It was partially purified and found to be a peptide of approximately 3800 daltons. Sakai et al. (1977) found a similar inhibitor in corpora lutea of porcine ovaries, but not in other ovarian tissues. These crude porcine luteal extracts inhibited basal and LH-stimulated progesterone secretion by cultured granulosa cells of large porcine follicles (Kumari et al., 1979). In another in vitro system using ovarian slices, Yang and his colleagues (1979) demonstrated the inhibition of LH-stimulated progesterone synthesis in presence of luteinized rat ovarian extracts. These observations, as well as the finding of increased extractable LHRBI activity in old porcine corpora lutea compared to those of early luteal phase (Kumari et al., 1979) pointed to the importance of LHRBI in regulating the follicular responsiveness to LH and regression of corpus luteum under normal physiological conditions.

A specific inhibitor of binding of LH to ovarian cells, thus, would provide new avenues of approach for regulation of ovarian processes in the female. It was therefore of interest to look for

---

*Department of Reproductive Biomedicine, National Institute of Health and Family Welfare, Munirka, New Delhi, India.
†Department of Zoology, University of Delhi, New Delhi, India.
‡Department of Obstetrics and Gynaecology, Lady Hardinge Medical College, New Delhi, India.

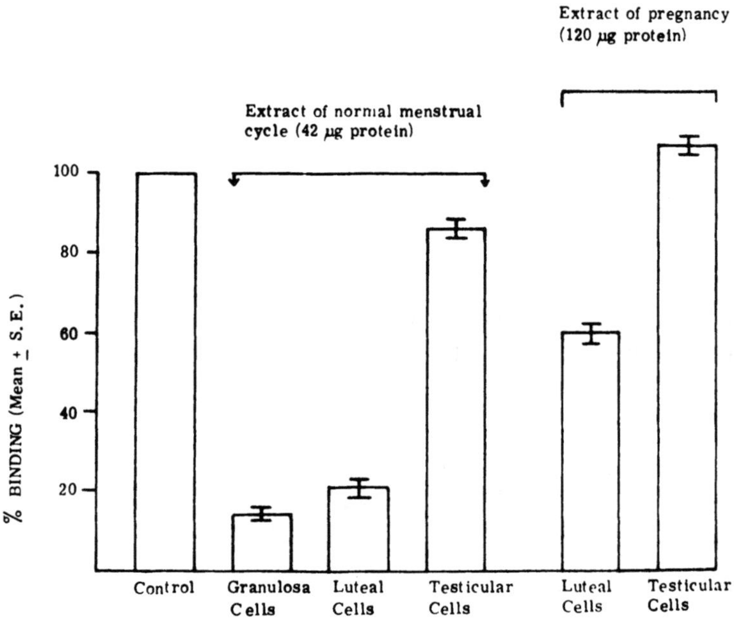

Fig. 1. Effect of addition of 30,000 g supernatant of luteal extracts of menstrual cycle and pregnancy on binding of ^{125}I-hCG to granulosa and luteal cells of sheep ovaries and rat testicular homogenates. The control binding was taken as 100% and the results are expressed in reference to the control binding as mean ± S.E. of five observations after deduction for non-specific binding.

the presence of LHRBI in human ovaries, especially in view of the potential for regulation of fertility that such a factor might offer. Characterization of the inhibitor from chemical and biological standpoints would require adequate amounts of source material. With this in view, purification of LHRBI from sheep ovaries was also undertaken, so that the methodology established in this species could be applied in the human.

I. LHRBI IN HUMAN CORPORA LUTEA

Human corpora lutea were collected from non-pregnant and pregnant women in the reproductive age group, who were undergoing abdominal surgery for hysterectomy or tubal ligation. The corpora lutea were dissected out from the ovary and kept frozen at -20°C for 12 to 20 weeks. These were thawed and eight each from normal menstrual cycle (4.7 g) and pregnancy (4.4 g) were minced and homogenized in six volumes of water following the method of Sakai et al. (1977). The details of methodology employed for the following experiments were described earlier (Kumari et al., 1980).

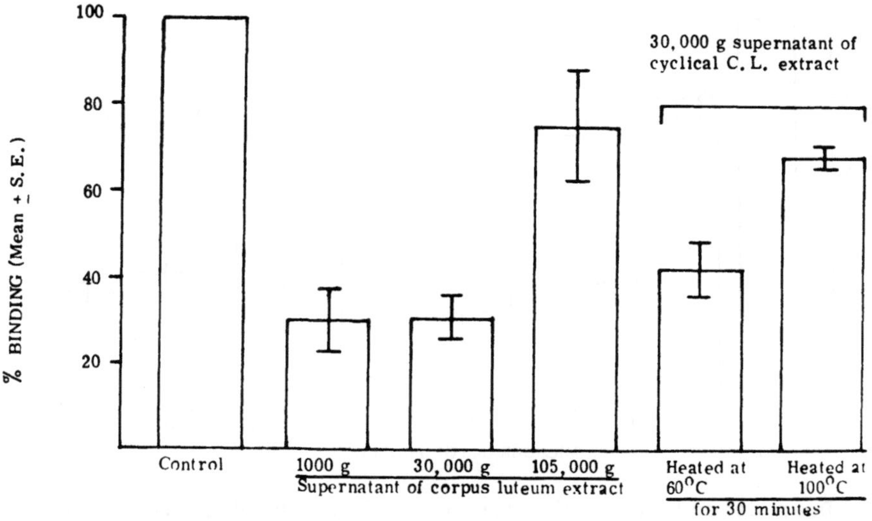

Fig. 2. The ability of 1000, 30,000, and 105,000 g supernatants of luteal extracts of normal menstrual cycle to inhibit binding of ^{125}I-hCG to sheep luteal cells and effect of heating the extract at 60 and 100°C respectively for 30 min on LHRBI activity. The rest is as in Fig. 1.

In earlier studies, Yang and his colleages (1976) found that rat LHRBI inhibited binding of LH specifically to ovarian cell homogenates, but not to testicular cells. In fact addition of rat LHRBI to in vitro incubations stimulated the steroidogenesis of testicular cells (Yang et al., 1979). Therefore, the crude extracts of human corpora lutea were tested for their ability to inhibit binding of ^{125}I-hCG to granulosa and luteal cells of sheep ovaries and to rat testicular homogenates. The 30,000 g supernatant of luteal extract of normal menstrual cycle (42 µg protein/3.6 mg tissue) affected 75 to 85% inhibition of binding of labeled hCG to granulosa and luteal cells, while that of pregnancy (120 µg protein/3.3 mg tissue) caused approximately 40% inhibition (Fig. 1). Extracts of cyclical corpora lutea had slight but significant effect on the binding of labeled hCG to rat testicular cells. Since the luteal extract of pregnancy had no effect, the inhibition observed using cyclical corpora lutea could be due to biologically active LH present in the corpus luteum itself.

As in the case of porcine luteal extracts, LHRBI activity is mainly present in the 1000 and 30,000 g supernatants of homogenates of human corpora lutea (Fig. 2). Heating the extract in a water bath at 60°C for 30 min had no effect on the binding of iodinated hCG to sheep luteal cells, but at 100°C, a major part of the inhibitory activity was abolished (Fig. 2).

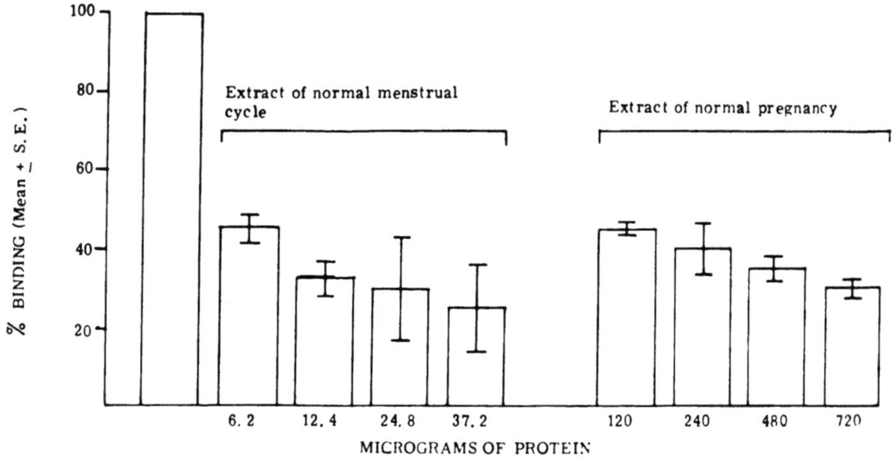

Fig. 3. Dose dependent inhibition of ^{125}I-hCG binding to sheep luteal cells on addition of 30,000 g supernatants of extracts of human corpora lutea. The rest as in Fig. 1.

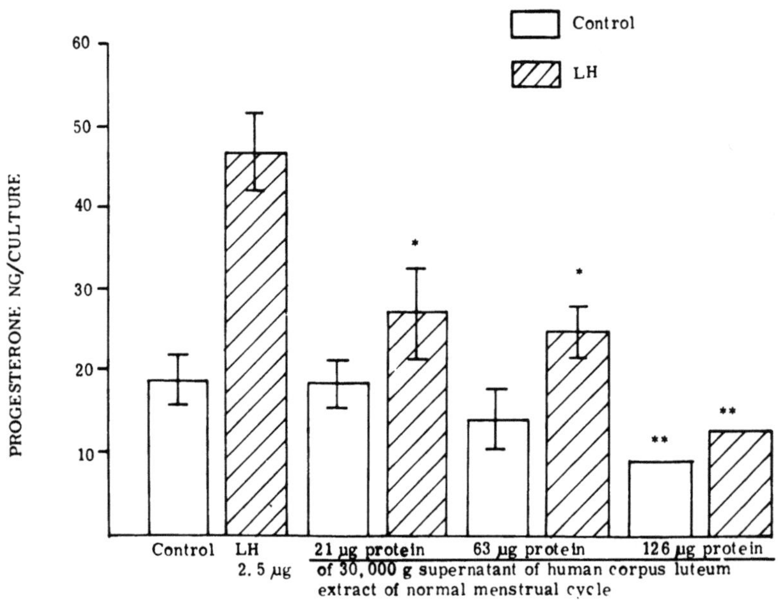

Fig. 4. Effect of addition of charcoal-treated (1%, w/v) 30,000 g supernatant of cyclical corpus luteum extract on progesterone secretion by granulosa cells of large sheep follicles in 0-2 day cultures. The results are expressed as mean ± S.E. of 4 individual experiments, except where 126 µg equivalent of protein was used. The latter represents results of one experiment. *$p < 0.05$; $p < 0.001$.

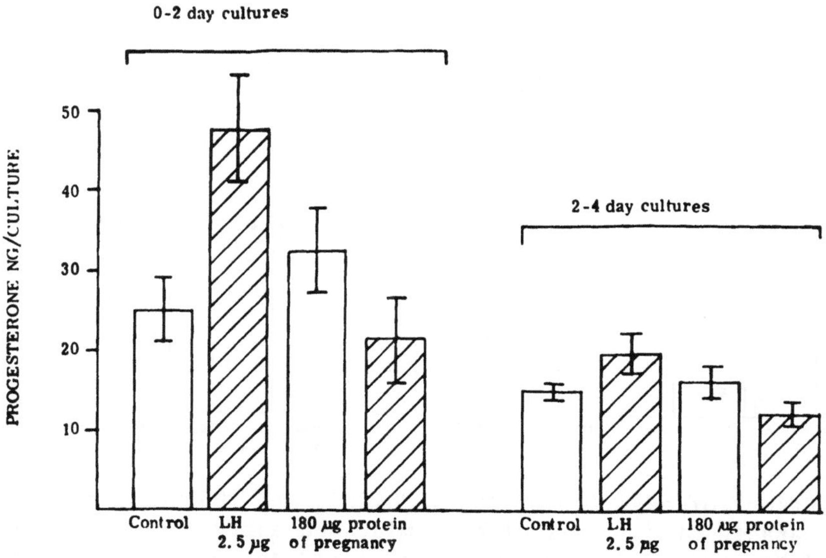

Fig. 5. Effect of addition of charcoal-treated (1%, w/v) 30,000 g supernatant of luteal extract of pregnancy on progesterone secretion by granulosa cells of large sheep follicles in 0-2 and 2-4 day cultures. The results are expressed as mean ± S.E. of three individual experiments.

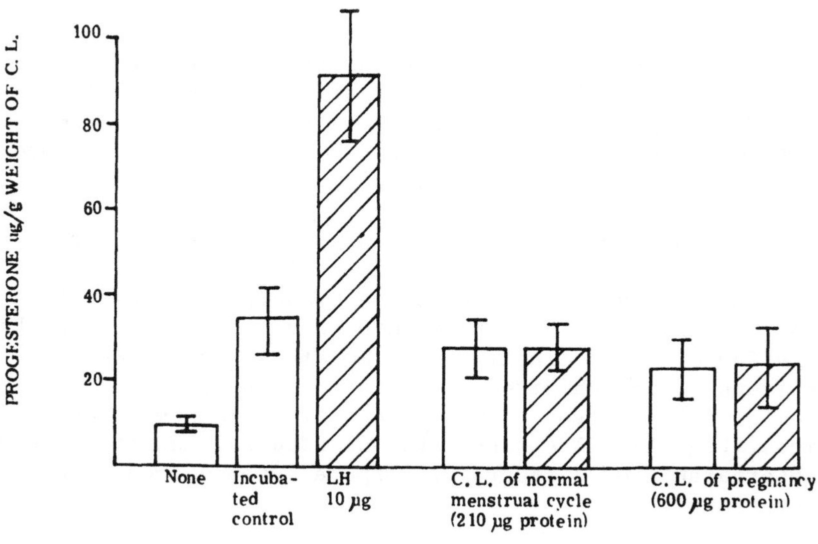

Fig. 6. Effect of charcoal-treated (1%, w/v) 30,000 g supernatant of corpora lutea extracts on LH-stimulated progesterone synthesis in sheep corpus luteum slices. The values represent mean ± S.E. of six observations.

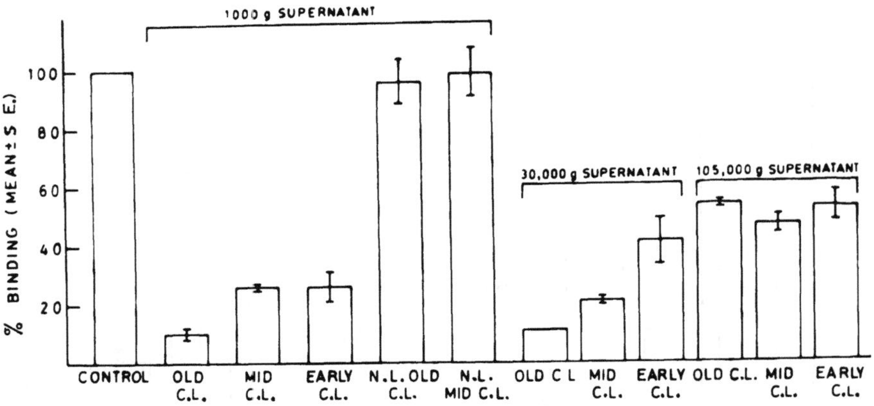

Fig. 7. Effect of addition of 1000, 30,000, and 105,000 g supernatants of extracts of old, mid, and early cycle corpora lutea and 1000 g supernatants of old and mid luteal phase non-luteal tissues on binding of ^{125}I-hCG to sheep luteal cells. The results are expressed as mean ± S.E. of five observations after deduction for non-specific binding.

Addition of increasing concentrations of the extract caused a gradual decrease in the binding capacity of sheep luteal cells to iodinated hCG (Fig. 3). With pregnant corpora lutea, more protein was needed to achieve inhibitory effects comparable to those obtained with extracts of non-pregnant corpora lutea.

While the binding of LH to target cells is considered to be the first step in the initiation of biological responses, increased synthesis of cAMP and progesterone can be measured for quantitation. An inhibitor of LH binding will therefore be expected to block these biological effects. Accordingly, LH-stimulated progesterone synthesis was found to be inhibited in cultured granulosa cells of large sheep follicles (Figs. 4 and 5) as well as in the sheep corpus luteum slices (Fig. 6). The test material used was charcoal-treated 30,000 g supernatants of extracts of corpora lutea of menstrual cycle and pregnancy.

These in vitro studies provide evidence for the presence of LHRBI in the human corpus luteum. Whether it also occurs in other ovarian tissue and whether it is specific for LH alone are questions awaiting more detailed investigation.

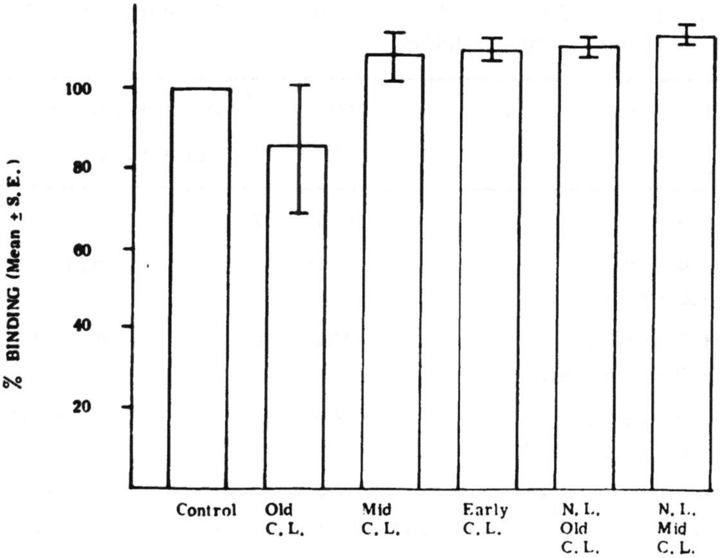

Fig. 8. Effect of addition of 1000 g supernatants of sheep luteal and non-luteal tissue extracts on the binding ability of rat testicular homogenates to iodinated hCG. The rest as in Fig. 7.

II. IDENTIFICATION AND CHARACTERIZATION OF LHRBI IN SHEEP OVARIES

The presence of LHRBI in sheep ovaries was earlier reported by Yang and his colleagues (1978). Our studies with crude extracts of sheep corpora lutea confirmed these findings (Kumari, 1981). As in the case of porcine corpora lutea (Kumari et al., 1979), the extracts of old and mid-cycle corpora lutea of sheep ovaries had more extractable LHRBI activity compared to the extracts of early luteal phase (Fig. 7). The activity, localized in the 30,000 g fraction was present in extracts of corpora lutea, but not in those of non-luteal ovarian tissue. None of these extracts had any effect on the binding of iodinated hCG to rat testicular homogenates (Fig. 8).

Differential pH Extraction of Old Corpora Lutea

Frozen old corpora lutea were thawed overnight at 4°C and approximately 60 g of tissue was homogenized with distilled water in Waring blendor in three bursts of 60 seconds each. The crude homogenate was divided into three equal portions which were adjusted to pH 4,6 and 8 respectively, following the addition of ethylene diamine citrate to make 0.1 M. Extraction with the buffer was carried out for 2 h at room temperature. The extracts were then centrifuged at 1000 g at 4°C; the supernatants were further centrifuged

Table 1. LHRBI Activity in Different pH Extracts of Old Sheep Corpora Lutea

S.No.	Additions	CPM of ^{125}I-hCG bound*	% Binding Mean ± S.E.
1.	None	5576 ± 97	-
2.	pH 4 extract (500 μg protein)	Nil	Nil
3.	pH 6 extract (500 μg protein)	194 ± 58	3.47 ± 1.03
4.	pH 8 extract (500 μg protein)	1086 ± 88	19.47 ± 1.58

* The binding of ^{125}I-hCG to rat ovarian homogenates (prepared as described by Yang et al, 1976) in presence of different luteal extracts, which were desalted after passing through Sephadex G-10 column equilibrated with distilled water. The results represent mean ± S.E. of four values after deduction for non-specific binding.

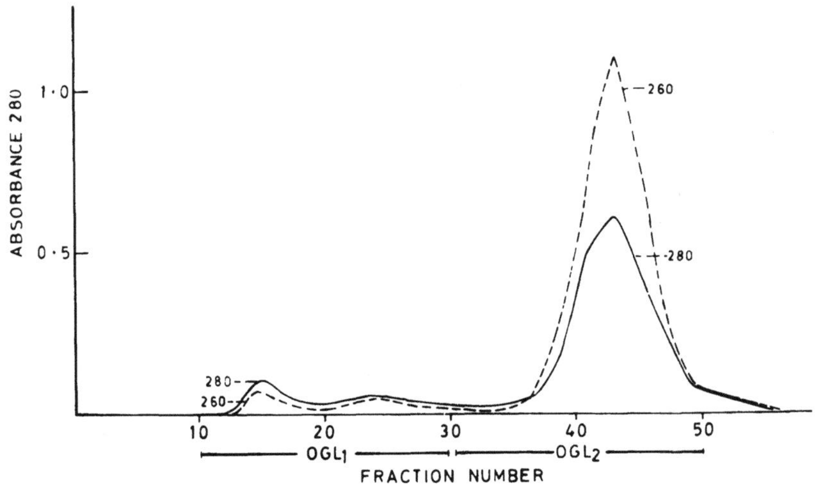

Fig. 9. Fractionation of pH 4 extract of old sheep corpora lutea on Sephadex G-100 column (100 × 2 cm) equilibrated with 0.1 M sodium acetate buffer, pH 5.2 at 4°C. Five milliliter fractions were collected.

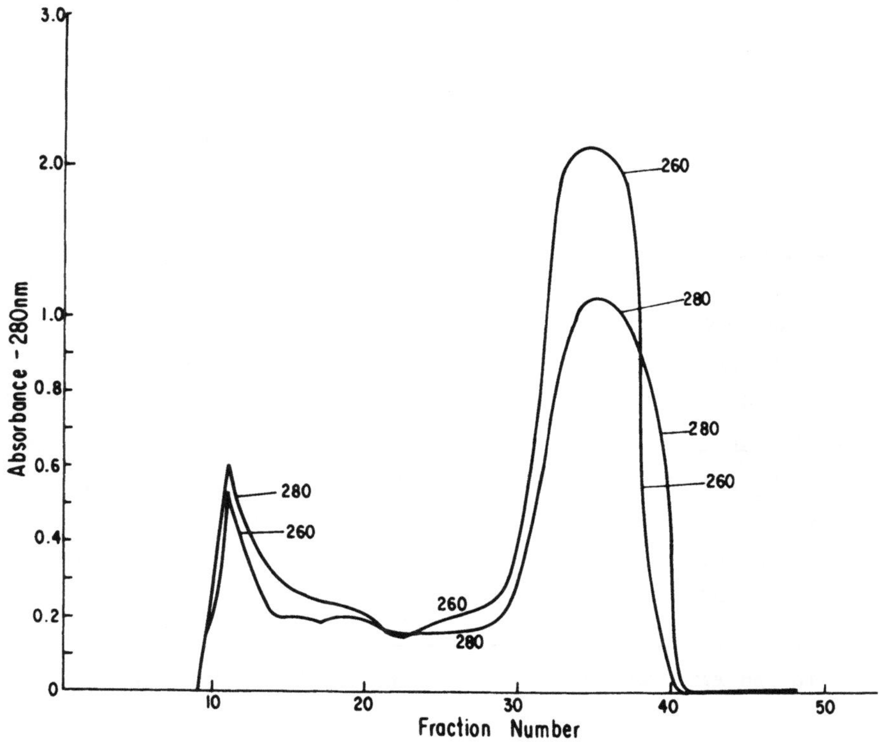

Fig. 10. Fractionation of pH 6 extract on Sephadex G-100 column. Conditions same as in Fig. 9.

at 30,000 g in the ultracentrifuge. The resulting supernatants were tested for their ability to inhibit binding of ^{125}I-hCG to rat ovarian homogenates, prepared as described by Yang et al. (1976). Extracts made at pH 4 and 6 had more LHRBI activity as compared to the pH 8 extract (Table 1).

<u>Fractionation of pH 4 and 6 Extracts on Sephadex G-100</u>

The pH 4 extract was fractionated on a Sephadex G-100 column (100 × 2 cm) equilibrated with 0.1 M sodium acetate buffer pH 5.2 at 4°C and eluted with the same buffer. It resolved into two components, a minor o-GL_1 and a major o-GL_2 fraction (Fig. 9). Chromatography of pH 6 extract under similar conditions gave a bigger peak of o-GL_1 compared to pH 4 extract (Fig. 10). Among the two fractions of pH 4 extract, o-GL_2 appeared to inhibit approximately 96% binding of ^{125}I-hCG to rat ovarian homogenates. The two fractions of pH 6 extract contained LHRBI activity, the o-GL_1 causing

Table 2. LHRBI Activity in Different Fractions of pH 4 and 6 Extracts of Old Sheep Corpora Lutea after Sephadex G-100 Chromatography

S.No.	Additions	CPM of ^{125}I-hCG bound*	% Binding Mean ± S.E.
1.	None	5576 ± 97	-
2.	pH 4 Extract		
2.	o-GL$_1$ fraction (50 µg protein)	4618 ± 12	82.83 ± 0.21
3.	o-GL$_2$ fraction (50 µg protein)	192 ± 98	3.45 ± 1.78
	pH 6 Extract		
4.	o-GL$_1$ fraction (50 µg protein)	1584 ± 134	28.41 ± 2.24
5.	o-GL$_2$ fraction (50 µg protein)	3424 ± 602	61.42 ±10.81

* Luteinized rat ovarian homogenates were used for binding.

the major part of inhibition in the binding system (Table 2). From these observations, it appeared that LHRBI might exist in two forms — a low molecular weight and a high molecular weight component. The latter could have been generated either by aggregation of the low molecular weight form or through coupling of the former to another high molecular weight molecule.

Fractionation on DEAE-C Column

The o-GL$_2$ fraction of the pH 4 extract was subjected to chromatography on DEAE-C column (60 × 2 cm) equilibrated with 0.05 M sodium acetate buffer at pH 5.2. A portion DE$_1$ was unadsorbed. DE$_2$ was eluted with 0.1 M sodium acetate buffer at pH 4 and DE$_3$ with 0.1 M sodium acetate plus 1 M sodium chloride buffer at pH 4 (Fig. 11). The fraction DE$_3$ was filtered through UM-2 and UM-05 Amicon filters successively. DE$_1$, DE$_2$, DE$_3$ as well as the UM-2 retentate of DE$_3$ and the UM-05 retentate of the UM-2 filtrate were tested for their ability to inhibit binding of ^{125}I-hCG to rat ovarian homogenates. The DE$_3$ fraction (at 20 µg equivalent of protein) inhibited 64% of binding of labelled hCG to ovarian cells. On comparing the data summarized in Table 3 with the LHRBI activity of o-CL$_2$ (96%

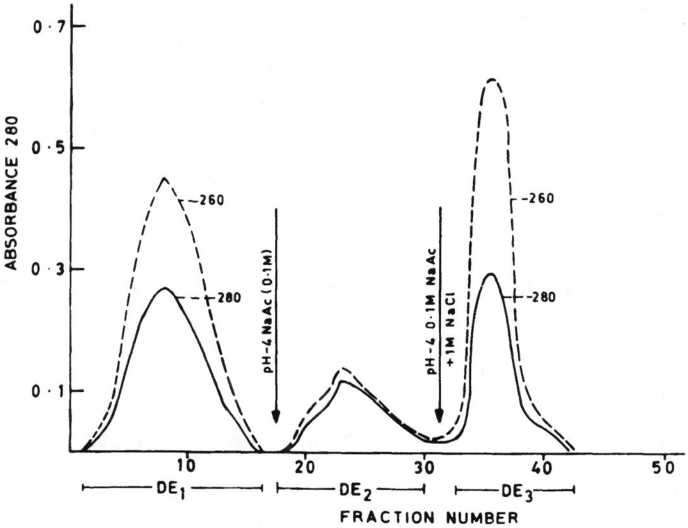

Fig. 11. Chromatography of o-GL_2 fraction of pH 4 extract on DEAE-C column (60 × 2 cm) equilibrated with 0.05 M sodium acetate buffer pH 5.2, eluted with 0.1 M sodium acetate buffer pH 4, followed by 0.1 M sodium acetate plus 1 M sodium chloride buffer pH 4.

inhibition using 50 µg protein in the assay) it appears that biological activity is lost following chromatography on DEAE-cellulose. This assumption was further supported by the finding that the o-GL_1 fraction of pH 6 extract resolved into two components on a DEAE-C column both of which had no LHRBI activity (Table 4).

Effect of Different Fractions on LH-Stimulated Progesterone Secretion by Cultured Granulosa Cells of Large Sheep Follicles

The pH 4,6 and 8 extracts and the o-GL_1 and o-GL_2 fractions of pH 4 extract inhibited LH-stimulated progesterone synthesis in the granulosa cell cultures. In addition, the extracts of pH 6 and 8 also affected the basal secretion of progesterone, probably due to presence of toxic components (Table 5). The fractions DE_2 and DE_3 inhibited LH-stimulated progesterone synthesis in granulosa cell cultures at a level of 20 µg protein (Table 6). These data are in agreement with the earlier binding studies (Table 3), in which DE_3 had some LHRBI activity.

Table 3. LHRBI Activity in Different Fractions of o-GL_2 (pH 4 Extract) after DEAE-C Chromatography and Amicon Filtration of DE_3 Fraction

S.No.	Additions	CPM of ^{125}I-hCG bound*	% Binding Mean ± S.E.
1.	None	5576 ± 97	-
2.	DE_1 fraction (20 μg protein)	5266 ± 194	94.46 ± 3.51
3.	DE_2 fraction (10 μg protein)	5406 ± 262	96.95 ± 4.72
4.	DE_3 fraction (20 μg protein)	2014 ± 48	36.12 ± 0.88
5.	UM-2 retentate of DE_3 (10 μg protein)	3494 ± 378	62.65 ± 6.79
6.	UM-05 retentate of UM-2 filtrate (10 μg protein)	2690 ± 700	47.16 ± 12.70

* Luteinized rat ovarian homogenates were used for binding.

Table 4. LHRBI Activity in Different Fractions of o-GL_1 (pH 6 Extract) after DEAE-C Chromatography

S.No.	Additions	CPM of ^{125}I-hCG bound*	% Binding Mean ± S.E.
1.	None	5576 ± 97	-
2.	DE_1 unadsorbed fraction (20 μg protein)	4730 ± 240	84.82 ± 4.33
3.	DE_2 adsorbed fraction (20 μg pretein)	4054 ± 74	72.55 ± 1.11

* Luteinized rat ovarian homogenates were used for binding.

Table 5. Effects of Addition of Different Extracts of Old Sheep Corpora Lutea before and after Chromatography on Sephadex G-100 Column, to Cultures of Sheep Granulosa Cells on Progesterone Secretion

S.No.	Additions	Progesterone ng/culture 0-2 day	2-4 day
1.	None	5.4	14.5
2.	2.5 μg o-LH	14.4	17.0
3.	pH 4 extract (100 μg protein)	5.2	6.2
4.	pH 4 extract (100 μg protein) plus 2.5 μg o-LH	4.5	2.9
5.	pH 6 extract (100 μg protein)	1.7	0.3
6.	pH 6 extract (100 μg protein) plus 2.5 μg o-LH	0.8	0.2
7.	pH 8 extract (100 μg protein)	2.6	0.6
8.	pH 8 extract (100 μg protein) plus 2.5 μg o-LH	1.8	1.0
9.	o-GL_1 fraction of pH 4 extract (50 μg protein)	5.4	3.6
10.	o-GL_1 fraction of pH 4 extract (50 μg protein) plus 2.5 μg o-LH	3.2	4.3
11.	o-GL_2 fraction of pH 4 extract (50 μg protein)	4.2	12.5
12.	o-GL_2 fraction of pH 4 extract (50 μg protein) plus 2.5 μg o-LH	4.8	8.5

The pH extracts were all passed through Sephadex G-10 equilibrated with distilled water to get rid of the salts. Similarly, the o-GL_1 and o-GL_2 fractions were filtered on UM-05 Amicon filter and washed two times before these were added to cultures. The results represent triplicate additions for each extract in an individual experiment.

Table 6. Effect of Addition of Different Fractions of o-GL_2 (pH 4 Extract) after DEAE-C Chromatography on Progesterone Secretion by Cultured Sheep Granulosa Cells

S.No.	Additions	Progesterone ng/culture 0-2 day	2-4 day
1.	None	10.1 ± 1.9*	7.1
2.	2.5 μg o-LH	16.5 ± 1.5	10.8
3.	DE_1 (20 μg protein)	15.2 ± 3.1	8.3
4.	DE_1 (20 μg protein) plus 2.5 μg o-LH	16.7 ± 1.0	12.1
5.	DE_2 (10 μg protein)	15.7 ± 0.8	11.3
6.	DE_2 (10 μg protein) plus 2.5 μg o-LH	12.8 ± 0.6	10.8
7.	DE_2 (20 μg protein)	8.8	5.3
8.	DE_2 (20 μg protein) plus 2.5 μg o-LH	9.2	6.9
9.	DE_3 (20 μg protein)	3.6 ± 0.6	2.6
10.	DE_3 (20 μg protein) plus 2.5 μg o-LH	3.4 ± 1.4	1.4

* These data represent mean ± S.E. of three individual experiments, while the rest express mean of two experiments.

III. EXPERIMENTAL STUDIES IN IMMATURE FEMALE RATS

The role of LHRBI in the regulation of follicular maturation and luteal function has remained speculative due to lack of availability of purified preparations. In view of the availability of the partially purified DE_3 fraction, the following in vivo studies were carried out.

Effects on Ovulation: Twenty seven day old immature female rats were treated with PMSG (40 I.U./S.C.) followed 55 h later by hCG (30 I.U./S.C.). A total of four injections of DE_3 (80 μg per injection/i.p.) was given, the first one being 1 h prior to PMSG administration. Two more injections were given 24 and 48 h after the first. The last dose was administered 1 h before injection of hCG. Control animals received saline. Seventy three hours after the first injection, the peripheral plasma was collected and the animals were killed. Ovarian weight and the number of ova released were also assessed. Plasma levels of progesterone and estradiol-17β were estimated by radioimmunoassay. Plasma levels of estradiol-

Table 7. Effect of Intraperitoneal Injections of DE_3 Fraction on Ovulation and Hormone Levels in PMSG and hCG Treated Immature Female Rats

S.No.	Parameters studied	Group I (saline treated)	Group II (DE_3 injected)
		Mean ± S.E.	
1.	Number of ova released	8.8 ± 0.3	6.6 ± 1.6
2.	Ovarian weight in mg	116.0 ± 4.3	124.0 ± 12.7
3.	Progesterone ng/ml plasma	80.5 ± 18.9	79.8 ± 4.3
4.	Estradiol-17β pg/ml plasma	166.5 ± 38.5	112.9 ± 27.8 ($p < 0.4$)

Twenty seven day old immature rats, six each in two groups were treated with PMSG and then 55 h later with hCG. The first group received saline and the second group was injected with four doses of DE_3 (80 μg per injection/i.p.). One was given 1 h prior to PMSG, the other three at 24,48 and 54 h after the first injection. The animals were killed at 73 h.

17β as well as the number of ova released were less in the treated group compared to the control group, but the differences were not statistically significant (Table 7).

Effects on Luteal Function: Immature female rats were treated with PMSG and hCG as described earlier. To one group of rats, DE_3 (80 μg/i.p.) was given 1 h before hCG injection and then daily on days 1, 2, 3, and 4 after hCG treatment. On the 5th day, peripheral plasma was collected and the animals were killed. The ovarian and uterine weights were taken and the plasma concentrations of progesterone and estradiol-17β were measured by radioimmunoassay. The ovarian weight and the levels of plasma progesterone were found to be low in the treated group. The results were again not statistically significant (Table 8).

In another experiment, a second dose of hCG was given on the 5th day after the initial hCG injection to PMSG primed rats. One of the groups received DE_3 fraction (80 μg/i.p.) daily on days 3, 4, and 5 after the first hCG injection. On the 6th day, the animals were sacrificed and same parameters as above were studied.

Table 8. Effect of Intraperitoneal Injections of DE_3 on Luteal Function in Immature Female Rats Primed with PMSG and hCG

S.No.	Parameters studied	Group I (saline treated)	Group II (DE_3 injected)
1.	Ovarian weight in mg	199.0 ± 10.9	170.8 ± 10.3*
2.	Uterine weight in mg	114.8 ± 9.7	129.1 ± 14.1
3.	Progesterone ng/ml plasma	137.0 ± 11.7	115.3 ± 7.8**
4.	Estradiol-17β pg/ml plasma	51.6 ± 4.0	59.7 ± 12.6

Twentyseven day old immature female rats, six each in two groups, were treated with PMSG and 55 h later with hCG. The control group received saline, while Group II was injected with DE_3 (80 μg per injection/i.p.) in five doses. The first injection was given at 54 h just prior to hCG injection. Rest of the doses were given on days 1,2,3 and 4 after hCG treatment. The animals were killed on the fifth day.

* $p < 0.2$; ** $p < 0.2$.

The weights of ovary and the uterus and the plasma levels of estradiol-17β showed stimulation due to second injection of hCG in the luteal phase compared to the values obtained earlier in Table 9. Plasma progesterone concentrations remained unchanged. In the treated group, the concentrations of both progesterone and estradiol-17β in plasma were low compared to the control group of rats. This decrease was found to be insignificant.

From these in vivo experiments it would appear that DE_3 consistently causes a decrease in one or other of the parameters studied, even though the results are not significant. Clearly further investigations are called for.

SUMMARY

Crude extracts of frozen human corpora lutea, obtained from normal menstrual cycle and pregnancy, were shown to inhibit ^{125}I-hCG binding to granulosa and luteal cells of sheep ovaries. The corpora lutea of pregnancy seemed to have less LHRBI activity com-

Table 9. Effect of Intraperitoneal Injections of DE_3 on hCG Stimulated Luteal Function in Immature Female Rats Primed with PMSG and Treated with hCG

S.No.	Parameters studied	Group I (saline treated)	Group II (DE_3 injected)
1.	Ovarian weight in mg	202.0 ± 11.9	209.2 ± 22.3
2.	Uterine weight in mg	155.6 ± 5.5	158.8 ± 10.6
3.	Progesterone ng/ml plasma	121.5 ± 4.4	103.5 ± 11.6*
4.	Estradiol-17β pg/ml plasma	158.0 ± 45.9	129.3 ± 29.6**

Twentyseven day old immature female rats, six each in two groups, were treated with PMSG and 55 h later with hCG. A second dose of hCG was given on the 5 th day after first hCG injection. The control group received saline and the Group II were injected with DE_3 (80 μg/i.p.) in three doses on days 3,4, and 5 after the first hCG injection. The rats were killed on the 6 th day. * $p < 0.2$; ** $p < 0.5$.

pared to the cyclical corpus luteum. These have either less or no effect on the binding of iodinated hCG to rat testicular homogenates. The charcoal-treated extracts further inhibited LH-stimulated progesterone synthesis in cultured granulosa cells and in incubations of corpus luteum slices. These studies provide evidence for the presence of LHRBI in human corpora lutea.

Extensive studies on the characterization of LHRBI were undertaken in sheep ovaries. Its presence in corpora lutea but not in other non-luteal tissues and the increased activity in old and midcycle corpora lutea compared to those of the early luteal phase have also been demonstrated. The effect of the crude extracts were specific in that it was inhibitory to ovarian cells, but not to testicular cells.

Differential pH extraction of old sheep corpora lutea and chromatography on Sephadex G-100 column demonstrated that at pH 4 and 6, different molecular weight fractions had the major part of LHRBI activity. It appeared to be a low molecular weight component at pH 4 and at higher pH, the major part of the activity was associated with the high molecular weight fraction. The latter may represent the product of aggregation of the low molecular weight components or else association with another molecule of high molecular weight.

Fractionation of both the components on DEAE-C column resulted in discernible loss of activity. The low molecular weight fraction (o-GL_2) was resolved into three components DE_1, DE_2, and DE_3 on DEAE-C; the DE_3 fraction had major LHRBI activity. Intraperitoneal injections of DE_3 to immature rats primed with PMSG and hCG resulted in a consistent decrease of either plasma estradiol-17β at ovulation or ovarian weight and plasma progesterone in the luteal phase, which were not statistically significant. The data suggest that LHRBI may affect ovulation and luteal function.

ACKNOWLEDGEMENTS

This work was supported by grants from World Health Organization (Contract No. 78049) and Indian Council of Medical Research. Ovine LH (Batch 21) and hCG (CR-119) were gifts from National Pituitary Agency, NIAMDD, Baltimore, Maryland. Specific antisera raised against BSA conjugates of 3-o-carboxy methyl ether derivative of estradiol-17β and 11α-hemiglutarate derivative of progesterone were generously supplied by Dr. P. N. Rao of Southwest Foundation for Research and Education, San Antonio, Texas. Our thanks are also due to Dr. A. R. Sheth, Institute for Research in Reproduction, Bombay, for providing iodinated hCG for the initial studies.

REFERENCES

Kumari, G. L., Tucker, S., and Channing, C. P., 1979, Changes in levels of a LH binding inhibitor in aqueous extracts of porcine corpora lutea as a function of aging of the corpus luteum, Biol. Reprod., 21:1043.

Kumari, G. L., Vohra, S., Joshi, L., and Roy, S., 1980, Demonstration of luteinizing hormone/human chorionic gonadotrophin receptor-binding inhibitor in aqueous extracts of frozen human corpus luteum, Hormone Res., 13:57.

Kumari, G. L., 1981, Presence of LH/hCG receptor binding inhibitor in sheep corpora lutea, Ind. J. Exptl. Biol., 19:16.

Saki, C. N., Engel, B., and Channing, C. P., 1977, Ability of an extract of pig corpus luteum to inhibit binding of ^{125}I-labeled human chorionic gonadotropin to porcine granulosa cells, Proc. Soc. Exp. Biol. Med., 155:373.

Yang, K. P., Samaan, N. A., and Ward, D. N., 1976, Characterization of an inhibitor for luteinizing hormone receptor site binding, Endocrinology, 98:233.

Yang, K. P., Samaan, N. A., and Ward, D. N., 1976, Lutropin receptors from male and female tissues: different responses to a lutropin receptor inhibitor, Proc. Soc. Exp. Biol. Med., 152: 605.

Yang, K. P., Gray, K. N., Jardine, J. H., Yan, H. L., N., Samaan, N. A., and Ward, D. N., 1978, LHRBI-an inhibitor of in vitro luteinizing hormone binding to ovarian receptors and LH-stimulated progesterone synthesis by ovary, in: "Novel Aspects of Reproductive Physiology," C. H. Spilman and J. W. Wilks, eds., pp. 61-80, S. P. Medical and Scientific Books, New York.
Yang, K. P., Samaan, N. A., and Ward, D. N., 1979, Effects of luteinizing hormone receptor binding inhibitor on the in vitro steroidogenesis by rat ovary and testis, Endocrinology, 104:552.

REGULATION OF CHOLESTEROL ACQUISITION AND UTILIZATION IN THE CORPUS LUTEUM

Jerome F. Strauss, III, Toshinobu Tanaka, Leslie MacGregor, and Richard W. Tureck

Departments of Obstetrics and Gynecology
and Physiology
University of Pennsylvania School of Medicine
Philadelphia, Pennsylvania

Introduction

The enzymatic reactions involved in the formation of progesterone in the corpus luteum have been known for some time, but the factors which control the synthesis and secretion of this hormone are not fully understood (Strauss et al., 1981). Two aspects of the steroidogenic process deserve attention as loci for regulation by gonadotropins or other substances; the acquisition of the substrate, cholesterol, and cholesterol side chain cleavage, which is generally accepted as a rate-determining step. Because functional luteal cells produce relatively large quantities of steroid, they must have an efficient means of obtaining cholesterol. Recent studies we have carried out demonstrate that lipoprotein-carried cholesterol is the primary substrate for rat luteal tissue and that specific, saturable binding sites mediate the uptake of high density lipoprotein sterol (Christie et al., 1979; Schuler et al., 1979; Schuler, Toaff and Strauss, 1981; Rosenblum et al., 1981; Schuler et al., 1981a; Christie et al., 1981). Just how the number of lipoprotein receptors on luteal cells is regulated has not yet been defined. There is, however, abundant evidence that gonadotropins, specifically LH, act to augment the conversion of cholesterol into pregnenolone in mitochondria, but the mechanism by which this is accomplished remains obscure (Strauss et al., 1981). In this paper we describe the gonadotropic stimulation of lipoprotein uptake by rat ovarian cells. We also provide evidence that the formation of pregnenolone in luteal mitochondria is modulated by alterations in the phospholipid milieu of the cholesterol side chain cleavage enzyme.

Lipoprotein-Carried Cholesterol as a Substrate for Luteal Cells

Most cells have the capacity to synthesize cholesterol de novo from acetyl coenzyme A, but luteal cells primarily utilize blood-borne cholesterol for steroidogenesis and cholesteryl ester storage. If the plasma cholesterol concentrations of immature PMSG-hCG-primed rats are reduced by treatment with 4-aminopyrazolopyrimidine (4-APP), blood progesterone levels decline within 24 hours and the ovaries fail to accumulate sterol esters (Christie et al., 1979; Schuler et al., 1981b). This occurs despite compensatory changes in ovarian sterol metabolism which increase de novo sterol synthesis and reduce sterol ester storage. There is a more than 10-fold rise in activity of 3-hydroxy-3-methylglutaryl coenzyme A reductase (HMG-CoA reductase), a rate-limiting enzyme in cholesterol biosynthesis, and a significant stimulation of [U-^{14}C] acetate incorporation into free cholesterol by the luteal cells, while activity of the cholesterol esterifying enzyme, acyl coenzyme A: cholesterol acyltransferase, falls. Intravenous injection of human lipoproteins rapidly restores plasma progesterone concentrations to normal in the 4-APP-treated animals, suppresses the elevated HMG-CoA reductase activity and promotes ovarian cholesteryl ester storage.

Cultured rat granulosa cells display a similar dependence upon exogenous cholesterol as a substrate for steroidogenesis and sterol ester storage (Schuler et al., 1979; Rosenblum et al., 1981). Cells cultured in lipoprotein-deficient medium produce about 1/10th as much progestin as do cells growing in medium containing lipoproteins, and cellular sterol ester stores are also reduced. The diminished steroid secretion and sterol ester storage also occur in the face of increased cellular HMG-CoA reductase activity and augmented [U-^{14}C] acetate incorporation into sterols, while sterol ester synthesis, as assessed by incorporation of [U-^{14}C] acetate or [1-^{14}C] oleate into the fatty acyl moieties of sterol esters, is reduced. Addition of human low density (LDL) or high density (HDL) lipoproteins to cells growing in lipoprotein-depleted media enhances progestin secretion in a dose-dependent, saturable fashion.

HDL, the class of lipoproteins which carry most of the cholesterol in rat blood, also stimulate progestin production and cellular sterol ester storage when presented to freshly dispersed cells prepared from ovaries of PMSG-hCG-primed rats, during short incubations (1-6h) (Schuler et al., 1981a). Human and rat HDL stimulate these processes in a dose-dependent manner. In contrast, free cholesterol provided in an ethanolic solution or dispersions of cholesterol in phospholipid do not produce similar effects unless added in exceedingly high concentrations or in the presence of apolipoproteins of HDL.

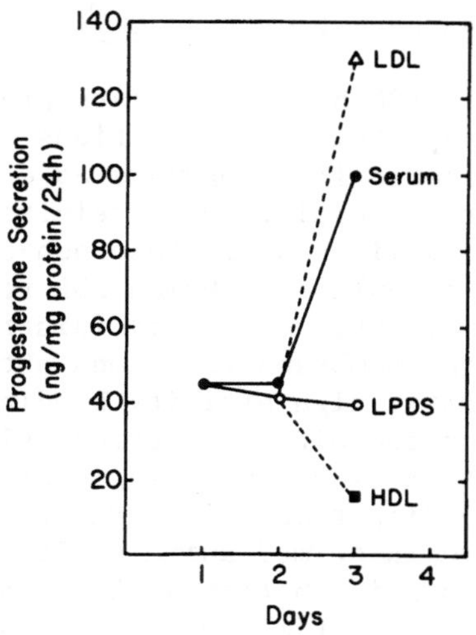

Fig. 1. Effect of lipoproteins on progesterone secretion by cultured human granulosa cells. Granulosa cells were collected from follicles of women undergoing laparoscopic examination. Cells were cultured in RPMI-1640 containing 25 mM Hepes, 4 mM glutamine and 50 μg/ml gentamicin supplemented with 20% human male serum. After 2 days in culture the experiment was begun; some cells were changed into medium supplemented with 20% lipoprotein-deficient human serum (LPDS). Twenty-four hours later some of these cultures received either 200 μg human LDL (density 1.019-1.063 g/ml) protein/ml or 200 μg human HDL (density 1.095-1.21 g/ml) protein/ml. Incubations were terminated 24 h later. Progesterone released into the incubation medium was quantitated by radioimmunoassay. Values presented are means from 2 separate experiments in which each treatment group consisted of duplicate cultures.

The utilization of lipoprotein-carried sterol as a substrate for steroid synthesis is not unique to rat ovarian cells. Human granulosa cells reduce progesterone secretion when cultured in lipoprotein-deficient medium (Fig. 1). Addition of human LDL stimulates progestin secretion, but unlike the situation with rat granulosa or lutein cells which utilize HDL-carried sterol, human HDL tends to inhibit it. The effects of HDL under these circumstances are of interest since HDL enters into follicular fluid while LDL does not pass through the blood-follicle barrier. Therefore, human granulosa cells in situ are deprived of substrate in the form of LDL as well as being exposed to HDL which may promote efflux of cellular cholesterol and, therefore, further limit steroidogenesis.

HDL Uptake and Metabolism by Rat Ovarian Cells

In vivo, ovaries of PMSG-hCG-primed rats pre-treated with 4-APP to reduce endogenous lipoprotein concentrations accumulate human ^{125}I-labeled HDL administered intravenously as a bolus (Christie et al., 1981). On a mg wet wt. basis, the luteinized ovaries take up substantially more labeled lipoprotein than non-steroid producing organs. Administration of unlabeled human HDL reduces the ovarian uptake of the ^{125}I-labeled HDL, whereas administration of unlabeled human LDL does not. Using membrane fractions obtained from ovaries of the PMSG-hCG-primed rats, high affinity binding of ^{125}I-labeled human HDL can be demonstrated with an apparent single class of binding sites having a K_d of 56 µg HDL protein/ml at 4°C (Christie et al., 1981). Collagenase-dispersed cells prepared from the ovaries of the immature gonadotropin-primed animals also readily take up human ^{125}I-labeled HDL upon incubation at 37°C (Schuler et al., 1981a). The uptake process is time and temperature dependent and half-maximal uptake at 37°C is achieved with human HDL concentrations of about 20 µg HDL protein/ml. Furthermore, the amount of HDL associated with the dispersed cells during a 1 h incubation under saturating conditions is sufficient to supply the cholesterol required to support the observed increased progestin synthesis and sterol ester storage.

It should be pointed out that the utilization of HDL sterol by rat ovarian cells does not involve the LDL receptor pathway described by Goldstein and Brown (1977); a discrete HDL receptor exists. Human ^{125}I-labeled HDL binding to both ovarian membranes and dispersed ovarian cells is competitively inhibited by human and rat HDL but not by human LDL. A number of other characteristics including the absence of a divalent cation requirement for binding distinguish HDL association with rat ovarian membranes from the binding of LDL to the LDL receptor. The metabolism of HDL by rat ovarian cells also seems to differ from the LDL pathway in that prior or simultaneous degradation of the HDL apolipoproteins is not required for utilization of the cholesterol carried by these particles (Schuler et al., 1981a; Gwynne and Hess, 1981), but further studies are required to clarify the exact fate of these particles following uptake by the ovarian cells.

Control of HDL Uptake by Rat Ovarian Cells

To determine whether the uptake of HDL by rat ovaries correlates with their steroidogenic capacity, we compared the accumulation of human ^{125}I-labeled HDL by ovaries of immature females and immature animals primed with PMSG-hCG. The rats were pre-treated with 4-APP for 3 days to lower endogenous lipoprotein levels before receiving a bolus of human ^{125}I-labeled HDL via a tail vein. The 4-APP treatment reduced plasma cholesterol levels from a normal value

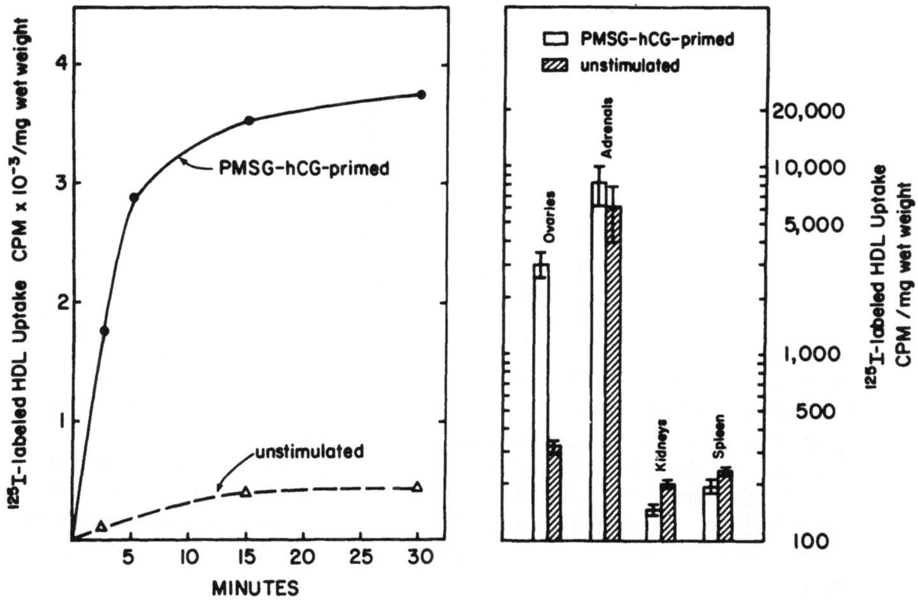

Fig. 2. In vivo uptake of human ^{125}I-labeled HDL by ovaries of unstimulated and PMSG-hCG-primed immature rats. Immature females, 23 days of age, were either unprimed or given 50 IU of PMSG followed 60 h by 25 IU of hCG. The animals received 3 days of treatment with 4-APP (10 mg/kg), starting on day 4 post-hCG in the gonadotropin-primed animals. Twenty-four hours after the last 4-APP treatment, day 7 post-hCG in the gonadotropin-primed animals, rats were anesthetized with ether and given a bolus of human ^{125}I-labeled HDL (density 1.125-1.21 g/ml; 40 μg, 1.5×10^7 cpm) in 0.9% saline by tail vein. Animals were exsanguinated at various times after injection of the labeled HDL and organs were removed, rinsed in chilled 0.9% saline, weighed and counted using a gamma spectrometer. Mean ovarian weights (mg/pair of ovaries) for PMSG-hCG primed rats and unprimed animals were 162 ± 21 (N = 8) and 11.8 ± 0.8 (N = 6), respectively. In left panel, points represent mean values for ^{125}I-labeled HDL uptake obtained from 2 animals, except for 2.5 min point which represents uptake in one rat. In right panel, values are means ± S.E. for uptake from 3 rats in each treatment group.

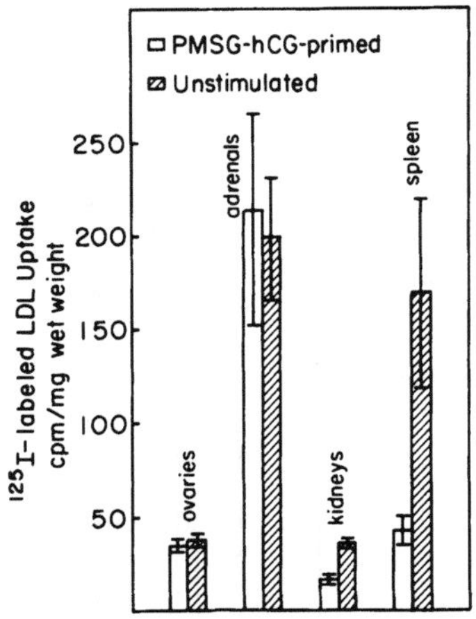

Fig. 3. Uptake of human ^{125}I-labeled LDL by organs of PMSG-hCG-primed immature and unstimulated immature rats. Animals were treated as described in the legend for Fig. 2. On the day of the experiment rats were given an intravenous bolus of human ^{125}I-labeled LDL (12 µg; 2 × 10^6 cpm) and were killed 15 min later. ^{125}I accumulated in various organs was measured. Values are means ± S.E. for 3 animals in each group.

of 50 mg/dl to 12 mg/dl. ^{125}I in the ovaries and other organs was determined at various times after administration of the labeled HDL by counting the organs in a gamma counter. The uptake of ^{125}I-labeled HDL by ovaries of PMSG-hCG-primed rats was 10-fold greater on a mg wt. basis than ovaries of unprimed immature animals (Fig. 2). Greater than 95% of the label taken up by the ovaries of the PMSG-hCG primed rats at 15 min was precipitable by 10% trichloroacetic acid. While the ovaries revealed a differential uptake, the adrenals, kidneys, and spleen of the gonadotropin-primed and unprimed rats accumulated similar amounts of labeled HDL at 15 min post-injection (Fig. 2). Unlike human ^{125}I-labeled HDL, the ovaries of PMSG-hCG-primed and unstimulated rats did not selectively accumulate human ^{125}I-labeled LDL (Fig. 3). These findings clearly demonstrate that the highly luteinized ovaries of immature gonadotropin-primed rats have an increased capacity to take up HDL, consistent with their greater steroidogenic activity as compared to ovaries of unprimed animals.

The observation that human ^{125}I-labeled LDL is not selectively accumulated by luteinized ovaries appears to be at odds with the findings that cultured rat granulosa cells utilize LDL-carried sterol for steroidogenesis and sterol ester storage (Schuler et al., 1979; Schrieber et al., 1980; Rosenblum et al., 1981). However, this discrepancy may be explained by the different expression of lipoprotein receptors under in vitro conditions, or the failure of human LDL to readily penetrate the endothelium and reach the lutein cells in vivo.

We next examined the ability of cells from ovaries of gonadotropin-treated hypophysectomized rats to take up HDL. In these experiments the animals were given a sequence of hormones to induce follicular maturation and luteinization as described by Richards and Williams (1976). The ovaries were dispersed with collagenase and cellular uptake of human ^{125}I-labeled HDL during a 2 h incubation was determined using methods which we have previously described (Schuler et al., 1981a). Cells were incubated with various concentrations of human HDL so that lipoprotein accumulation under saturating conditions could be measured. Concentrations of lipoprotein allowing half-maximal accumulation were estimated from Eadie-Hofstee plots of the data for specific lipoprotein uptake as a function of HDL concentration.

Cells prepared from ovaries of hypophysectomized immature females took up relatively little HDL, as did cells from ovaries of rats treated for 4 days with 1 mg estradiol (Table 1). The plasma progesterone and 20α-hydroxypregn-4-en-3-one levels, determined by specific radioimmunoassays, were also relatively low. Treatment of rats for 4 days with estradiol followed by 2 days of oFSH produced a modest increase in HDL uptake and plasma progestins. However, within 48 h of administration of a single dose of oLH to induce luteinization of estrogen-FSH-primed ovaries, HDL uptake per μg DNA was increased more than 6-fold and HDL uptake per ovary more than 14-fold above that observed with unprimed hypophysectomized rats, and plasma progestin concentrations were substantially increased. Addition of prolactin to the hormonal regimen further enhanced HDL uptake and raised plasma progestin levels as well as shifting the ratio of progesterone to 20α-hydroxypregn-4-en-3-one in favor of progesterone. Four days after the injection of LH, HDL uptake by cells was augmented and plasma progestin concentrations had also risen. Those rats receiving prolactin for the 4 days had the highest progestin concentrations and dispersed cells from their ovaries displayed the greatest HDL uptake. The plasma cholesterol levels in animals receiving the various hormonal regimens remained relatively constant around 50 mg/dl. The concentrations of HDL permitting half-maximal lipoprotein uptake during the 2 h incubations increased from approximately 7-14 μg HDL protein/ml for cells from ovaries of hypophysectomized estrogen-treated and untreated animals, respectively, to values of 30-49 μg HDL protein/ml for cells from luteinized ovaries. Thus, the enhanced lipoprotein uptake by luteinized cells cannot be attributed to an increase in affinity of the uptake system for HDL.

Our previous studies in which uptake of human ^{125}I-labeled HDL by enzymatically dispersed rat lutein cells was studied revealed that there was little cell specific degradation of the apolipopro-

Table 1. Effect of Various Gonadotropic Hormones on Uptake of ^{125}I-Labeled HDL by Ovarian Cells, and Plasma Sterol and Progestin Concentrations

Treatment	Ovarian wt mg/ovary	^{125}I-labeled HDL uptake ng/μg DNA/2h	^{125}I-labeled HDL uptake ng/ovary/2h	K 1/2 μg HDL protein/ml	Plasma Cholesterol mg/dl	Plasma Progestins ng/ml Progesterone	Plasma Progestins ng/ml 20α-OH-P
Unprimed	4.6 ± 0.4	1.79 ± 0.75	37.8 ± 17.6	14.6 ± 8.9	64 ± 5	1.7 ± 0.7	0.5 ± 0.2
E_4	6.2 ± 1.2	1.77 ± 0.17	39.9 ± 8.8	6.98 ± 3.2	52 ± 7	1.1 ± 0.3	1.2 ± 0.9
E_4F_2	16.4 ± 1.5	3.87 ± 1.81	249 ± 13	25 ± 8.4	54 ± 8	9.7 ± 3.5	4.3 ± 2.8
$E_4F_2L_2$	24.7 ± 1.8	12.17 ± 1.05	565 ± 61	19.7 ± 3.8	50 ± 5	19.3 ± 4.7	45 ± 9.8
$E_4F_2LP_2$	40.0 ± 5.3	32.74 ± 9.40	1596 ± 192	23.2 ± 12.7	46 ± 9	70.3 ± 11.5	82 ± 8.6
$E_4F_2L_4$	50.7 ± 8.0	32.79 ± 8.56	3400 ± 1156	48.5 ± 5.7	58 ± 7	54.6 ± 27.7	62.3 ± 7.6
$E_4F_2LP_4$	54.0 ± 9.2	51.38 ± 12.50	6064 ± 2021	30.4 ± 16.7	51 ± 6	198 ± 40	72.9 ± 7.3

Immature female rats, hypophysectomized on Day 21 of life received the following treatments starting 4 days after surgery:

Unprimed:	propylene glycol
E_4:	1 mg estradiol in propylene glycol for 4 days
E_4F_2:	1 mg estradiol for 4 days followed by oFSH (1 μg B.I.D.) for 2 days
$E_4F_2L_2$:	1 mg estradiol for 4 days, 2 days oFSH followed by oLH (60 μg) with killing 48 h after LH treatment

$E_4F_2LP_2$: 1 mg estradiol for 4 days, 2 days of oFSH followed by oLH (60 μg) and prolactin (125 μg B.I.D) for 2 days, starting on the day of LH treatment

$E_4F_2L_4$: Same as $E_4F_2L_2$ except killing 96 h after LH treatment

$E_4F_2LP_4$: Same as $E_4F_2LP_2$ except prolactin (125 μg B.I.D.) was administered for 4 days starting on the day of LH treatment

Animals were killed 12-24 h after the last hormone treatment. Ovaries were removed and plasma collected for sterol and progestin assays. The ovaries were dispersed with collagenase and uptake of human ^{125}I-labeled HDL (density 1.125-1.21 g/ml) during a 2 h incubation was determined. Uptake of ^{125}I-labeled HDL/ovary was estimated from the value for maximal specific uptake per aliquot of cells X the number of aliquots from the ovarian dispersion. Complete dispersion of the tissue has been assumed. HDL concentrations permitting half-maximal uptake (K 1/2) were estimated from Eadie-Hofstee plots of the specific binding data as a function of HDL concentration in the incubation medium. The following gonadotropins were used NIH-FSH-S11, NIH-LH-S22, and NIH-P-S12. Values presented are means ± S.E. for 3 separate experiments.

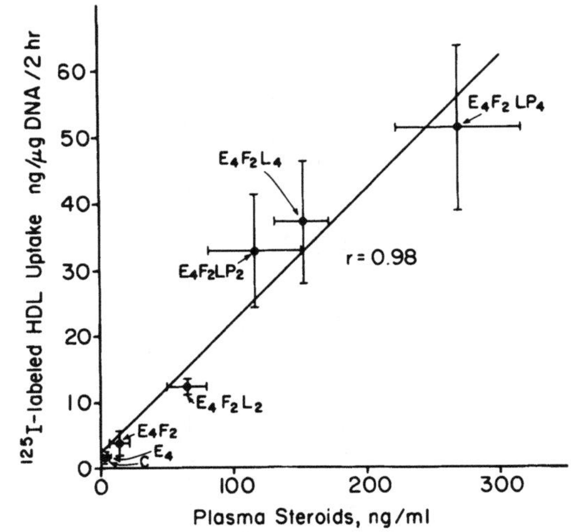

Fig. 4. Relationship between cellular uptake of human ^{125}I-labeled HDL and plasma progestin concentrations. Data for uptake of human ^{125}I-labeled HDL and plasma progestins (progesterone + 20α-hydroxypregn-4-en-3-one) presented in Table 1 have been plotted.

teins during the several hours of incubation (Schuler et al., 1981a). The apolipoprotein degradation which was observed could mostly be accounted for by proteolytic activity carried over from the dispersing medium. Therefore, it was not possible in these experiments to critically assess whether the various treatments affected rates of lipoprotein metabolism. Whether gonadotropin treatment specifically affected the internalization of lipoprotein sterol, as distinct from HDL binding, also remains to be determined. None the less, our observations show that luteinization is attended by a significant increase in the capacity of rat ovarian cells to take up human ^{125}I-labeled HDL and this occurs concomitantly with enhancement of steroidogenic activity as reflected by elevated plasma progestin concentrations. When maximal lipoprotein uptake by cells/µg DNA/2h is plotted against total plasma $C21\ \Delta^4$-3 ketosteroid concentrations (progesterone + 20α-hydroxypregn-4-en-3-one), a linear relationship is apparent (Fig. 4). The increased lipoprotein uptake presumably reflects greater numbers of cell surface lipoprotein receptors and LH, acting in concert with prolactin, appears to trigger this response. It is not yet known if the effect of prolactin on lipoprotein uptake and steroidogenesis is permissive, or if this hormone has a direct influence on these processes. It is evident that LH alone is sufficient to induce an increase in lipoprotein uptake, so that we currently favor a permissive action for prolactin in this regard.

Control of Cholesterol Side Chain Cleavage in Rat Luteal Mitochondrial

Cholesterol side chain cleavage occurs in the inner mitochondrial membrane and is catalyzed by cytochrome P-450scc supported by its soluble electron transport system (Strauss et al., 1981). Although the mitochondrial content of cytochrome P-450scc is one major determinant of the steroidogenic capacity of luteal tissue, conversion of cholesterol to pregnenolone can be acutely stimulated by LH without an increase in enzyme protein. This stimulatory effect of LH requires the intermediacy of cyclic nucleotides; but in addition, an undefined factor appears to transmit a signal to the mitochondria resulting in enhanced cholesterol side chain cleavage. Since inhibitors of protein synthesis, such as cycloheximide and puromycin, rapidly and reversibly block steroidogenesis (Toaff et al., 1979), this factor has been assumed to be a rapidly turning over protein.

Treatment of immature PMSG-hCG primed rats with cycloheximide results within 1 h in a significant reduction in the synthesis of steroids by mitochondria isolated from the ovaries (Toaff et al., 1979). The reduction in steroid formation is not due to a deficiency in substrate since mitochondrial free cholesterol contents are increased by cycloheximide. Moreover, steroid production by mitochondria prepared from ovaries of cycloheximide-treated rats is not stimulated by addition of exogenous substrate in the form of cholesterol-phospholipid dispersions, whereas steroid synthesis by mitochondria from ovaries of vehicle-treated rats is (Toaff et al., 1979).

Cycloheximide treatment does not immediately reduce mitochondrial contents of cytochrome P-450 as determined by spectral analysis (Toaff et al., 1979). Furthermore, mitochondria from ovaries of cycloheximide-treated rats produce as much steroid as those from ovaries of vehicle-treated rats when sonicated and incubated with NADPH as a source of reducing equivalents, or when the intact organelles are incubated with oxygenated sterols such as 5-cholesten-3β, 25-diol, or 26-nor-5-cholesten-3β-ol-25-one. These findings suggest that utilization of mitochondrial cholesterol for steroidogenesis is restrained by cycloheximide treatment, and that enzyme levels are not affected. This restraint is presumably due to a deficiency of a rapidly turning over protein whose synthesis is stimulated by LH but inhibited by cycloheximide. Disrupting the mitochondrial architecture by sonication or presenting a substrate which more readily penetrates the mitochondrial membranes and achieves a higher concentration near the cytochrome P-450scc overcomes the restraint. However, attempts to isolate the putative

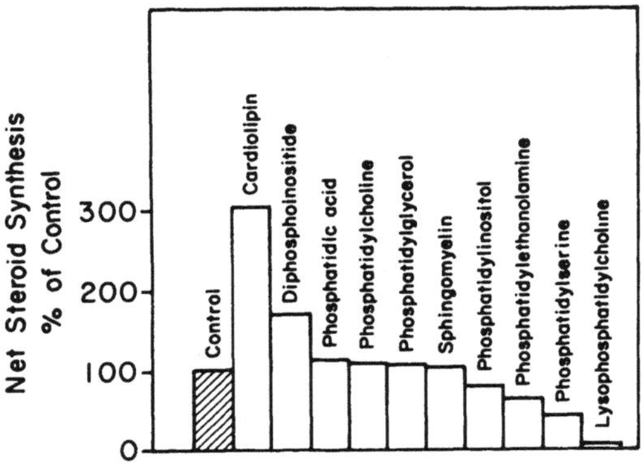

Fig. 5. Effect of various phospholipids on steroid synthesis by mitochondria from ovaries of cycloheximide-treated rats. Mitochondria were prepared from ovaries of PMSG-hCG primed rats which received 5 mg of cycloheximide, i.p., 30 min before killing on day 8 post-hCG. Mitochondria were prepared and incubated using our previously described methods with some modifications. The mitochondria were incubated with phospholipids dispersed by sonication in buffer containing 10% propylene glycol with 10 mM isocitrate serving as a source of reducing equivalents. Net steroid synthesis was determined, representing the sum of pregnenolone + progesterone formed minus the contents of these steroids at the initiation of the incubation. All phospholipids, except bovine brain diphosphoinositide, were added at a final concentration of 800 μM. Diphosphoinositide was added at a concentration of 200 μM. Net steroid synthesis as a percentage of that produced by mitochondria incubated with the vehicle (2.7 nmol/mg protein/10 min) is presented. Values are means from 2 separate experiments.

rapidly turning over protein from steroidogenic tissues have been unsuccessful (Toaff et al., 1980). Alternatively, the factor whose formation or activity is affected by cycloheximide might be a substance other than a protein.

Modulation of Luteal Mitochondrial Cholesterol Side Chain Cleavage by Phospholipids

Phospholipids are known to have a significant effect upon membrane-associated enzymes and certain mitochondrial enzymes have absolute requirements for phospholipid, particularly cardiolipin (e.g., Fry and Green, 1981). We had previously presented evidence for active turnover of phospholipids in dispersed rat luteal cells (Strauss and Flickinger, 1977), and found some acute effects of LH on incorporation of ^{14}C-acetate into polar lipids. Although the LH-induced

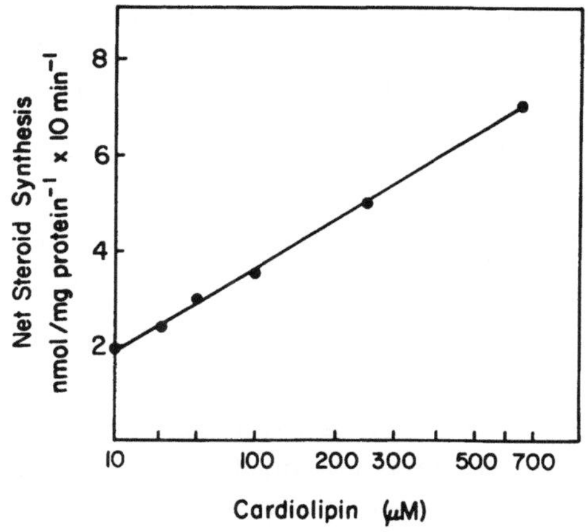

Fig. 6. Effect of different concentrations of cardiolipin on steroid synthesis by mitochondria from ovaries of cycloheximide-treated rats. Mitochondria were prepared from ovaries of PMS-hCG-primed rats treated with 5 mg cycloheximide 30 min before killing as described in the legend for Fig. 5. The organelles were incubated in the presence of the indicated concentrations of cardiolipin and net steroid synthesis during a 10 min incubation was determined.

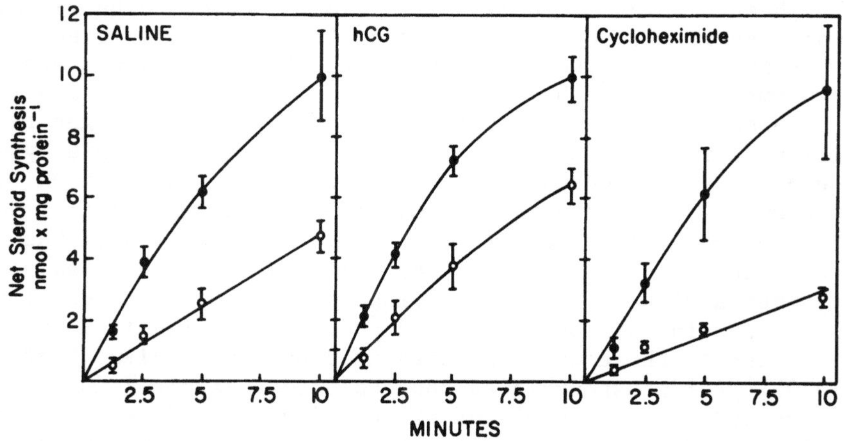

Fig. 7. Effect of acute treatment of rats with hCG or cycloheximide on the response of luteal mitochondria to exogenous cardiolipin. Mitochondria were prepared from ovaries of PMSG-hCG-primed rats (day 8 post-hCG) which received either 25 IU hCG, 5 mg cycloheximide or the 0.9% saline vehicle, i.p., 30 min before killing. Mitochondria were incubated in the presence of 800 μM cardiolipin or vehicle with 10 mM isocitrate serving as a source of reducing equivalents. Net steroid synthesis was determined. Values are means ± S.E. for 3-5 separate experiments.

Table 2. Effect of Cardiolipin and Phosphatidylcholine on Cholesterol Side Chain Cleavage Carried out by Acetone Powders of Luteal Mitochondria

	% of Control
cardiolipin:	585 ± 94
phosphatidylcholine:	3]8 ± 96

Acetone powders prepared from luteal mitochondria were incubated in our routine assay medium containing 1 mM NADPH, 2 mM glucose-6-phosphate, 2.5 mIU glucose-6-phosphate dehydrogenase, 100 µM cholesterol in ethanol and phospholipid dispersed by sonication in a vehicle of incubation medium and 10% propylene glycol. Bovine liver cardiolipin (800 µM) or egg phosphatidylcholine (800 µM) served as the polar lipids. Steroid synthesis during a 30 min. incubation was determined for 3 separate preparations. Mean net steroid synthesis by the acetone powders in the presence of vehicle (control) was 0.38 nmol/mg protein/30min. Values presented are means ± S.E.

changes in phospholipids in whole cell extracts seemed to be slight, alterations occurring specifically in the mitochondrial membranes might either facilitate movement of cholesterol to the side chain cleavage enzyme or affect the enzyme directly, altering its catalytic activity. These effects of gonadotropins on polar lipids might require ongoing protein synthesis (Silbert, 1975). To test this hypothesis, we obtained mitochondria from ovaries of PMSG-hCG primed rats which had been pre-treated on day 8 post-hCG with 5 mg cycloheximide, i.p., 30 min before killing and incubated these organelles in the presence of dispersions of various phospholipids and determined net steroid synthesis with isocitrate serving as the source of reducing equivalents. The incubation system and analyses of steroidal products are similar to those which we used in previous studies (Toaff et al., 1979). Polyphosphorylated phospholipids, cardiolipin and diphosphoinositide, were capable of stimulating steroidogenesis under these in vitro conditions at concentrations of 800 µM and 200 µM, respectively. Other polar lipids at concentrtions of 800 µM, including precursors of cardiolipin (phosphatidylglycerol) and diphosphoinositides (phosphatidylinositol), either had no effect or were inhibitory, as was lysophosphatidylcholine (Fig. 5). The effects of cardiolipin on steroidogenesis by mitochondria from ovaries of cycloheximide-treated rats were dose-related, yielding a linear log-dose response curve (Fig. 6). The relatively high concentrations of cardiolipin required for stimulation of steroidogenesis presumably reflect the limited entry of the lipid onto the organelles when presented as a sonicate in our buffer.

The steroidogenic response to cardiolipin depended upon the status of the luteal mitochondria (Fig. 7). Mitochondria prepared from ovaries of immature PMSG-hCG-primed rats treated with 25 IU hCG and killed 30 min later produced 1.5-fold as much steroid as mitochondria prepared from ovaries of saline-treated animals, while mitochondria from ovaries of rats treated with 5 mg cycloheximide synthesized half as much steroid as those from saline-treated animals. The addition of 800 μM cardiolipin increased steroid production by mitochondria from all 3 groups to similar values. Thus, steroid synthesis by organelles from cycloheximide-treated animals was increased 4-fold; steroid production by mitochondria from saline-treated rats 2-fold; and steroid production by mitochondria from hCG-treated rats 1.4-fold.

Cardiolipin also increased cholesterol side chain cleavage by acetone powders of luteal mitochondria incubated with cholesterol and NADPH as a source of reducing equivalents (Table 2). Cardiolipin (800 μM) elevated steroid synthesis 6-fold above vehicle-containing preparations, while egg phosphatidylcholine dispersions (800 μM) increased steroid production 3-fold. The cardiolipin effect was significantly greater than that of phosphatidylcholine ($p < 0.02$).

It should be mentioned that ACTH acutely increases synthesis of polyphosphoinositides and their precursors, phosphatidic acid and phosphatidylinositol, in rat adrenal tissue in a temporal pattern coinciding with increased corticosteroid formation (Farese et al., 1980). These effects of ACTH are blocked by cycloheximide. Farese and his colleages have also observed stimulation of cholesterol side chain cleavage by polyphosphorylated phospholipids using rat adrenal cortex mitochondria (Farese and Sabir, 1980). The exact mechanism by which this stimulation is affected is uncertain. The phospholipid milieu of purified bovine adrenal cortex cytochrome P-450scc influences the Km of the enzyme for cholesterol (Lambeth et al., 1980) and such an alteration (e.g., a lowered Km for cholesterol) might well account for the increased side chain cleavage activity produced by tropic hormones or polyphosphorylated polar lipids since saturating concentrations of free sterol may never be achieved in the inner membrane of the intact organelles.

One might speculate on the basis of our findings, and those of Farese and his colleagues, that polyphosphorylated phospholipids serve as intracellular messengers mediating the steroidogenic response to LH. In such a scheme LH is envisioned to stimulate the synthesis of polyphosphorylated lipids with the translocation of these lipids to the mitochondria resulting in enhanced pregnenolone formation. The rapid changes in turnover of specific phospholipids, particularly phosphatidylinositol, following exposure of a variety of target tissues to hormones or neurotransmitters (Michell, 1975) is consistent with such a notion. We have found diphosphoinositides, triphosoinositides and cardiolipins in our mitochondria at levels of 0.67, 1.26, and 16 nmol/mg protein, respectively. However, we have not yet determined if the contents of any of these lipids increase following hCG treatment and are reduced by cycloheximide.

Summary

In this report we have focused on two aspects of control of corpus luteum function; the regulation of lipoprotein uptake and the modulation of cholesterol side chain cleavage through alterations in the phospholipid milieu. Until recently, a role of gonadotropins in the acquisition of lipoprotein-carried cholesterol had not been generally appreciated. However, it is likely that the regulation of expression of cell surface lipoprotein receptors will prove to be an important function of tropic hormones, particularly LH, since luteal cells generate progestins at a substantial rate and must, therefore, have an equally substantial supply of cholesterol. Factors other than gonadotropins may have a function in control of lipoprotein receptors and it would not be surprising if luteolytic factors, such as prostaglandin $F_{2\alpha}$, are found to diminish lipoprotein-uptake as part of their action in reducing steroidogenesis.

The control of mitochondrial pregnenolone synthesis by gonadotropin-induced alterations in phospholipid composition of mitochondrial membranes represents a novel concept. Whether certain polar lipids will emerge with the status of intracellular messengers remains to be seen. In any event, our observations clearly demonstrate the possibility for such a mechanism of control.

Acknowledgements

This work was supported by USPHS Grant HD-12314. Dr. Tanaka is supported by a fellowship from the Rockefeller Foundation. Ms. MacGregor is a trainee of the Veterinary Medical Scientist Training Program of the University of Pennsylvania.

We thank Ms. Charlene Smith for her help in preparation of this manuscript.

References

Christie, M. H., Gwynne, J. T., and Strauss, J. F., III, 1981, Binding of human high density lipoproteins to membranes of luteinized rat ovaries, J. Steroid Biochem., 14:671.

Christie, M. H., Strauss, J. F., III, and Flickinger, G. L., 1979, Effect of reduced blood cholesterol on sterol and steroid metabolism by rat luteal tissue, Endocrinology, 105:92.

Farese, R. V. and Sabir, A. M., 1980, Polyphosphoinositides: Stimulator of mitochondrial cholesterol side chain cleavage and possible identification as an adrenocorticotropin-induced, cycloheximide-sensitive, cytosolic, steroidogenic factor, Endocrinology, 106:1869.

Farese, R. V., Sabir, M. A., Vandor, S. L., and Larson, R. E., 1980, Are polyphosphoinositides the cycloheximide-sensitive mediator in the steroidogenic actions of adrenocorticotropin and adenosine-3',5'-monophosphate?, J. Biol. Chem., 255:5728.

Fry, M. and Green, D. E., 1981, Cardiolipin requirement for electron transfer complex I and III of the mitochondrial respiratory chain, J. Biol. Chem., 256:1874.

Goldstein, J. L. and Brown, M. S., 1977, The low-density lipoprotein pathway and its relation to atherosclerosis, Annu. Rev. Biochem., 46:897.

Gwynne ,J. T. and Hess, B., 1980, The role of high density lipoproteins (HDL) in rat adrenal cholesterol metabolism and steroidogenesis, J. Biol. Chem., 254:10875.

Lambeth, J. D., Kamin, H., and Seybert, D. W., 1980, Phosphatidylcholine vesicle-reconstituted cytochrome P-450 scc. Role of the membrane in control of activity and spin state of the cytochrome, J. Biol. Chem., 255:8282.

Michell, R. H., 1975, Inositol phospholipids and cell surface receptor function, Biochim. Biophys. Acta, 415:81.

Richards, J. S. and Williams, J. J., 1976, Luteal cell receptor content for prolactin (PRL) and luteinizing hormone (LH): Regulation by LH and PRL, Endocrinology, 99:1571.

Rosenblum, M. F., Huttler, C. R., and Strauss, J. F., III, 1981, Control of sterol metabolism in cultured rat granulosa cells, Endocrinology, 109:1518.

Schuler, L. A., Scavo, L., Kirsch, T. M., Flickinger, G. L., and Strauss, J. F., III, 1979, Regulation of de novo biosynthesis of cholesterol and progestins and formation of cholesteryl ester in rat corpus luteum by exogenous sterol, J. Biol. Chem., 254:8662.

Schuler, L. A., Langenberg, K. K., Gwynee, J. T., and Strauss, J. F., III, 1981a, High density lipoprotein utilization by dispersed rat luteal cells, Biochim. Biophys. Acta, 664:583.

Schuler, L. A., Toaff, M. E., and Strauss, J. F., III, 1981b, Regulation of ovarian cholesterol metabolism: Control of 3-hydroxy-3-methylglutaryl coenzyme A reductase and acyl coenzyme A: cholesterol actyltransferase, Endocrinology, 108:1476.

Silbert, D. F., 1975, Genetic modification of membrane lipid, Annu. Rev. Biochem., 44:315.

Strauss, J. F., III and Flickinger, G. L., 1977, Phospholipid metabolism in cells from highly luteinized rat ovaries, Endocrinology, 101:883.

Strauss, J. F., III, Schuler, L. A., Rosenblum, M. F., and Tanaka, T., 1981, Cholesterol metabolism by ovarian tissue, Adv. Lipid Res., 18:99.

Toaff, M. E., Schleyer, H., and Strauss, J. F., III, 1980, Presence of a low molecular weight inhibitor of succinate-supported cholesterol side chain cleavage in rat ovaries, Biochim. Biophys. Acta, 617:291.

Toaff, M. E., Strauss, J. F., III, Flickinger, G. L., and Shattil, S. J., 1979, Relationship of cholesterol supply to luteal mitochondrial steroid synthesis, J. Biol. Chem., 254:3977.

SOME OBSERVATIONS THAT MAY PERTAIN TO STRUCTURE-FUNCTION RELATIONSHIPS OF hCGβ*

Robert Ryan, Patrick McIlroy,
Kavi Ratanabanangkoon, and Henry Keutmann†

Department of Cell Biology
Mayo Medical School
Rochester, Minn. 55901

Atassi (1977) in a series of elegant experiments defined the three antigenic determinants on the disulfide-containing protein lysozyme. They are all conformationally dependent sites and each consists of 5 or 6 topographically adjacent amino acids across the surface of the molecule. As surface residues, the amino acids involved are primarily hydrophilic. It seems likely to us that the receptor binding site(s) on hCG and other glycoprotein hormones will be similar.

Ward has reviewed the studies from his laboratory and others concerning chemical modification of glycoprotein hormones and their effects on subunit recombination and biologic activity (Ward, 1978). One of the major road blocks to determining the receptor binding site(s) and subunit interacting sites is the lack of a crystal structure for any of the glycoprotein hormones.

The present report relates a series of preliminary observations that may pertain to the structure-function relationships of the β subunit of hCG.

---

*Supported by funds from the National Institutes of Health (HD 12851, 9140, 7108), The World Health Organization and the Mayo Foundation.
†The Endocrine Unit, Massachusetts General Hospital, Boston, Mass. 02114.

1. The Cys-Ala-Gly-Tyr (GACY) Sequence is Not Directly Involved in Subunit Attachment

There are two tyrosine residues in LHβ and hCGβ and these are located at residues 37 and 59. These tyrosines cannot be iodinated, nitrated or perturbed by solvent in native LH or hCG, although they can be altered readily in the isolated β subunits (Ward, 1978; Bishop and Ryan, 1975). The above suggests that they are in a hydrophobic environment in the native molecule and that this environment may be provided directly or indirectly by attachment of the α subunit.

The amino acid sequences around both tyrosine residues are conserved in the glycoprotein hormones with the degree of conservation being greater for tyr 37. The Cys-Ala-Gly-Tyr-sequence at residues 34-37 is present in the β subunits of all the glycoprotein hormones as well as many serine proteases and the β subunit of cholera toxin. This suggested that it may play a fundamental role in subunit recombination, receptor binding or adenylate cyclase activation.

Two synthetic peptides were prepared by solid phase synthesis and their homogeneity was confirmed by gel filtration, amino acid analysis and Edman degradation:

1. CAGY-8: Thr-Asp-Ile-Cys(SH)-Ala-Gly-Tyr-His-NH_2

2. CAGY-11: Cys-Val-Asn-Thr-Asp-Ile-Cys-Ala-Gly-Tyr-His-NH_2 (Cys¹ and Cys⁷ joined by S—S)

The CAGY-11 peptide was prepared with the closed disulfide as illustrated and as the half cystine acetamidomethyl protected open ring structure (CAGY-11-ACM).

These peptides were incubated in molar excess (50 to 500 fold) with hCGα and hCGβ (both from Dr. Robert Canfield, Columbia Univ.). Formation of hCG was assessed by receptor binding assays (Lee and Ryan, 1973) and activation of rat ovarian adenylate cyclase (McIlroy et al., 1980).

The data in Table 1 were obtained using a 50-fold molar excess of CAGY 11. Similar results were obtained with a 500 molar excess of CAGY 11 and with CAGY 8 and CAGY 11-ACM.

These data suggest that the amino terminal sequence adjacent to tyrosine 37 in the β subunit is not directly involved in the contact surface with the α subunit. Whether the hydrophobic environment in this region of the β subunit in the native molecule is provided by

Table 1. Effect of CAGY-11 on Subunit Recombination

Incubation Mixture	Radioreceptor ID-50 ng/ml	Radioreceptor RP	Adenylate Cyclase ½max μg/ml	Adenylate Cyclase RP
hCG native (CR 123)	5.5	1.0	.105	1.00
CR 123α + CR 123β	8.5	.65	.175	.60
CR 123α + CR 123β + CAGY 11	11.2	.49	.208	.51

Table 2. Potency of DAPA and Its Derivatives in Inhibiting ^{125}I[iodo]hCG Binding to Ovarian Receptors

Compound	ID_{50}[M]
hCG native	1×10^{-10}
2-N-dansylarginyl-N-[4 ethyl]piperidine (DAPA)	7×10^{-5}
2-N-dansylarginyl-N-piperidine	1×10^{-3}
2-N-dansylarginyl-N-[3 methyl]piperidine	5×10^{-4}
2-N-dansylarginyl-N-[4 methyl]piperidine	2×10^{-4}
2-N-dansylarginyl-N-[4 propyl]piperidine	2.1×10^{-5}
2-N-dansylarginyl-N-[4 isobutyl]piperidine	1.8×10^{-5}
2-N-dansylarginine	$>1 \times 10^{-3}$

the residues of the α subunit or whether the α subunit induces a conformational change in this region of the β subunit to provide a hydrophobic "pocket" remains to be seen. Alternatively, these β subunit tyrosines may become involved in intrachain hydrogen bonding after recombination with the α subunit.

2. 2-N-Dansylarginyl-N-[4-ethyl] Piperidine (DAPA) Inhibition of hCG Binding

Several years ago, while investigating the effects of proteinase inhibitors on the adenylate cyclase enzyme in rat luteal cell membranes, it was noted that DAPA, a highly potent thrombin inhibitor, blocked activation of this enzyme by hCG (McIlroy et al., 1980). Unlike the other proteinase inhibitors, it also blocked hCG binding at the same doses required to block cyclase activation.

DAPA inhibited ^{125}I[iodo]hCG and ^{125}I[iodo]LH binding to rat luteal membranes (method of Lee and Ryan, 1973) in a dose related manner. The competition curve was parallel to that of unlabeled hCG (or LH) but the ID_{50} was 7×10^{-5} as opposed to 1×10^{-10} for hCG. A series of DAPA derivatives, obtained from Dr. Michael Nesheim, Hematology Research Unit, Mayo Clinic, were examined and their relative potencies are listed in Table 2.

Analysis of the DAPA-effect by Scatchard plot indicated a lowering of the affinity constant without a significant effect on the number of binding sites. It thus appears to act as a competitive inhibitor of hCG (LH) binding.

To investigate the specificity of this effect ^{125}I[iodo]hFSH binding to rat testicular membranes was studied using the method of Abou-Issa and Reichert (1976). DAPA did not inhibit FSH binding until mM concentrations were used, 1 to 2 log orders more than required to inhibit hCG binding. At these concentrations, the incubation tubes became cloudy and it was thought that the effect was nonspecific.

How is this DAPA effect mediate? The structure of DAPA shows homology with an arginine-hydrophobic sequence in a peptide. Several such sequence exist in LHβ and hCGβ, that differ from the homologous regions in FSHβ or TSHβ. One is the-Leu-Arg-Pro sequence at residues 5-7 in hCGβ. This sequence is also found at the C terminal region of GNRH. We examined the effects of GNRH and the synthetic peptide-Leu-Arg-Pro-Glycine amide (both provided through the courtesy of Dr. Roger Guilleman) on hCG binding and they were less potent than DAPA.

Another sequence unique to hCG and LH, is the-Arg-Val-Leu at residues 43 to 45. The increased potency of DAPA with greater hydrophobic substitution at position 4 favors this peptide as opposed to the-Leu-Arg-Pro-sequence. The Arg-Val-Leu sequence is found in ovine LHβ between two methionine residues at positions 42 and 53 and can be obtained as a cyanogen bromide cleavage product. Dr. Harold Papkof (Univ. California, San Francisco) provided the -Lys-Arg-Val-Leu-Val-Ile-Leu-Pro-Pro-Homoserine peptide from ovine LHβ. When this was examined, it failed to inhibit hCG binding at doses up to 3 mM.

We would like to speculate that DAPA acts by mimicking one of the sequences involved in the biologically active site(s) on hCGβ or LHβ. However, we cannot as yet define this site. From a study of molecular models of the arginyl containing regions unique to hCGβ and hLHβ it was found that [4-isobutyl] DAPA most closely resembled the Arg-Gly-Val sequence at residues 74-76) however, in ovine, bovine, and porcine LHβ there is a proline substituted for argining at position 74. It should also be noted that the ovine, bovine, and porcine hormones have a lower affinity for the rat receptor than hLH or hCG (Leidenberger and Reichert, 1973). It is possible that this arginine for proline substitution may contribute to the higher affinity of the human hormones.

3. Deglycosylated hCGβ

The definition of the functional role of the carbohydrate side chains of the glycoprotein hormones remains incomplete. Removal of terminal sialic acid residues has been shown to accelerate the rate

Table 3. Carbohydrate Analysis of Deglycosylated hCGβ (mean of 3 preparations)

Sugar	degly β g%	hCGβ* g%
Mannose	0.41	3.0
Galactose	1.89	6.3
Glucose	0.34	--
N-acetylgalactosamine	4.62	3.4
N-acetylglucosamine	2.68	7.0
Sialic acid	2.04	8.3

*from Canfield et al. 1976.

of disappearance of hCG from the circulation (Van Hall et al., 1971) but not receptor binding of hCG (Moyle et al., 1975). Removal of additional carbohydrates by sequential enzymatic hydrolysis has resulted in reduced in vitro biologic responses (Moyle et al., 1975). These latter studies, however, have not excluded an internal peptide cleavage and, therefore, the results cannot be clearly attributed to carbohydrate removal.

The presence of carbohydrate has been thought to inhibit the crystallization of glycoproteins. We, therefore, decided to deglycosylate hCGβ (degly β) and attempt to form crystals. This effort failed. While wondering what to do with the degly β we saw Sairam's and Schiller's report (1979) that deglycosylated whole ovine LH retained receptor binding activity but lost biologic activity. We, therefore, investigated degly β for the same effects.

Thoroughly dried hCGβ was treated with anhdydrous HF and 20% anisole in an Ohno HF distillation apparatus for 1 h at 0°C (Mort and Lamport, 1977). The product was extracted with ether (to remove anisole) and filtered on a column of Sephadex G100. As expected, it eluted from the gel filtration column as a single species of smaller size than intact hCGβ.

Amino acid analysis of the deglycosylated hCGβ, with or without reduction and carboxymethylation (RCXM) revealed a composition that agreed with the theoretical composition of hCGβ. Gel filtration of RCXM degly β revealed a single peak with a constant amino acid composition across the peak. Edman degradation of the RCXM material confirmed that the product was a single peptide chain.

The carbohydrate analysis of degly β is shown in Table 3. Carbohydrate was reduced by 60% but complete removal of sugar was not achieved. This may account for the failure to crystallize. The results are consistent with the removal of peripheral carbohydrates (sialic acid, mannose, galactose, N-acetylglucosamine) from the long asparinine-linked side chains.

Table 4. Relative Potencies of Recombined Degly β-hCGα in Receptor Binding and Adenylate Cyclase Assays

Preparation	RRA ID^{50} ng/ml	RRA RP	Adenylate cyclase ½max ng/ml	Adenylate cyclase RP
incubated hCG CR 123	8.2	1.00	80	1.00
CR 123 hCGα-degly β (peak A)	8.0	1.03	>1000	<.08

Table 5. Effects of hCGα-Degly β (A) on the Rat Ovarian Response to Native hCG (CR 117)

Preincubation (30 min 20°C)	assay (20 min 30°C)	pMoles cAMP/mg/20 min mean ± SE
None	Buffer	106 ± 5
None	CR 117 (1 μg/ml)	290 ± 25
None	CR 117 (1 μg/ml) + A (0.3 μg/ml)	258 ± 23
None	CR 117 (1 μg/ml) + A (1.0 μg/ml)	222 ± 20
Buffer	Buffer	146 ± 5
Buffer	CR 117 (1 μg/ml)	374 ± 11
A(.3 μg/ml)	CR 117 (1 μg/ml)	231 ± 7
A(1.0 μg/ml)	CR 117 (1 μg/ml)	210 ± 15

The residual N-acetylgalactosamine and sialic acid suggest that the short chain serine-linked (C terminal region) carbohydrates and the N-acetylgalactosamine residues immediately adjacent to the asparagines were not removed by HF treatment.

Deglycosylated hCGβ was incubated with a 1.5-fold molar excess of CR 123 hCGα in 0.1 M bicarbonate at 25°C for 24 h. The incubation mixture was then passed through a Sephadex G100 column and two peaks of protein were resolved. The larger species (A) emerged as a slightly smaller material than native hCG while the smaller species (B) had the same elution position as hCGα. Both peaks were lyophylized. On amino acid analysis peak A was deglycosylate hCGβ recombined with hCGα while peak B was free α subunit.

The results of radioreceptor and rat ovarian adenylate cyclase assays of the Peak A material are given in Table 4.

Peak A was not devoid of cyclase stimulating activity as the table suggests, but the activity was quite low (the approximate relative potency was .05 × CR 123). This is in sharp contrast to the very good potency in the receptor binding assay and suggests that the deglycosylated hormone might be an inhibitor of intact hCG. To this end further experiments were carried out and the results are shown in Table 5.

Additional experiments are in progress and these preliminary results indicate that the degly β hCG is an antagonist of the native hormone. The fact that preincubation is more effective than concurrent incubation suggests that a difference in binding affinity exists, yet the competition for receptor binding showed no difference. This apparent paradox could be explained by the 'spare receptor' notion if only a small fraction of the total receptors were actually coupled to the cyclase system. How the carbohydrate chains, presumably those that are asparginine linked, activate the cyclase system is unknown.

Other recent experiments indicate that deglycosylated hCGα recombines readily with intact hCGβ and is fully active in a radioreceptor assay but is also relatively impotent in stimulating adenylate cyclase activity.

4. Phosphate Accepting Sequence

Recent data indicates that cAMP-dependent protein kinase induces phosphorylation of a serine or threonine residue in a particular amino acid sequence (Yeaman et al., 1977; Matsuo et al., 1978, Feramisco et al., 1980). In general, this is an -Arg-X-Ser- sequence but the reaction rate is enhanced if it is an -Arg-Arg-X-Ser- sequence. The rate of phosphorylation is also more rapid if X is a hydrophobic residue like leucine. These sequences exist in β turn regions of the protein substrate (Small et al., 1977).

Examination of the amino acid sequence of hCG-β and hLHβ reveals an -Arg-Arg-Ser-Thr-Thr-Asp- sequence at residues 94-99. Computer analysis using the Chou-Fasman rules suggests that this is a β turn region (Ryan, unpublished). Garnier (1978) also found this to be a β turn region using a different predictive method. The turn is further enforced by the disulfide bridge between residues 93 and 100 (Guidice and Pierce, 1978).

As indicated residues 93-100 in hCGβ and hLHβ are identical. Ovine, bovine, and porcine LHβ differ by substitution of leucine for arginine at position 95. In contrast, this region has significantly different sequences in FSHβ and TSHβ and has been discussed with respect to charge differences among the glycoprotein hormones and the dual activity of PMSG (Ward and Moore, 1979; Moore et al., 1980).

Various glycoprotein hormones and related materials were tested in vitro as substrates for cAMP-dependent protein kinase. Substrate

Table 6. cAMP-Dependent in vitro Phosphorylation of Histone and Glycoprotein Hormone Related Materials

Substrate (1.25 nMole)	Phosphate incorporation pMole/30 min
hCG	8.8
hCGβ	109.6
RCXM hCGβ	234.3
hLH	17.1
hFSH	0
hTSH	2.0
Histone IIA	50.0

(0.125 mM) was incubated at 30°C for 30 min with 300 μg of rabbit muscle cAMP-dependent protein kinase (Sigma) and 0.25 mM ^{32}P-γ-ATP (233 dpm/pmole, New England Nuclear) in a total volume of 80 μl. The buffer was 62.5 mM MES containing 12.5 mM $Mg(OAc)_2$, 0.25 mM EGTA, 9 μM cAMP and 0.5 mM DTT. Essentially the method is that described by Fermisco et al. (1980).

Results are given in Table 6. The radioactivity incorporated into hCG and hCGβ migrated with β subunit during gel filtration and SDS disc gel electrophoresis and was precipitable with antibody against hCGβ. A synthetic peptide representing hCGβ 93-101 with a C terminal tyrosine and disulfide bridge between 93 and 100 was made and it phosphylated as well as RCXM-hCGβ. Digestion and amino acid analysis indicated that both serine and threonine were phosphorylated.

The facts that RCXM-hCGβ phosphorylated better than hCGβ, which in turn was a better substrate than hCG suggests that the 93-100 region of the molecule may be masked in the native state. Since LH was a better substrate than hCG the masking may be due in part to the C terminal region of hCGβ as well as the α subunit. Ovine LH could also be phosphorylated (data not shown). It will be important to determine whether hCG or LH become phosphorylated after binding to receptor.

5. Concluding Remarks

These preliminary results offer some hints about structural features of LH and hCG that warrant further study. They also suggest two potential intraovarian regulatory mechanisms. The requirement for asparginine linked carbohydrate for full activation of ovarian adenylate cyclase activity suggests that there may be a membrane lectin involved in the regulation of this enzyme.

The fact that hCG and LH can be phosphorylated in vitro raises the possibility that it may be phosphorylate in vivo. In general protein phosphorylation is an important regulatory mechanism and in this circumstance it could be related to the phenomena of down regulation, desensitization, or internalization. It should also be pointed out that phosphorylation would protect the -Arg-Arg-peptide sequence from tryptic cleavage.

These studies also suggest some cautions about intraovarian inhibitors of various activities. If our speculation about the topographic nature of the biologic site on gonadotrophins is correct then peptide fragments that mimic parts of this site could inhibit hormone binding to receptor. These fragments could arise from metabolism of the gonadotrophins, but also could arise from other proteins. They would, however, need to be present in relatively high concentrations with respect to the gonadotrophins and, therefore, the gonadotrophins per se would be an unlikely source unless these fragments were concentrated by some purification procedure.

References

Abou-Issa, H., and Reichert, E. R., Jr., 1976, Properties of the follitropin (RSH)-receptor interaction: characterization of the interaction of follitropin receptors in purified membranes isolated from mature rat testes tubules, J. Biol. Chem., 251: 3326.

Atassi, M. Z., ed., 1977, Immunochemistry of Proteins, Volume 2, Plenum Press, New York.

Bishop, W. H., and Ryan, R. J., 1975, Solvent exposure of aromantic groups in human luteinizing hormone and its subunits. Measurement by solvent perturbation difference spectroscopy, Biochemistry, 14:4083.

Canfield, R. E., Birken, S., Morse, J. H., and Morgan, F. J., 1976, Human chorionic gonadotropin, in: "Peptide Hormones," J. A. Parsons, ed., McMillan Press Ltd., London, pp. 299-315.

Fermisco, J. R., Glass, D. B., and Krebs, E. G., 1980, Optimal spatial requirements for the location of basic residues in peptide substrates for cyclic AMP-dependent protein kinase, J. Biol. Chem., 255:4240.

Garnier, J., 1978, Molecular aspects of the subunit assembly of glycoprotein hormones, in: "Structure and Function of the Gonadotropins," K. W. McKerns, ed., Plenum Press, New York, pp. 81-109.

Guidice, L. C., and Pierce, J. G., 1978, Glycoprotein hormones: some aspects of studies of secondary and tertiary structure, in: "Structure and Function of the Gonadotropins," K. W. McKerns, ed., Plenum Press, New York, pp. 81-109.

Lee, C. Y., and Ryan, R. J., 1973, Interaction of ovarian receptors with human luteinizing hormone and human chorionic gonadotropin, Biochemistry, 12:4609.

Matsuo, M., Huang, C.-H., and Huang, L. C., 1978, Evidence for an essential arginine recognition site on adenosine 3':5' cyclic monophosphate-dependent protein kinase of rabbit skeletal muscle, Biochem. J., 173:441.

McIlroy, P. J., Richert, N. D., and Ryan, R. J., 1980, Effects of proteinase inhibitors on adenylate cyclase, Biochem. J., 188: 423.

Moore, W. T., Burleigh, B. C., and Ward, D. N., 1980, Chorionic gonadotropin: comparative studies and comments on relationships to other glycoprotein hormones, in: "Chorionic Gonadotropin," S. Segal, ed., Plenum Press, New York, pp. 89-126.

Mort, A. J., and Lamport, D. T. A., 1977, Anhydrous hydrogen fluoride deglycosylates glycoproteins, Anal. Biochem., 82:289.

Moyle, W. R., Bahl, O. P., and Marz, L., 1975, Role of the carbohydrate of human chorionic gonadotropin in the mechanism of hormone actions, J. Biol. Chem., 250:9163.

Leindenberg, F., and Reichert, L. E., Jr., 1973, Species differences in luteinizing hormone as infered from slope variations in a radioligand receptor assay, J. Clin. Endocr. and Metab., 92:646.

Sairam, M. R., and Schiller, P. W., 1979, Receptor binding, biological, and immunological properties of chemically deglycosylated pituitary lutropin, Arch. Biochem. Biophys., 197:294.

Small, D., Chou, P. Y., and Fasman, G. D., 1977, Occurrence of phosphorylated residues in predicted β-turns; implication for β turn participation in control mechanism, Biochem. Biophys. Res. Comm., 79:341.

Van Hall, E. V., Vaitukaitis, J. L., Ross, G. T., Hickman, J. W., and Ashwell, G., 1971, Effects of progressive desialylation on the rate of disappearance of immunoreactive hCG from plasma in rats, Endocrinology, 89:11.

Ward, D. N., 1978, Chemical approaches to the structure-function relationships of luteinizing hormone (lutropin), in: "Structure and Function of the Gonadotropins," K. W. Kerns, ed., Plenum Press, New York, pp. 31-36.

Ward, D. N., and Moore, W. T., 1979, Comparative study of mammalian glycoprotein hormones, in: "Animal Models for Research on Fertility and Contraception," N. J. Alexander, ed., Harper and Row, Baltimore, Maryland, pp. 151-164.

Yeaman, S., Cohen, P., Watson, D. C., and Dixon, G. H., 1977, The substrate specfcity of adenosine 3':5'-cyclic monophosphate dependent protein kinase of rabbit skeletal muscle, Biochem. J., 162:411.

STIMULATION OF GRANULOSA CELL MATURATION BY FOLLICULAR FLUID

Florence Ledwitz-Rigby, Brian W. Rigby,
Su Yu Ling, Luanne Stewart,
and Mark McLean

Department of Biological Sciences
Northern Illinois University
Dekalb, Illinois 60115

Introduction

Antral follicle granulosa cells experience a series of biochemical and morphological changes as the follicles enlarge and mature. These include increased steroidogenic activity (Channing, 1970), increased cellular content of steroidogenic organelles (Rigby et al., 1980a), enhanced responsiveness to the gonadotrophins, and enhanced adenyl cyclase activity (Channing and Ledwitz-Rigby, 1974). Concomitant with follicular enlargement granulosa cell dependence on hormones to stimulate luteinization declines. Granulosa cells from large (6-12 mm) preovulatory, porcine follicles luteinize in vitro without requiring gonadotrophins in the culture media; medium (3-5 mm) follicles require either FSH or LH (Channing, 1970). However, even with estrogen, FSH and LH in the media the granulosa cells of small follicles do not secrete as much progesterone as cells obtained from more mature follicles (Thanki and Channing, 1976). The question remains as to which factors control in vivo maturation of granulosa cells and especially what determines whether an individual follicle will mature or become atretic.

We have examined the influence of porcine follicular fluid on porcine granulosa cell function in vitro. In this paper we will describe experiments with granulosa cells from small immature follicles in which we have been able to enhance basal and gonadotrophin stimulated progesterone secretion, basal and FSH stimulated estrogen secretion and morphological maturation by incubating the cells in the presence of follicular fluids. We have observed that fluid from small follicules, 50-80% of which are atretic (Ryan, 1981) is only weakly stimulatory of steroidogenesis in granulosa cells from small follicles (Ledwitz-Rigby and Rigby, 1979) and inhibitory of several functions of granulosa cells from larger follicles (Ledwitz-Rigby et al., 1977; Alexander et al., 1978; Rigby et al., 1980b; Ledwitz-Rigby, 1980). Fluid from large follicles, 80% of which will ovulate (Ryan, 1981), is highly stimulatory of granulosa cells from small follicles (Ledwitz-Rigby and Rigby, 1979) and slightly stimulatory of granulosa cells from large follicles (Alexander et al., 1978). There appears to be a correlation between the percentage of atretic follicles in the sample that the fluid is obtained from and the ratio of maturation stimulator: luteinization inhibitor activity in the fluid. We have hypothesized that this ratio may determine which follicles become atretic and which mature. Thus only a small percentage of small follicles would have a relatively high content of maturation stimulator. The stimulator would enhance granulosa cell responsiveness to gonadotrophins and estrogen secretion. The estrogen produced within the follicle would further enhance gonadotrophin responsiveness, mitosis of the granulosa cells and growth of the follicle. Those follicles containing a high ratio of inhibitor to stimulator would be less responsive to the gonadotrophins and become atretic.

General Methods and Materials

In all the experiments described in this paper we have used porcine follicular fluids and granulosa cells obtained from slaughterhous material returned to the lab and collected within a few hours after slaughter. Adult porcine serum was either collected by ourselves at local slaughterhouses or purchased from Flow laboratories. All sera batches are tested for cytotoxicity and only those showing no effect were used. Follicular fluids and sera were charcoal treated (CT) to remove 99% of the steroids and prostaglandins (Ledwitz-Rigby et al., 1977; Batta and Channing, 1979) and filtered through 0.2 micron filters to remove microorganisms.

Incubations lasting longer than 24 h were done as monolayer cultures. Shorter incubations were performed in glass culture tubes. Estrogen and progesterone content of media and cells were determined by radioimmuno assays (Ledwitz-Rigby, 1980). The endogenous estrogen and progesterone in the unincubated media were subtracted from the values assayed in the incubated media.

Stimulation of Progesterone Secretion

Incubation of granulosa cells from small follicles in media composed of equal parts of TC199 and either fluid from small follicles (SFFl), fluid from large follicles (LFFl), or adult porcine serum from 2, 4, or 6 days indicated that the follicular fluids enhanced both basal and gonadotrophin stimulated progesterone secretion (Fig. 1) (Ledwitz-Rigby and Rigby, 1979). The follicular

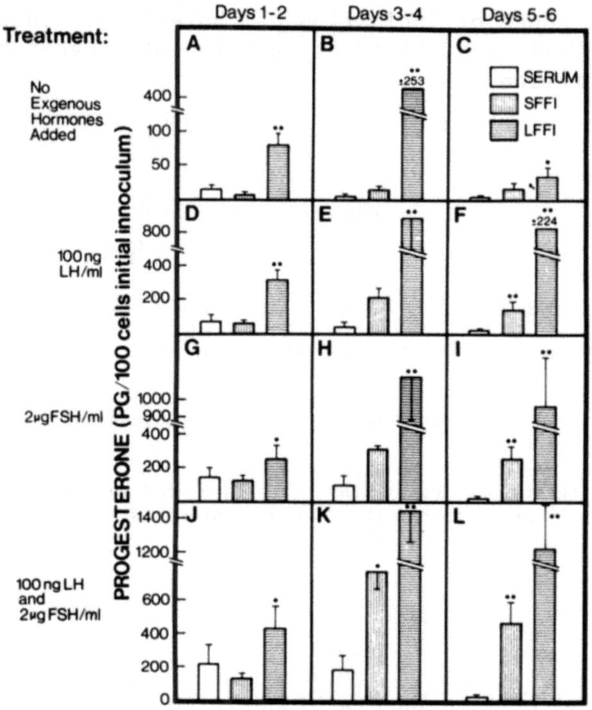

Fig. 1. Influence of follicular fluid on basal and gonadotrophin-stimulated progesterone secretion by immature porcine granulosa cells. Progesterone secreted during 48 h culture periods by porcine granulosa cells from 1-2 mm follicles incubated in culture media containing 50% serum, 50% fluid from small follicles (SFFl) or 50% fluid from large follicles (LFFl) without exogenous hormones (A, B, and C) or in the presence of 100 ng LH/ml (D, E, and F), 2 µg FSH/ml (G, H, and I) or 100 ng LH and 2 µg FSH/ml (J, K, and L). * and ** indicate that the progesterone secreted by the cells cultured in a follicular fluid was significantly greater ($P < 0.05$ and 0.1 respectively), than that secreted by cells cultured in serum under identical conditions. From: Ledwitz-Rigby and Rigby (1979), with permission of the publishers.

fluids and gonadotrophins were synergistic in their action. Of particular interest was the observation that the granulosa cells were completely unresponsive to LH when incubated in serum but highly responsive to LH in the presence of fluid from large follicles. LFFl stimulated progesterone secretion earlier (within the first 2 days) and to a greater extent than SFFl. Since those experiments, we have concentrated our studies on the ability of LFFl to enhance progesterone secretion.

Table 1. Influence of Inhibition of Cholesterol Synthesis on Follicular Fluid Stimulation of Progesterone Secretion

	Progesterone secreted ng/culture/48 hr. (n=3)			
	Concentration of AY9944*			
Culture medium	0	10^{-5}M	10^{-4}M	5 x 10^{-4}M
50% CT Serum	7.3 ± 0.2	5.5 ± 0.2	2.5 ± 0.1	1.4 ± 0.1
50% CT LFF1	17.9 ± 0.8***	14.2 ± 0.9**	10.3 ± 0.5**	7.7 ± 0.7**

*AY9944 inhibitor of the conversion of 17 dehydrocholesterol to cholesterol (see text)

**P<0.01 that progesterone secreted by cells incubated in LFF1 was significantly greater than that secreted by cells in serum with the same concentration of AY9944

Time Course

We have found LFF1 to have a biphasic influence on progesterone secretion. The first very rapid effect is to stimulate the release of a small amount of progesterone within the time it takes to mix the follicular fluid with the cells and centrifuge the cells at 1000 × g to end the experiment. Assays of freshly collected cells, sonicated in saline have failed to detect any preformed progesterone in the cells (<50 pg/5 × 10^5 cells). Following the initial release, there is little additional secretion of progesterone into the media until 24-48 h. By 48 h, the cells incubated in LFF1 have secreted 2 to 10-fold more progesterone than those incubated in serum.

Dose Effect

Our initial observations of follicular fluid stimulation of progesterone secretion used media containing 50% follicular fluid. More recently we have tested whether lower concentrations of follicular fluid can enhance progesterone secretion.

Preliminary dose response curves have indicated that 10% LFF1 did not stimulate progesterone secretion by cells from small follicles, but 30% LFF1 was equally as stimulatory as 50% LFF1. The addition of 0.7 μg testosterone/ml did not alter the dose response curve. In the presence of 1.0 μg oFSH, however, 10% LFF1 was as stimulatory of progesterone secretion as 30% and 50% LFF1.

Table 2. Influence of Follicular Fluids on Estrogen Secretion by Granulosa Cells from Small, Medium, and Large Follicles

Size of follicles granulosa cells were obtained from	Culture Medium	Estrogen Secreted/Culture % stimulation over control cultures in 50% serum $\bar{x} \pm$ s.e.; number of experiments=( )
small (1-2 mm)	50% SFF1*	132 ± 31.7 (3)
	50% LFF1	568 ± 171 (7)
medium (3-5 mm)	50% SFF1	173 ± 86 (3)
	50% LFF1	261 ± 38 (6)
large (6-12 mm)	50% SFF1	156 ± 46 (4)
	50% LFF1	331 ± 96 (4)

*All media contained 2 μg ovine FSH and 0.14 μg testosterone/ml.

Effect of Inhibition of Cholesterol Synthesis

We have examined whether de novo synthesis of cholesterol is required for follicular fluid stimulation of progesterone secretion. Granulosa cells were incubated for 48 h with 50% LFF1 or 50% serum with 0, 10^{-5}, 10^{-4}, and 5×10^{-4} M AY9944 (trans 1-, 4-bis {2 chlorobenzyl/aminomethyl} cyclohexane dihydrochloride) (an inhibitor of the Δ7 reductase enzyme catalyzing the conversion of 17 dehydrocholesterol to cholesterol, supplied by Dr. Dvornik of Ayerest Laboratories). Progesterone secretion by cells incubated in both LFF1 and serum was decreased in a dose related manner by AY9944. However, even at the highest concentration of AY9944, LFF1 was clearly stimulatory of progesterone secretion (Table 1). These studies suggest that LFF1 may stimulate progesterone secretion by enhancing both cholesterol synthesis and the conversion of cholesterol to progesterone (Stewart, 1980).

Stimulation of Estrogen Secretion

The initial experiments examining the influence of follicular fluids on estrogen secretion by granulosa cells were performed in the presence of 0.14 μg testosterone and 2 μg ovine FSH/ml (Table 2). Incubations were for 2 or 3 days. Fluid from large follicles stimulated estrogen secretion by cells from small, medium, and large follicles during the 3 day incubations. The two-day incubations with LFF1 stimulated estrogen secretion by cells from medium and large follicles but not from small follicles.

Fluid from small follicles did not reproducibly stimulate estrogen secretion although stimulation was occasionally observed.

Table 3. Time Course for Follicular Fluid Stimulation of Estrogen Secretion by Granulosa Cells from Medium Sized Follicles

Culture medium: Length of Incubation	50% C.T. Serum +0.7μg Testosterone +1μg o FSH/ml	50% C.T. LFF1 +0.7μg Testosterone +1μg o FSH/ml
	Estrogen secreted pg/culture(n=3)	
3 hr.	16 ± 1.7	26 ± 3
24 hr.	65 ± 3	158 ± 42
48 hr.	89 ± 19	236 ± 96

Table 4. Influence of Preincubation with Follicular Fluid on FSH Stimultion of Estrogen Secretion

Estrogen secreted during second 48 hr incubation Pg/culture (n=3)

2nd Incubation medium (48 hrs)

Preincubation medium (48 hours)	10% CT Serum	10% CT Serum +0.7μg Testosterone/ml	10% CT Serum +0.7μg Testosterone/ml +1μg FSH/ml
50% CT Serum	25 ± 4	75 ± 14	58 ± 17
50% CT LFF1	30 ± 3	54 ± 1	132 ± 19

In each experiment where SFF1 was stimulatory it was less stimulatory than LFF1.

Time Course

Preliminary time course studies for follicular fluid stimulation of estrogen secretion have indicated that cells from medium sized follicles can be stimulated within 24 h but not within 3 h by LFF1 (Table 3). Cells from small follicles have only been stimulated after 3-day incubations. We have not yet examined the time course for stimulation of large cells. We were unable to detect estrogen in freshly collected granulosa cells from all size follicles (≤ 25 pg/5 × 10^5 cells). Our preliminary data, indicate that follicular fluid stimulation of estrogen secretion takes longer than stimulation of progesterone secretion.

Influence of FSH

We have more recently found that the presence of FSH has no consistent significant influence on estrogen secretion by cells incubated with either serum or follicular fluids. FSH did stimulate progesterone secretion in these same experiments. We had originally included FSH in the media as we expected one significant difference between serum and the follicular fluids would be a higher FSH content in the fluids from non-atretic follicles (McNatty and Baird, 1978). Thus, the enhancement of estrogen secretion by the follicular fluids might have been due to the FSH content. The lack of effect of FSH on antral porcine granulosa cell estrogen secretion observed by ourselves and others (Anderson et al., 1979) suggests this is not the case. Stimulation of estrogen secretion by follicular fluid appears to be different from stimulation of progesterone secretion where LFF1 and FSH are synergistic.

We next questioned whether FSH and follicular fluid might influence each other's actions on granulosa cell estrogen secretion. Three or five day preincubations of granulosa cells from small follicles with 2 μg ovine FSH in 10% serum had no consistent influence on the amount of estrogen secreted by these cells when they were incubated for 3 days in either 50% serum or 50% LFF1. However, following 2 day preincubations of cells from small follicles with 50% LFF1 or 50% serum, FSH did stimulate estrogen secretion by the cells preincubated in LFF1 but not those preincubated in serum (Table 4). These experiments suggest that LFF1 may enhance FSH responsiveness of granulosa cells from small follicles. Other experiments done several years ago by Ledwitz-Rigby and Channing indicate that FSH stimulation of cyclic AMP accumulation in cells from small follicles was enhanced by 2 day preincubations in 50% LFF1 as compared to a preincubation with 50% serum (unpublished). We have not yet examined whether these two phenomenon are related.

Influence of Follicular Fluid on the Morphology of Granulosa Cells from Small Follicles

Granulosa cells from small follicles incubated for 5 days in monolayer culture in the presence of either SFF1 or LFF1 appeared primarily epitheliod-like whereas cells incubated with serum were predominantly fibroblast-like (Ledwitz-Rigby and Rigby, 1979). More recently we have performed a morphometric analysis of the ultrastructure of freshly collected granulosa cells and of cells incubated for 2 and 5 days in 50% serum or 50% LFF1 (McLean, 1981). The cultures were first innoculated in 10% serum. After allowing 2 days for plating, the media was changed to either 50% serum or 50% LFF1 in TC199. To allow a direct comparison to the 'spherical' in vivo granulosa cell, the cells were released from the culture plate by scraping. This process results in rounded-up cells. These were then pelleted gently and embedded in agarose for processing for electron microscopy.

Micrographs were analyzed for comparative volume of the whole cells and percent volumes of nucleii, cytoplasm, mitochondria, lipids, lysosomes, dark staining granules and whorled endoplasmic reticulum.

The most dramatic difference observed was an increase in lysosome volume with time amongst the cells incubated in serum. The lysosome content of cells incubated in LFF1 did not differ from that of freshly collected cells. One hypothesis to explain our data is that a factor in follicular fluid not present in serum prevents atresia. The increasing lysosomal content of cells incubated in serum may reflect atretic changes that normally occur in the majority of small follicle granulosa cells in vivo. The lack of an increase in lysosomal content seen in cells incubated in LFF1 may reflect a decreased rate of atresia. Another major difference amongst the cultures was a rise in the content of dark staining granules after 2 days in LFF1 which declined sharply by 5 days. The dark staining granule content of cells in serum rose during the 5 day incubation but not to a statistically significant extent. Dark staining granules have been postulated to contain steroids. Our data may reflect an initial build up of steroids followed by secretion by the cells incubated in LFF1. The depletion of the granules from the cells after 5 days of incubation in LFF1 parallels our previous findings that progesterone secretion was highest during the first 4 days of incubation in LFF1 and markedly reduced by 5 and 6 days of culture (Ledwitz-Rigby and Rigby, 1979). The lipid content of the cultured cells declined steadily with time and was most reduced in the cells incubated in LFF1.

Characterization of Stimulatory Activity

As all follicular fluids assayed for stimulation of steroid secretion have been charcoal extracted, their native steroid or prostaglandin content is not responsible for their activity. The activity has survived repeated freeze-thawing and long-term storage at -20°C. Based on ultrafiltration the activity is greater than 10,000 daltons. Activity is consistently lost when column fractionation or salting-out procedures are utilized. Preliminary experiments indicate that some restoration of activity can be achieved by recombining separated fractions raising the possibility of multiple factors being involved in the activity.

Fractions have been examined by SDS-gel electrophoresis for the presence of follicular fluid specific proteinaceous material. We have observed a 35,000 dalton species in follicular fluid but not in similarly treated porcine serum or plasma. The fluids must be freed of albumin on cibacron-blue sepharose and concentrated before this band can be demonstrated. The biological activity of the fraction containing this unique protein has not yet been determined.

Summary and Conclusions

In vitro incubation of granulosa cells from small immature and atretic follicles with follicular fluids enhances both basal estrogen and progesterone secretion. FSH stimulation of estrogen secretion and LH stimulation of progesterone secretion are seen only after incubation of the granulosa cells with the follicular fluids. Fluid from large follicles is always more stimulatory than fluid from small follicles. At the ultrastructure level an increase of dark staining granules is observed in cells incubated in LFF1 prior to the greatest stimulation of progesterone secretion. After 5 days of culture in LFF1, the cell content of dark staining granules is markedly reduced. Another major difference between cells incubated with serum and LFF1 is a steadily increasing lysosome content in cells incubated in serum. The increase in lysosomes has been suggested as a component of atresia. Follicular fluid may therefore contain a factor(s) which prevents atresia from occurring.

We have hypothesized that differences in the follicular fluid content of "maturation stimulating molecules" may determine which follicles are able to produce adequate estrogen and respond to the gonadotrophins maximally and which are not. Maximal response to gonadotrophins and the actions of the estrogen produced within the follicles may prevent atresia. We are currently investigating whether one or several different factors act to exert the several observed effects on estrogen and progesterone production, FSH and LH responsiveness, and morphological changes.

Acknowledgements

The work presented in these papers was supported in part by BRSG Grant RR 07176 awarded by the Biomedical Research Support Grant Program, Division of Research Resources, National Institutes of Health and by the Graduate School of NIU. The authors gratefully acknowledge the technical assistance of Joan Kumpulla and Victor McCullum.

References

Alexander, J. S., Rigby, B. W., and Ledwitz-Rigby, F., 1978, Biol. Reprod., 19:693-700.
Anderson, L. D., Schaerf, F. W., and Channing, C. P., 1979, in: Ovarian Follicular and Corpus Luteum Function," C. P. Channing and J. M. March, eds., pp. 187-196, Plenum Press, New York.
Batta, S. K., and Channing, C. P., 1980, in: "Advances en obstetricia y ginecologia," J. Gonzales-Merlo and J. Gviv, eds. (in Spanish), Salvat Publisher, Barcelona, pp. 125-134.
Channing, C. P., and Ledwitz-Rigby, F., 1974, in: "The Physiology and Genetics of Reproduction," E. Coutinho and F. Fuchs, eds., pp. 353-370, Plenum Press, New York.

Channing, C. P., 1970, Endocrinol., 87:156-164.
Ledwitz-Rigby, F., 1980, Biol. Reprod., 23:324-330.
Ledwitz-Rigby, F., and Rigby, B. W., 1979, Mol. Cell. Endocr., 14: 73-79.
Ledwitz-Rigby, F., Rigby, B. W., Gay, V. L., Stetson, M., Young, J., and Channing, C. P., 1977, J. Endocrinol., 74:175-184.
McLean, M., 1981, Master's Thesis, Northern Illinois University (in preparation), directed by F. Ledwitz-Rigby and B. W. Rigby.
McNatty, K., and Baird, D. T., 1978, J. Endocrinol., 76: 527-531.
Rigby, B. W., Ledwitz-Rigby, F., and Hunzicker-Dunn, M., 1980, Mol. Cell. Endo., 17:17-24.
Rigby, B. W., Workman, J., McLean, M., and Ledwitz-Rigby, F., 1980a, Biol. Reprod. (Suppl. #1).
Ryan, R. J., 1981, in: "Dynamics of ovarian function," N. B. Schwartz and M. Hunzicker-Dunn, eds., Raven Press, New York. pp. 1-12.
Stewart, L. E., 1980, Master's Thesis, Northern Illinois University, directed by F. Ledwitz-Rigby.
Thanki, K. H., and Channing, C. P., 1976, Endocrine Res. Commun., 3:319-333.

INTRAOVARIAN REGULATION OF GRANULOSA-CELL REPLICATION

James M. Hammond, Johannes D. Veldhuis,
Thomas W. Seale,* and Matthew M. Rechler†

Division of Endocrinology
Deparment of Medicine
The Milton S. Hershey Medical Center
The Pennsylvania State University
Hershey, Pennsylvania 17033

INTRODUCTION

From the reinitiation of growth by primordial follicles until terminal differentiation (luteinization), the number of granulosa cells per follicle may increase from approximately 50 (Lintern-Moore et al., 1974) to more than 5×10^7 (McNatty et al., 1979) in the human. Although estimates of cell number vary among species and observers, granulosa cell replication is unquestionably a dominant feature of ovarian physiology. The earliest phase of follicular growth occurs in fetal or immature animals, during seasonal diapause, during pregnancy, and after hypophysectomy (Review, Richards, 1980) with no apparent relationship to the hormonal milieu. Although a case can be made fo a permissive role of hormones during this phase of follicular growth (Ross and Schreiber, 1978), the proximate regulators of this process must be chemical messages generated within the ovary. After antrum formation, there is a further 100- to 1000-fold increase in granulosa cell number per follicle (McNatty et al., 1979). During this period, FSH and estradiol cause a synergistic enhancement of granulosa cell multiplication in vivo (Rao et al., 1978). In addition, both estradiol and FSH (McNatty and Sawers, 1975; Thanki and Channing, 1976, 1978) have previously

*Department of Pediatrics, University of Oklahoma School of Medicine, Oklahoma City, Oklahoma 73190.
†Laboratory of Biochemical Pharmacology, NIAMDD, National Institute of Health, Bethesda, Maryland 20014.

been found to be mitogenic for cultured granulosa cells. However, these data do not exclude the participation of "non-hormonal" mitogens in the regulation of this second phase of follicular growth.

By far, the most potent stimulators of granulosa cell replication evaluated in vitro are the low molecular weight peptides epidermal growth factor (EGF) and fibroblast growth factor (FGF) (Gospodarowicz et al., 1977a, 1977b; Gospodarowicz and Bialecki, 1979). In addition, requirements for insulin (Channing et al., 1976; Orly et al., 1980), transferrin (Orly et al., 1980), as well as steroids and thyroxine (Channing et al., 1976) have been defined for optimal maintenance of granulosa cells under in vitro conditions. However, a physiological role for these compounds in vivo has remained uncertain. Recently, Sertoli cells, which function in a fashion analogous to granulosa cells, have been found to contain a polypeptide mitogen (Feig et al., 1980). In addition, recent recombination experiments have suggested that thecal elements secrete substances which enhance granulosa cell proliferation in vitro (McNatty et al., 1980). Thus, a regulatory role for endogenous mitogens, perhaps produced within the follicle seemed plausible. The present studies are concerned with a search for such mitogens in porcine follicular fluid and spent culture medium from porcine granulosa cells. In part, we have employed the enzyme ornithine decarboxylase (ODC) as an easily quantified albeit indirect assay for potential mitogens. In addition, somatomedins, one type of potential growth regulator, have been identified by specific assays. The possibility that ODC and polyamines may serve as intracellular regulators in follicular growth is an additional hypothesis of these studies.

Regulation of Granulosa Cell Replication by Hormones, Serum, and Follicular Fluid

As discussed above, the influence of estradiol on granulosa cell replication in vivo and in vitro seems established. Although present evidence favors a direct mitogenic action for FSH in vitro, this effect has not been confirmed by all authors (Gospodorowicz et al., 1977a). The interpretation of FSH action is further confused by the finding that some gonadotropin preparations are contaminated with potent mitogens (Gospodarowicz, 1975), and by the knowledge that cAMP, the presumed mediator of FSH action causes a cessation of ovarian cell replication (Gospodarowicz and Gospodarowicz, 1975). Our own studies have supported a direct effect of FSH on granulosa cell replication which is additive to the effects of estrogens (Fig. 1). However, the possibility remains that such effects could be caused by potent growth promoting contaminants or by stimulation of granulosa cells to produce mitogens.

The mitogenic effects of serum appear additive to those of hormones in this system (Thanki and Channing, 1976; Fig. 1). We

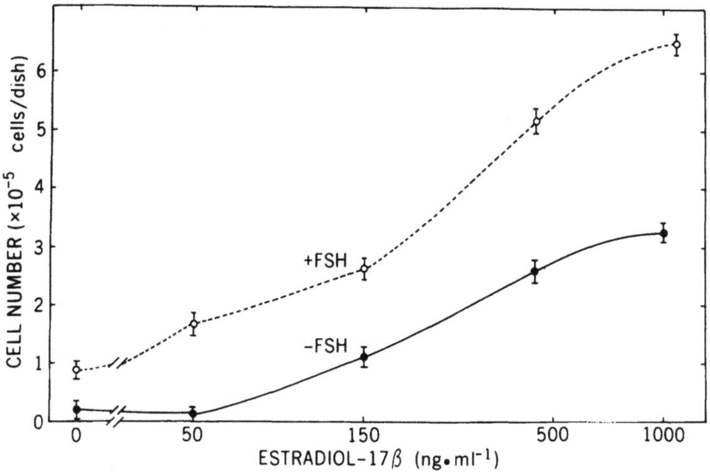

Fig. 1. Effect of estradiol-17β and oFSH (200 ng/ml) on replication of granulosa cells from 1-2 mm porcine follicles. Cells were grown in medium 199 with 10% FCS and enumerated by coulter counting after trypsinization.

have routinely employed fetal calf-serum (FCS) at 10% (v/v) for granulosa cell cultures. Peak mitogenic activity is exerted at approximately 20-30% (Hammond et al., unpublished). Porcine serum is roughly eqivalent to FCS in growth sustaining capabilities, and, in turn, follicular fluid from immature follicles appears similar in this regard (Fig. 2). In general, we have not employed charcoal-stripped serum since activated charcoal is known to bind insulin-like growth factors (Moses et al., 1979). However, the measured concentrations of estrogens in these supplements (Veldhius et al., 1979) have been low compared to the doses of estrogens required to stimulate cell growth, and the effects of estrogens and serum have been additive. To fully understand the effects of growth factors on granulosa cells, it will be necessary to maintain replicating cultures for some time under serum-free conditions. As shown in Fig. 2, we have been unable to achieve this goal to date with porcine cells. In part, because of this failure, we have turned to the assay of growth-related enzyme, ODC, to screen potential mitogens for effects on granulosa cells under serum-free conditions.

ACTIVATION OF GRANULOSA CELL ODC BY PUTATIVE MITOGENS

As indicated by Maudsley (1979), "An increase in ornithine decarboxylase is an invariant component of the integrated response of a cell stimulated to proliferate." Under most circumstances, ODC controls the synthesis of polyamines, compounds which have been linked biochemically to numerous aspects of the growth response

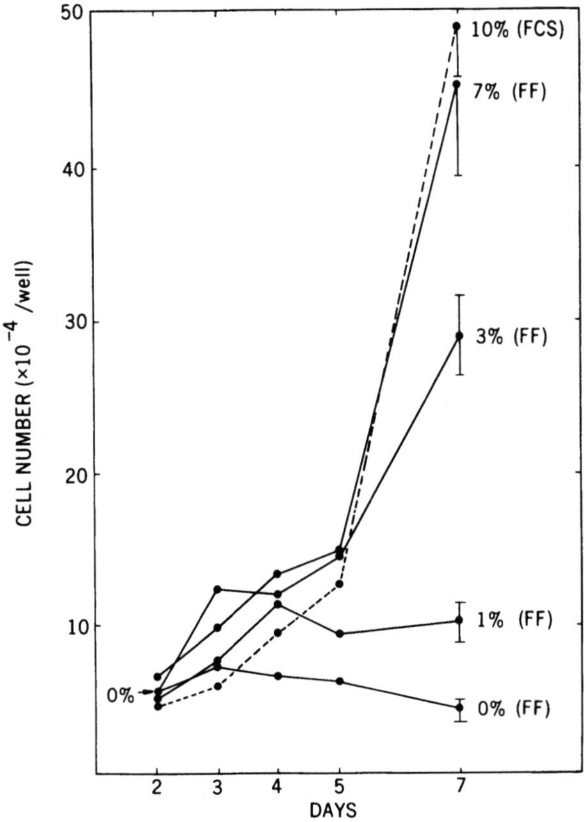

Fig. 2. Comparative effects of fetal calf serum and follicular fluid on growth of granulosa cells from small (1-2 mm) porcine follicles. Cell counts as in Fig. 1.

ranging from DNA to protein synthesis (Review — Cohen, 1971; Janne et al., 1978). The use of ODC as an assay for suspected granulosa cell mitogens is supported by our own in vitro data (reviewed below). However, the possibility that ODC may also be involved in aspects of granulosa cell function other than replication has not been excluded.

The results derived from numerous experiments and several publications concerning the effects of potential mitogens on ODC have been summarized in Fig. 3. Each of the agonists depicted has saturable dose-dependent effects on enzyme activity. In addition, each of the mitogens appears to have a discrete mode of action in that maximally effective concentrations of each are additive or supra-additive when combined. MSA and follicular fluid (FF1) form an exception to this rule since this combination was not additive in sev-

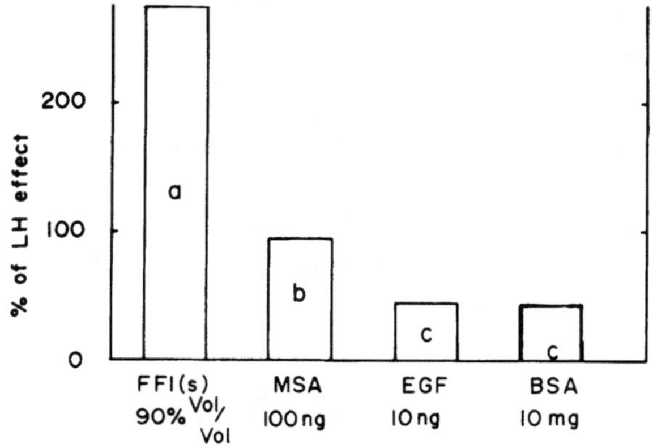

Fig. 3. Average effects of "non-hormonal" factors on ornithine decarboxylase in granulosa cells from small follicles. Results have been normalized in terms of effects of LH which was used as an internal control. For details regarding various agonists see the following references: a. (Veldhuis et al., 1979): b. (Veldhuis and Hammond, 1979b); c. (Osterman and Hammond, 1979). Osterman and Hammond (1979) found that the activity of bovine albuminum was due to low molecular weight contaminants.

eral experiments (Veldhuis and Hammond, unpublished), possibly indicating some overlap in function between the factors in FFl and MSA (see below). Porcine serum (data not shown) is approximately equipotent with FFl in stimulating ODC, but at submaximal concentrations the two are more than additive (Veldhuis et al., 1979) indicating some divergence in the active principles contained in the two mixtures. Finally, each of the compounds or mixtures depicted in Fig. 3 has been shown to have additive or synergistic effects when used in combination with cyclic nucleotides or the cAMP-dependent stimulators of ODC, LH, FSH, catecholamines, and prostaglandins (discussed in the final section). These data indicate that the effects of the putative mitogens in Fig. 3 are independent of cAMP and of these hormonal agents. Additional support for this conclusion is shown in Fig. 4 which contrasts the effects of prostaglandin E_2 and follicular fluid on ODC and cAMP. FFl from small follicles was unable to stimulate granulosa cell cAMP levels in agreement with the studies of Rigby et al. (1979).

Ultrafiltration studies of the ODC-stimulating activity in follicular fluid revealed that the active principles were restricted to 2 of the 6 fractions generated, one nominally <10,000 daltons and the other nominally >300,000. In the light of more recent studies (discussed below), it is tempting to relate these two fractions to

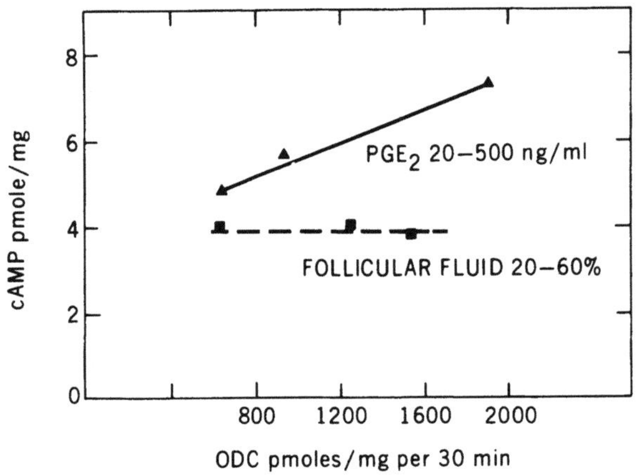

Fig. 4. Dose-response curves for a cAMP-dependent (prostaglandin E_2) and independent (follicular fluid) stimulator of ODC. Cells were maintained in suspension culture for 4 h and assayed by methods for cAMP and ODC descrived by Osterman et al. (1978) and Osterman and Hammond (1977) respectively.

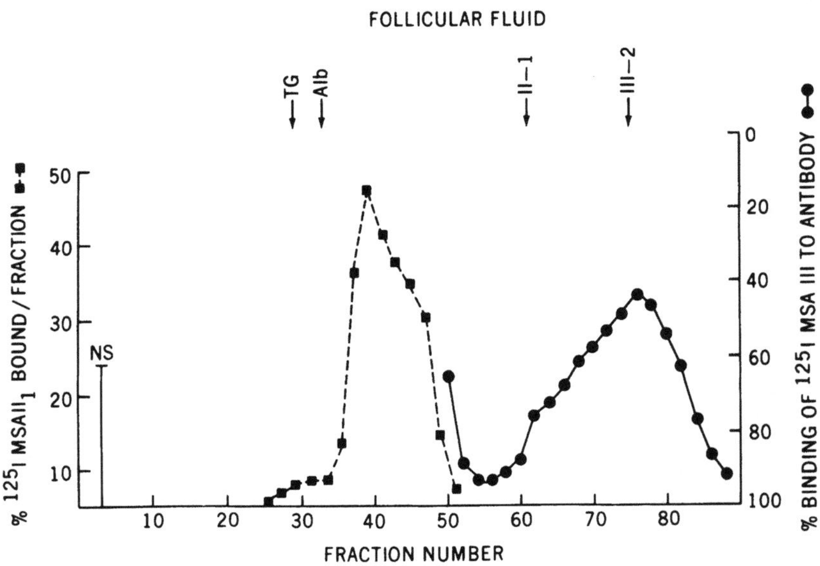

Fig. 5. Chromatography of FF1 from 1-2 mm porcine follicules on Sephadex G-100. A high molecular weight somatomedin binding protein — was assayed by direct binding of ^{125}I MSA II (method of Moses et al., 1979), and free MSA-like immunoreactivity was detected in the low molecular weight region (method of Moses et al., 1980a). Figure reproduced from Hammond (1981) with permission.

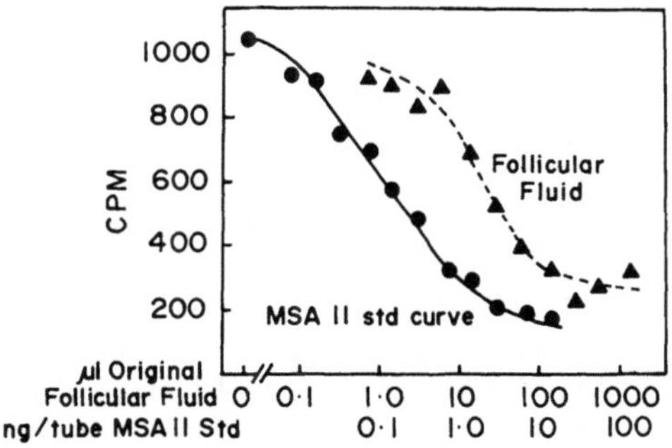

Fig. 6. Displacement of ^{125}I MSA from rat liver membrane somatomedin receptors by acid extracted low molecular weight fraction of follicular fluid. By comparison with dose-response of MSA II standards. FF1 from small follicles contained 660 ng of somatomedin/ml. (Method of Rechler et al., 1977). Figure reproduced from Hammond (1981) with permission.

the somatomedins and their binding proteins. However, our studies suggest discrete dose-response curves for the effect of the two fractions on ODC, and no precursor-product relationship has been demonstrated.

SOMATOMEDINS IN FOLLICULAR GROWTH

As discussed above, the characterization of stimulators of ODC in porcine FF1 disclosed a non-cyclic AMP-dependent stimulator in the <10,000 dalton eluate from ultrafiltration studies. We investigated the possibility that this activity might be conferred by a previously described mitogen in this molecular weight range. MSA was one contender, especially since we had shown this somatomedin to be active in stimulating granulosa cell ODC at exceptionally low concentrations (Veldhuis and Hammond, 1979b). Somatomedins, in general, and MSA in particular, have been found to stimulate replication in other culture systems (Temin et al., 1974). In 3T3 cells (Nissley et al., 1976) and chicken embryo fibroblast (Smith and Stange, 1980), MSA has been found to stimulate ODC, polyamine synthesis and DNA replication in a coordinated time-dependent fashion. We have recently shown MSA to be mitogenic for granulosa cells (Yoshida, Veldhuis, and Hammond, unpublished) and several laboratories (above) have noted a requirement for high concentrations of insulin for maintenance of granulosa cell cultures. At the concentrations utilized, insulin effects could be mediated via somatomedin receptors (Dulack and Temin, 1973; Nissley and Rechler, 1978).

Gel filtration studies on FF1 (Fig. 5) showed a compound reacting with anti-MSA antiserum which co-eluted with MSA II. The broad elution peak may indicate some heterogeneity; however, MSA itself is now recognized to be a family of peptides with several bands on polyacrylamide gels electrophoresis and a molecular weight range from 7,100 to 16,000 (Moses et al., 1980b). Although the reacting species was discriminated from MSA by nonparallel radioimmunoassay dose-response curves, the degree of cross reactivity encountered would be consistent with only a few recognized molecular species including other somatomedins but not insulin, proinsulin, EGF, FGF, or nerve growth factor (Moses et al., 1980a). Eluting prior to the MSA-peptides in Fig. 5 is a somatomedin binding protein detected by the direct binding of ^{125}I-MSA (Moses et al., 1979). Such binding proteins have been encountered almost uniformly in association with circulating somatomedins (Moses et al., 1979). The concentration of somatomedins in FFl from small follicles was assayed at 660 ng/ml using a rat liver plasma membrane MSA receptor preparation (Fig. 6). These levels approximate those in serum. These concentrations of somatomedin are well above the 1 ng/ml minimally effective dose for MSA effects on ODC or replication in granulosa cells (Veldhuis and Hammond, 1979b; Yoshida et al., unpublished).

The data above suggest a potential effect of such growth factors on granulosa cells in vivo. Such effects could be exerted in a permissive fashion by somatomedins which have filtered into the follicle from serum. However, they may also serve as a potential control point for intraovarian regulation of follicular growth. Preliminary evidence in favor of the latter hypothesis has been gleaned from easily detectible levels of somatomedins measured in medium from granulosa cells cultured for several days under serum-free conditions. Furthermore, two experiments have shown higher levels of somatomedin activity in medium from cultures treated with FSH and estradiol, suggesting a potential role for somatomedins in the mitogenic actions of these hormones. However, the levels of somatomedins in medium are low, and the release of previously absorbed or internalized somatomedins by these cultured cells has not been rigorously excluded. Culture of porcine granulosa cells under serum-free conditions long enough to eliminate this possibility has not been feasible to date.

Potential Regulatory Functions of ODC and Polyamines in the Ovarian Follicle

The data summarized in a previous section indicating stimulation of ODC in granulosa cells by known mitogens was consistent with the purported role of this enzyme in the action of EGF (Stastny and Cohen, 1970) and MSA (discussed above). In addition data from numerous rapidly proliferating cell types has shown a temporal relationship of peaks of ODC activity and/or polyamine levels with cell replication (Russel, 1973), and inhibitors of polyamine biosynthesis have almost uniformly blocked the response to mitogenic stimuli (Janne et al., 1978). By these criteria, ODC can also be linked to granulosa cell replication. The provision of serum to granulosa

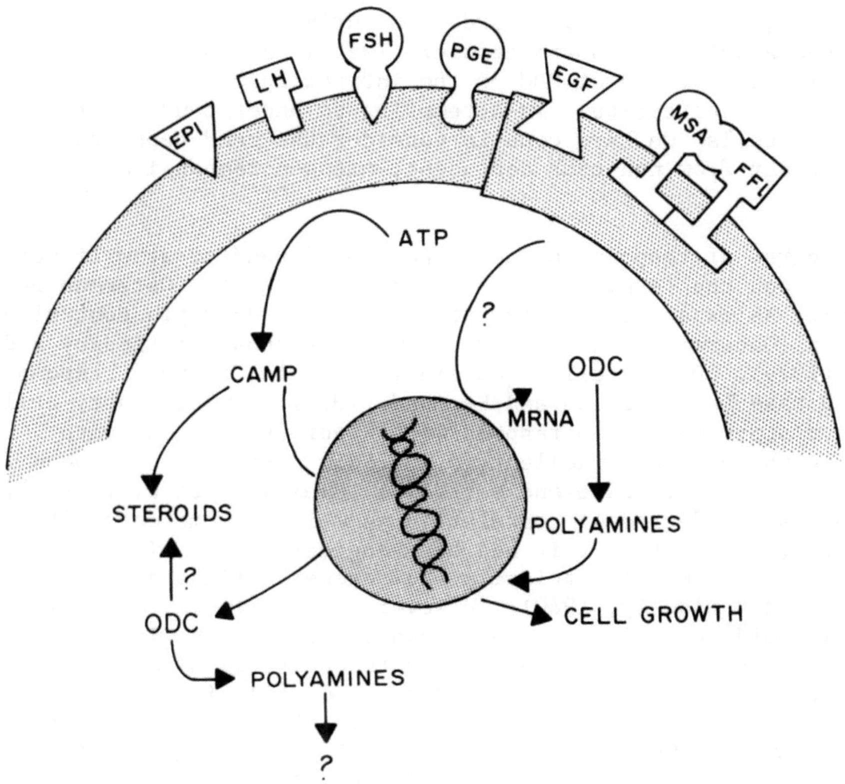

Fig. 7. Schematic representation of the control of ODC in the granulosa cell. cAMP dependent and independent pathways are contrasted. See text for details.

cells in monolayer is a condition of their ODC response and replication (Veldhuis and Hammond, unpublished), and blockade of the ODC response under these conditions with the specific non-competitive inhibitor, alpha methyl di-fluoroornithine markedly compromises the ability of granulosa cells to replicate (Veldhuis and Hammond, 1979a). In the latter studies the provision of putrescine, the compound distal to the blocked step, reversed the antimitotic effect of the inhibitor, arguing against a non-specific or toxic effect of the drug.

Nonetheless, it is also clear, at least in the ovary, that ODC activation is not specific for growth promoting stimuli. As shown schematically in Fig. 7 several cAMP-dependent hormonal agonists also stimulate ODC in the granulosa cell: FSH and LH (Osterman and Hammond, 1977), prostaglandins (Osterman and Hammond, 1978) and catecholamines (Veldhuis et al., 1980). With the exception of FSH, these factors, are generally regarded as antimitotic for ovarian cells (Gospodarowicz and Gospodarowicz, 1975), and this data prohibits a glib acceptance of ODC as the intracellular mediator of granulosa cell replication. In attempting to understand the role of ODC in these cells, we have come to consider the cAMP-dependent pathway as distinct from non-cAMP dependent, mitogen-responsive mechanisms.

The degree of response amongst the cAMP dependent stimuli seems predicated upon the levels of cAMP generated (Fig. 4) (Osterman et al., 1978). As expected from our studies of gonadotropin control of cAMP levels (Osterman et al., 1978) as well as those of Lindsey and Channing (1979), ODC in cells from small follicles is more responsive to FSH than LH (Osterman and Hammond, 1977). In contrast, cells from preovulatory follicles respond more vigorously to LH; with these cells there is an excellent (f = +.981) correlation between ODC generation and progesterone secretion (Sweinberg et al., 1981). Based on a similar degree of correlation as well as some data with indirect inhibitors of ODC, it had been suggested that, in the rat ovary in vivo, ODC might regulate steroidogenesis (Guha and Janne, 1977; Murthy and Midgely, 1980). However, in our in vitro systems, in which steroid secretion can be precisely correlated with ODC activity, a dissociation of these two events can be obtained under several circumstances: poorly differentiated cells in suspension respond to gonadotropins with an increase in ODC but no stimulation of steroidogenesis (Osterman and Hammond, 1977); in highly differentiated cells the responses of ODC and steroidogenesis are temporally dissociated, and short term inhibition of ODC with either direct or indirect inhibitors fails to inhibit steroidogenesis (Sweinberg et al., 1981); inhibition of ODC for 48 h produces a block in cell multiplication but enhances steroidogenesis (Veldhuis and Hammond, 1979). In short we feel that control of steroidogenesis in granulosa cells by ODC is unlikely.

We have further characterized the cAMP-dependent and independent pathways for ODC regulation through the use of inhibitors of protein and RNA synthesis (Veldhuis et al., 1979; Veldhuis et al., 1981, unpublished), microtubular and microfilament modifiers (Veldhuis and Hammond, submitted 1981), and inhibitors of calcium transport (Veldhuis and Hammond, 1981). In each instance, these inhibitors blocked ODC stimulation by both cAMP-dependent and independent mechanisms and failed to disclose differences in the manner in which these two pathways control ODC. An even more challenging question regards the function subserved by the two mechanisms. In part the latter question may be answered by more detailed measurements of the polyamines themselves. Present data suggests that LH stimulation of the rat ovary results in increased putrescine levels but little or no change in spermine or spermidine concentrations (Anderson and Henningsson, 1980; Fozard et al., 1980). Since the latter, more distal, polyamine have been found critical for mitogenic activity in several

Table 1

Concentrations of Polyamines were determined in Pooled Samples of Follicular Fluid from Porcine Follicles of Various Sizes: FFl(s), 1-2 mm; FFl(m), 3-5 mm; and FFl(l), 6-10 mm. The Analyses were performed on Durham Amino Acid Analyzer Modified for Polyamine Determinations.

	POLYAMINE CONCENTRATIONS (M x 10^6)		
	PUT.[A]	SP.[B]	SPD.[C]
PIG SERUM	4	2	Tr
FFl(s)	12	19	16.5
FFl(m)	12	11	8.5
FFl(l)	17	9.5	5.0

(PUTRESCINE[A], SPERMINE[B], SPERMIDINE[C])

systems (Janne et al., 1978; Maudsley, 1979), the inability for LH to generate these compounds is consistent with its lack of mitogenic activity. It also follows from these data that ODC cannot be regarded as rate-limiting for polyamine synthesis under these conditions. In this regard we (Hammond and Pegg, unpublished) and others (Maudsley and Kobayashi, 1974) have found S-adenosyl methionine decarboxylase, a critical enzyme for spermine and spermidine synthesis, to be unresponsive to LH. In contrast, rapid tissue growth has generally been associated with a uniform and coordinated increase in polyamine concentrations (Janne et al., 1978).

As an indication of the overall activity of the polyamine pathway in vivo, we measured polyamine levels in FFl. Follicular fluid from small porcine follicles (Table 1) contains high levels of all the polyamines with a predominance of spermidine and spermine. With increasing follicular size and differentiation (and the emergence of LH control) an apparent block between putrescine and the distal polyamines emerges. Parenthetically, these levels are extremely high values for biological fluids, and they may exert a substantial regulatory influence in and of themselves. In the light of the data above we currently hypothesize that growth promoting stimuli lead to a uniform increase in polyamine synthesis which influences cell replication. By contrast cAMP-dependent hormonal agents may stimulate putrescine levels alone. The significance of this increase in putrescine is unknown. The level of control of the two biosynthetic mechanisms seems distal to ODC. Studies testing this possibility are underway. However, even if this hypothesis is substantiated, critical information will remain unknown. Few details concerning the biochemistry of polyamine action are certain. No consistent receptor mechanism has been delineated, and it seems likely that their chemical effects derive from their structure as strong polycations. Any specificity in their actions is probably dependent on their ephemeral nature and the precise temporal (and/or spatial) control of their liberation in relation to other biochemical processes. This "compartmentalization" may continue to impede comprehension of their biological effects.

SUMMARY

A possible role for polypeptide growth factors in granulosa cell replication has been investigated. Follicular fluid was found to be comparable to serum in stimulating replication of porcine granulosa cells in vitro. To characterize the growth promoting activity in follicular fluid and to screen other compounds, the growth-related enzyme ornithine decarboxylase (ODC) was used as an indirect indicator of potential mitogenic activity. FFl was found to have ODC-stimulating activity in two molecular weight ranges by ultrafiltration — one <10,000 daltons and one >300,000. The effects of FFl were additive to those of the recognized hormonal stimulators of ODC as well as to the activity of the cyclic nucleotides. The previously characterized mitogen epidermal growth factor and the somatomedin, multiplication stimulating activity, also stimulated ODC. Specific assays for somatomedins in follicular fluid showed substantial activity with molecular size <10,000 daltons. It would appear that the concentrations of somatomedins in these fractions are adequate to account for the ODC-stimulating activity. Preliminary studies suggest that somatomedins are secreted by cultured granulosa cells in vitro. These latter observations are consistent with a critical role for these peptides in follicular growth. However, at present, the mitogenic activity, ODC stimulating activity, and MSA-like activity we have identified in follicular fluid have not been shown to be identical. The availability of potentially blocking antisera to MSA may permit an approach to this question. Until these studies have been completed, the inter-relationships suggested in the present studies will remain speculative. Finally the potential role of ODC and polyamines as intracellular regulators of granulosa cell replication was reviewed. The data from the granulosa cell cultures is similar to that in other systems in which polyamines have been assigned an essential role. However, the multiplicity of substances which stimulate ODC has prohibited a simple or consistent interpretation. More detailed studies of polyamine concentratons may clarify this area, but a comprehensive understanding will require information regarding the molecular mechanism of polyamine action. This information is presently unavailable in any cell type.

ACKNOWLEDGEMENTS

The studies reported here were supported by NIH Grant HD10122, and Fogarty Senior International Fellowship (F06 TW 00528) to James M. Hammond and National Research Service Award (#F32HD05755) to Johannes D. Veldhuis. Some of the studies were completed during the research training experience of Sharon Sweinberg, a second year medical student at the Hershey Medical Center. Dr. Kohji Yoshida performed the experiment shown in Fig. 2. Patricia Klase and Eileen Wolck have provided excellent technical support. Jill Volksberg, Marlene Thompson and Kay Cassel have prepared the manuscript.

REFERENCES

Andersson, A. C., and Henningsson, S., 1980, Biosynthesis and accumulation of cadaverine and putrescine in rat ovary after administration of human chorionic gonadotrophin, Acta Endocrinol., 95:237.

Channing, C. P., Tsai, V., and Sachs, D., 1976, Role of insulin, thyroxin and cortisol in luteinization of porcine granulosa cells grown in chemically defined media, Biol. Reprod., 15:233.

Cohen, S. S., 1971, Introduction to the Polyamines, Prentice-Hall, Englewood Cliffs, New Jersey.

Dulack, N. C., and Temin, H. M., 1973, A partially purified polypeptide fraction from rat liver cell conditioned medium with mulitiplication-stimulating activity for embryo fibroblasts, J. Cell Physiol., 81:153.

Feig, L. A., Bellve, A. R., Erickson, N. H., and Klagsbrun, M., 1980, Sertoli cells contain a mitogenic polypeptide, Proc. Natl. Acad. Sci. (U.S.A.), 74:4774.

Fozard, J. R., Nellikunja, J. P., and Grove, J., 1980, Ovarian function in the rat following irreversible inhibition of L-ornithine decarboxylase, Life Sciences, 27:2277.

Gospodarowicz, D., 1975, Purification of a fibroblast growth factor from bovine pituitary, J. Biol. Chem., 250:2515.

Gospodarowicz, D., and Bialecki, H., 1979, Fibroblast and epidermal growth factors are mitogenic agents for cultured granulosa cells of rodent, porcine, and human origin, Endocrinology, 104:757.

Gospodarowicz, D., and Gospodarowicz, F., The morphological transformation and inhibition of growth of bovine luteal cells in tissue culture induced by luteinizing hormone and dibutyryl cyclic AMP, Endocrinology, 96:438.

Gospodarowicz, D., Ill, C.R., and Birdwell, C. R., 1977a, Effects of fibroblast and epidermal growth factors on ovarian cell proliferation in vitro. I. Characterization of the response of granulosa cells to FGF and EGF, Endocrinology, 100:1108.

Gospodarowicz, D., Ill, C.R., and Birdwell, C. R., 1977b, Effects of fibroblast and epidermal growth factors on ovarian cell proliferation in vitro. II. Proliferative response of luteal cells to FGF but not EGF, Endocrinology, 100:1121.

Guha, S. K., and Janne, J., 1977, Inhibition of ornithine decarboxylase in vivo in rat ovary, Biochem. Biophys. Res. Commun., 75: 136.

Hammond, J. M., 1981, Peptide regulators in the ovarian follicle, Australian J. Biol. Sci., 34:491.

Janne, J., Poso, H., and Raina, A., 1978, Polyamines in rapid growth and cancer, Biochim. Biophys. Acta, 473:241.

Lindsey, A. M., and Channing, C. P., 1979, Influence of follicular maturation upon the effect of ovine follicle stimulating hormone and luteinizing hormone upon cyclic AMP accumulation by isolated porcine granulosa cells, Biol. Reprod., 20:473.

Lintern-Moore, S., Peters, H., Moore, G. P. M., and Faber, M., 1974, Follicular development in the infant human ovary, J. Reprod. Fert., 39:53.

Maudsley, D. V., 1979, Regulation of polyamine biosynthesis, Biochem. Pharmacol., 28:153.

Maudsley, D. V., and Kobayashi, 1974, Induction of ornithine decarboxylase in rat ovary after administration of LH or hCG, Biochem. Pharmacol., 23:2697.

McNatty, K. P., Makris, A., De Grazia, C., Osathanondh, R., and Ryan, K. J., 1980, Steroidogenesis by recombined follicular cells from the human ovary in vitro. J. Clin. Endocrinol. Metab., 51:1286.

McNatty, K. P., Moore, Smith, D., Makris, A., Osathanondh, R., and Ryan, K. J., 1979, The microenvironment of the human antral follicle: inter-relationships among the steroid levels in antral fluid, the population of granulosa cells and the status of the oocyte in vivo and in vitro, J. Clin. Endocrinol. Metab., 59:851.

Mcnatty, K. P., and Sawers, R. S., 1975, Relationship between the endocrine environment within the Graafian follicle and the subsequent rate of progesterone secretion by human granulosa cells in vitro. J. Endocrinol., 66:391.

Moses, A. C., Nissley, S. P., Passamani, J., and White, R. M., 1979, Further characterization of growth hormone-dependent somatomedin-binding proteins in rat serum and demonstration of somatomedin-binding proteins produced by rat liver cells in culture, Endocrinology, 104:536.

Moses, A. C., Nissley, S. P., Short, P. A., and Rechler, M. M., 1980a, Immunological cross-reactivity of multiplication-stimulating activity polypeptides, Eur. J. Biochem., 103:401.

Moses, A. C., Nissley, S. P., Short, P. A., Rechler, N. M., and Podskalny, J. M., 1980b, Purification and characterization of multiplication-stimulating activity, Eur. J. Biochem., 103:387.

Murthy, V., and Midgely, A. R., Jr., 1980, Polyamine mediation of hCG-stimulated progesterone production, 62nd Annual Meeting of the Endocrine Society (Abstract #12).

Nissley, S. P., Passamani, J., and Short, P., 1976, Stimulation of DNA synthesis, cell multiplication and ornithine decarboxylase in 3T3 cells by multiplication stimulating activity (MSA), J. Cell Physiol., 89:393.

Nissley, S. P., and Rechler, M. M., 1978, Multiplication-stimulating activity (MSA): A somatomedin-like polypeptide from cultured rat liver cells, Natl. Cancer Inst. Monogr., 48:167.
Orly, J., Sato, G., and Erickson, G. F., 1980, Serum suppresses the expression of hormonally induced functions in cultured granulosa cells, Cell, 20:817.
Osterman, J., Demers, L. M., and Hammond, J. M., 1978, Gonadotropin stimulation of porcine ovarian ornithine decarboxylase in vitro: The role of cyclic adenosine monophosphate, Endocrinology, 103:1718.
Osterman, J., and Hammond, J. M., 1977, FSH and LH stimulation of ornithine decarboxylase activity: studies with porcine granulosa cells in vitro, Endocrinology, 101:1335.
Osterman, J., and Hammond, J. M., 1978, Prostaglandin stimulation of ovarian ornithine decarboxylase in vitro, Biochem. Biophys. Res. Comm., 83:794.
Osterman, J., and Hammond, J. M., 1979, Effects of epidermal growth factor, fibroblast growth factor, and bovine albumin on ornithine decarboxylase activity of porcine granulosa cells, Horm. Metab. Res., 11:484.
Rao, M. C., Richards, J. S., and Midgley, A. R., Jr., 1978, Hormonal regulation of cell proliferation in the ovary, Cell, 14:71.
Rechler, M. M., Poskalny, J. M., and Kissley, S. P., 1977, Characterization of the binding of multiplication stimulating activity to a receptor for growth polypeptides in chick embryo fibroblasts, J. Biol. Chem., 252:3898.
Richards, J. S., 1980, Maturation of ovarian follicles: interaction of pituitary and ovarian hormones on follicle cell differentiation, Physiol. Rev., 60:51.
Rigby, B. W., Ledwitz-Rigby, F., and Hunzicker-Dunn, M., 1980, Follicular fluid depression of ovarian granulosa cell adenylyl cyclase activity, Mol. Cell. Endocrinol., 17:17.
Ross, G. T., and Schrieber, J. R., 1978, The ovary, in: "Reproductive Endocrinology," S. S. C. Yen and R. B. Jaffe, eds., Saunders, Philadelphia.
Russel, D. N., 1973, Polyamines in growth — normal and neoplastic, in: "Polyamines in normal and neoplastic growth," D. H. Russel, ed., pp. 1-13, Raven Press, New York.
Schomberg, D. W., Stouffer, R. L., and Tyrey, L., 1976, Modulation of progestin secretion in ovarian cells by 17β-hydroxy-5α-androstan-3 one (dihydrotestosterone): a direct demonstration in monolayer culture, Biochem. Biophys. Res. Commun., 68:77.
Smith, G. L., and Stange, A. W., 1978, Polyamines and the stimulation of DNA synthesis in chicken embryo fibroblasts by purified multiplication-stimulating activity, Life Sciences, 23: 1871.
Stastny, M., and Cohen, S., 1970, Epidermal growth factor. IV. The induction of ornithine decarboxylase, Biochim. Biophys. Acta, 204:578.

Sweinberg, S., Veldhuis, J. D., Klase, P. A., and Hammond, J. M., 1981, Evidence for dissociated hormonal regulation of cellular progesterone and cytosalic ornithine decarboxylase (ODC) activity in isolated, highly differentiated granulosa cells in vitro, Program of the Endocrine Society, Abstract #503.

Temin, H. M., Smith, G. L., and Dulak, N. C., 1974, Control of multiplication of normal and Rous sarcoma virus-transformed chicken embryo fibroblasts by purified multiplication-stimulating activity with nonsuppressible insulin-like and sulfation factor activities, in: "Control of Proliferation in Animal Cells," B. Clarkson and R. Baserga, eds., pp. 19-26, Cold Spring Harbor Laboratory, New York.

Thanki, K. H., and Channing, C. P., 1976, Influence of serum, estrogen, and gonadotropin upon growth and progesterone secretion by cultures of granulosa cells from small porcine follicles, Endocrine Res. Commun., 3:319.

Thanki, K. H., and Channing, C. P., 1978, Effects of follicle-stimulating hormone and estradiol upon progesterone secretion by porcine granulosa cells in tissue culture, Endocrinology, 103: 74.

Veldhuis, J. D., Demers, L. M., and Hammond, H. J., 1979, Regulation of ornithine decarboxylase in isolated granulosa cells in vitro by constituents of follicular fluid, Endocrinology, 105:1143.

Veldhuis, J. D., and Hammond, J. M., 1979a, Role of ornithine decarboxylase in granulosa-cell replication and steroidogenesis in vitro, Biochem. Biophys. Res. Commun., 91:770.

Veldhuis, J. D., and Hammond, J. M., 1979b, Multiplication stimulating activity regulates ornithine decarboxylase in isolated porcine granulosa cells in vitro, Endocrine Res. Commun., 6:299.

Veldhuis, J. D., and Hammond, J. M., 1981, Role of calcium in the modulation of ornithine decarboxylase activity in isolated porcine granulosa cells in vitro, Biochem. J., 196:795.

Veldhuis, J. D., Harrison, T. S., and Hammond, J. M., 1980, β-2-adrenergic stimulation of ornithine decarboxylase activity in porcine granulosa cells in vitro, Biochim. Biophys. Acta, 627: 123.

COMPARISON OF GRANULOSA AND SERTOLI CELLS AT VARIOUS STAGES OF MATURATION: SIMILARITIES AND DIFFERENCES

Irving B. Fritz

Banting and Best Department
of Medical Research
University of Toronto
112 College Street
Toronto, Ontario, M5G 1L6, Canada

I. INTRODUCTION

Homologies between granulosa cells of the ovary and Sertoli cells of the seminiferous tubule have often been noted (for review, see Ross and Lipsett, 1978). Sertoli cells and granulosa cells are thought to share a common embryological origin (Gilman, 1948), most probably from the surface epithelium of the gonadal ridge. If this is valid, the indifferent gonad contains pre-granulosa or pre-Sertoli cells which have a common precursor cell. The difficulties in ascribing specific origins to cells during differentiation of the gonads have been reviewed by others (Jost et al., 1970, 1973, 1974; Merchant, 1975; Gondos, 1977).

We have been impressed by several biochemical similarities between granulosa cells and Sertoli cells, which are particularly evident before these cells undergo terminal differentiation. Some of the similarities include the following: (1) they each have specific receptors for FSH and androgens; (2) each cell type produces and secretes several apparently identical proteins (including plasminogen activator, proteoglycans, and inhibin); (3) each cell type synthesizes estrogens from testosterone, dependent on FSH stimulation of aromatase activity; and (4) both types of gonadal somatic cells influence the development of adjacent germinal cells.

In this paper, we shall focus upon the properties of Sertoli cells, with emphasis on information recently acquired in our laboratory on the modulation of the production and secretion of plasminogen activator, and on the nature of glucose metabolism by Sertoli cells in culture. We shall then compare properties in Sertoli cells with those of granulosa cells reported in the literature, and we shall discuss these observations in relation to the nature of somatic cell-germ cell interactions in testis and in ovary.

II. CONTROL OF THE SYNTHESIS AND SECRETION OF PLASMINOGEN ACTIVATOR BY SERTOLI CELLS

A. Influence of Various Hormones and Drugs on the Production of Plasminogen Activator Activity by Sertoli Cells in Culture

Primary cultures of Sertoli cells, prepared from testes of 10 to 30-day-old rats, produce plasminogen activator, and release this protease into the medium. Plasminogen activator activity is increased in cells stimulated by FSH, dibutyryl cyclic AMP (dbcAMP), or cholera toxin (Lacroix et al., 1977, 1979, 1980). However, we have not detected any enhancement in cells treated with insulin (up to 10 mU/ml); hCG (2 μg/ml); testosterone (up to 5×10^{-7} M); prostaglandin E_1, E_2, or $F_{1\alpha}$ (up to 10^{-6} M); epidermal growth factor (up to 10 ng/ml) or dibutyryl cyclic GMP (up to 0.1 mM) (Lacroix et al., 1979 and unpublished observations). These results in Sertoli cell preparations are similar to those reported for granulosa cells (Strickland and Beers, 1976; Strickland, this volume), with the exceptions that prostaglandins and hCG did stimulate the formation of plasminogen activator activity by granulosa cells.

Plasminogen activator activity has been measured quantitatively by following the degradation of ^{125}I-fibrin, dependent upon the addition of plasminogen (Strickland and Beers, 1976), and it has been assessed qualitatively by the fibrin-agar overlay technique (Beers et al., 1975). The secretion of plasminogen activator activity by Sertoli cells but not by peritubular myoid cells, and the stimulation of activity by FSH, are readily observed with the latter technique (Fig. 1).

Both basal and stimulated rates of production are dependent upon de novo synthesis of RNA and protein, since the addition of actinomycin D, cycloheximide, or puromycin inhibits plasminogen activator production by Sertoli cells. Removal of these inhibitors is followed by normal rates of formation of plasminogen activator activity by Sertoli cells during subsequent periods of culture in the presence of dbcAMP, indicating that cells remain viable after treatment with inhibitors of RNA and protein synthesis at concentrations employed (Lacroix et al., 1979). The ED_{50} values for maximal stimulation of plasminogen activator activity by FSH (Sairam 1528C2) is 8 ng/ml; that for dbcAMP is 0.08 mM; and the ED_{50} for cholera toxin is 4.4 pg/ml (Lacroix et al., 1979, 1980). Addition of the phosphodiesterase inhibitor methylisobutyl xanthine also augments plasminogen activator activity formation.

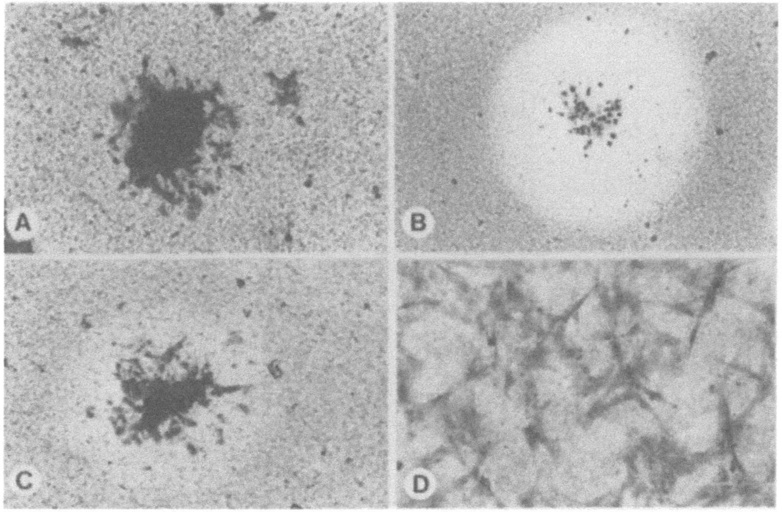

Fig. 1. Assay of plasminogen activator activity by the fibrin-agar overlay technique in Sertoli and myoid cell cultures. Plasminogen was preent in the fibrin-agar overlay medium in all preparations shown except for that in panel A. Magnification ×190. A) Sertoli cell aggregates previously cultured for 48 h in the presence of FSH (5 μg/ml) and then overlaid with fibrin-agar medium. Cells were cultured for 20 h before staining with Giemsa and Coomassie blue. B) Sertoli cell aggregates previously cultured for 48 h in the presence of FSH (5 μg/ml) and then overlaid with the fibrin-agar medium containing plasminogen. Cells were cultured for 20 h before staining. C) Sertoli cell aggregates previously cultured for 48 h in MEM alone and then overlaid with fibrin-agar medium containing plasminogen. Cells were incubated for 20 h before staining. D) Subcultured peritubular myoid cells cultured for 48 h in MEM containing 10% calf serum and then overlaid with the fibrin-agar medium containing plasminogen. Cells were incubated for 24 h before staining. From Lacroix et al., 1977. Reprinted with permission.

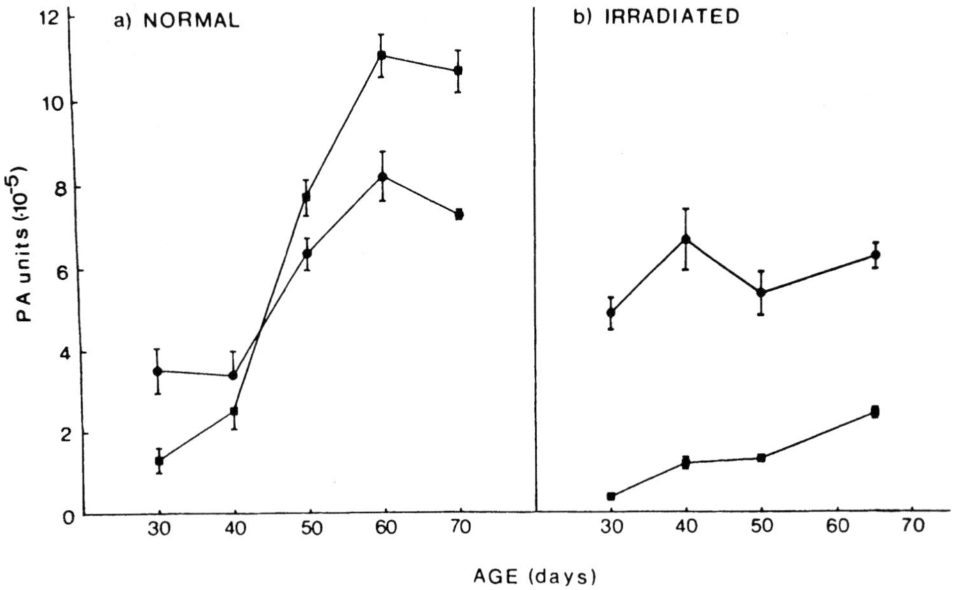

Fig. 2. Germ-cell-depleted animals were obtained by irradiating pregnant rats on their 19th day of gestation. The males from the ensuing litters were raised to the appropriate age required in the experiments. The degree of germ-cell depletion was monitored by histological examination. The testes were perfused, decapsulated and the tissue homogenized. Extracts were snap frozen in liquid nitrogen and kept at -60°C until the day of the plasminogen activator assay. Plasminogen activator activity/testis (■) and plasminogen activator activity/gram of testicular tissue (●) are shown. Each point is the average activity (± SEM) calculated from four testes. From Lacroix et al., 1979. Reprinted with permission.

B. Influence of Cellular Association Patterns in the Seminiferous Tubule on Plasminogen Activator Production by Sertoli Cells

Sertoli cells have been shown to be the primary source of testicular plasminogen activator activity (Lacroix et al., 1979). Changes observed in levels of testicular plasminogen activator during development, however, are not completely consistent with this interpretation. The greatest increases in total activity per testis and in specific activity occurs during the period between 40 and 60 days of age in testes obtained from normal Wistar rats (Fig. 2a). This is the time during which the considerable increase in testis

Table 1. Relative Levels of Plasminogen Activator Activity in Testicular Cells*

Peritubular Myoid Cells	1
Leydig Cells	1
Germinal Cells	50-200
Sertoli Cells (non-stimulated)	2,000
Sertoli Cells (maximally stimulated)	40,000

*Data are expressed in relative plasminogen activator activity units per mg proteins (Lacroix and Fritz, in preparation).

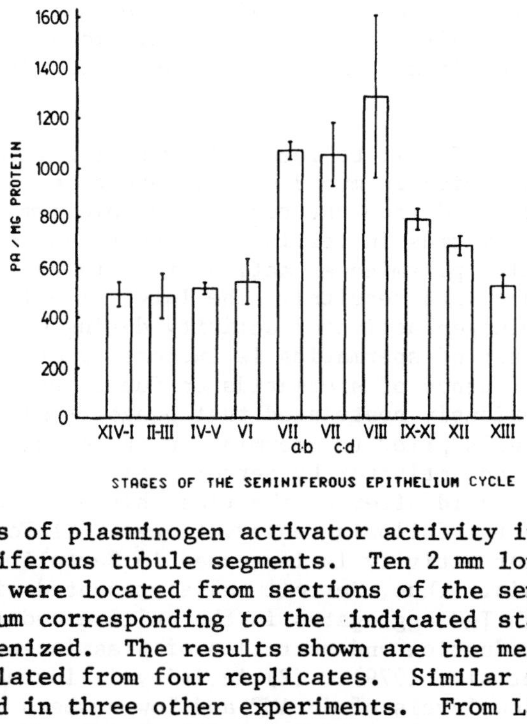

Fig. 3. Levels of plasminogen activator activity in extracts of seminiferous tubule segments. Ten 2 mm long tubule segments were located from sections of the seminiferous epithelium corresponding to the indicated stages and were homogenized. The results shown are the mean values (± S.D.) calculated from four replicates. Similar results were obtained in three other experiments. From Lacroix et al., 1981. Reprinted with permission.

mass is associated primarily with an increase in the number of spermatids (de Jong and Sharpe, 1977). The volume of Sertoli cells increases during this same period, but the number of Sertoli cells per testis remains relatively constant. In animals rendered germ-cell-deficient by irradiation in utero, the specific activity of plasminogen activator in testes from 30-day-old rats was higher than that in testes from normal animals of the same age (Fig. 2b). This seems reasonable if Sertoli cells are the primary source of testicular plasminogen activator activity, since Sertoli cells would constitute a higher percentage of the total population of cells in germ-cell depleted testes. The testicular plasminogen activator specific activity does not change appreciably during subsequent periods of development in irradiated rats (Fig. 2b).

Total amounts per testis increase during development in germ-cell-depleted animals from less than $3 \cdot 10^4$ units at 30 days of age to approximately $2 \cdot 10^5$ units per testis in 65-day-old rats. In contrast, total amounts per testis in normal rats increase from approximately $1 \cdot 10^5$ plasminogen activator activity units per testis at age 30 days of age to over $11 \cdot 10^5$ units in testes from 60-day old rats (Fig. 2). In summary, normal testes from adult rats have a plasminogen activator specific activity twice that of germ-cell-depleted testes from adult rats, and the total amounts per normal testis are about five times higher than those in irradiated testes (Fig. 2).

These data are difficult to reconcile with low levels of plasminogen activator activity in myoid cells, Leydig cells and germinal cell populations isolated from the testis and epididymis (Table 1). We estimate that spermatids in testis could contribute no more than 5% of total testicular plasminogen activator activity (Lacroix et al., 1979, and unpublished results). The large rise in total levels of testicular plasminogen activator activity during development at a time when the number of spermatids is increasing dramatically could reflect an influence of spermatids on functions of Sertoli cells. We have previously postulated that phagocytosis of residual bodies by Sertoli cells prior to spermiation may enhance the production of plasminogen activator by Sertoli cells. The only supporting evidence we could offer at the time this hypothesis was advanced consisted of the following observations: (1) Formation of plasminogen activator activity is increased in Sertoli cells engaged in phagocytosis. Thus, Sertoli cells presented with either dead _E. coli_ or with IgG aggregates in the culture medium have increased rates of production and secretion of plasminogen activator activity (Lacroix et al., 1979). (2) Activity is increased synergistically when low levels of dbcAMP and low numbers of autoclaved _E. coli_ or low levels of IgG aggregates are added to Sertoli cells in culture (Lacroix et al., 1979, and unpublished data).

We have recently measured plasminogen activator activity produced and released into the medium by defined segments of seminiferous tubules maintained in organ culture. Segments from various stages of the cycle of the seminiferous epithelium have been prepared by the transillumination procedures of Parvinen and Vanha-Perttula (1972). Levels of plasminogen activator activity are highest in segments at stages VII and VIII (Fig. 3). These are the stages at which spermiation occurs (Lebond and Clermont, 1952), and

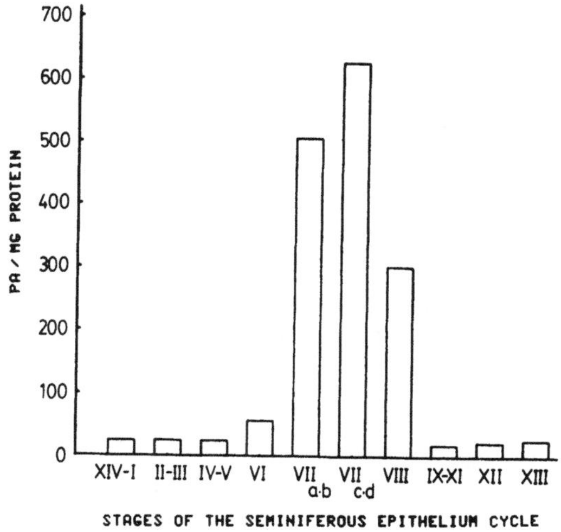

Fig. 4. Levels of plasminogen activator activity released by seminiferous tubule segments into culture medium. Five 2 mm long seminiferous tubule segments of the indicated stages were isolated and incubated in 0.2 ml MEM for 20 h. The results shown are the mean values from two experiments run in parallel (variation between experiments was never larger than 20%), and are expressed as units of plasminogen activator activity released into the medium per mg tissue/protein. From Lacroix et al., 1981. Reprinted with permission.

at which pre-leptotene spermatocytes committed to meiosis are enveloped by Sertoli cell cytoplasmic extensions prior to translocation of spermatocytes into the adluminal compartment (Russell, 1977). The increase in specific activity in extracts from segments at stages VII and VIII is modest (just over two-fold). However, the amounts of plasminogen activator activity produced and released into the medium during organ culture by segments at stages VII and VIII are more than an order of magnitude greater than amounts formed by segments from any other stage of the cycle (Fig. 4). Both basal and stimulated rates of formation of plasminogen activator activity are dependent upon de novo protein synthesis (Fritz et al., 1981; Lacroix et al., 1981).

These data demonstrate that cell association patterns in the seminiferous tubule greatly influence the rates of production of plasminogen activator activity by Sertoli cells, with highest levels formed by cells in stages VII and VIII. Localized proteolysis elicited by plasminogen activator secreted by Sertoli cells therefore appears to be modulated not only by hormones, but also by in-

Table 2. Rates of Conversion of Glucose to Metabolites by Sertoli Cells in Culture+

Additions to Basal Medium	Incorporation of U-[^{14}C]-Glucose Into:					Incorporation of 5[^{3}H]-Glucose into [^{3}H]-H_2O
	CO_2	Total Anions	Lipids	Glycogen	Sum of Products	
Basal	14.2 ± 0.7	474 ± 8	4.5 ± 0.3	2.3 ± 0.1	496	459 ± 4.2
Insulin	13.6 ± 0.2	462 ± 11	6.4 ± 0.1	2.1 ± 0.06	484	436 ± 6.2
oFSH	13.9 ± 1.3	457 ± 12	5.4 ± 0.6	2.1 ± 0.02	478	449 ± 14.9
dbcAMP	13.0 ± 0.3	469 ± 12	5.6 ± 0.2	2.1 ± 0.02	489	436 ± 6.2

+ Cells, cultured for 48 h in the presence or absence of insulin (10 mU/ml); oFSH (300 ng/ml of Sairam's S-1528C2); or dbcAMP (0.1 mM), were incubated for 60 min in Eagle's minimal essential medium containing U-[^{14}C]-glucose or 5-[^{3}H]-glucose (5.5 mM). Values are expressed as means of nmol/mg cell protein/h ± SEM for triplicate determinations in 4 separate experiments. For experimental details, the original publication should be consulted (Robinson and Fritz, 1981). (Reprinted with permission).

teractions with neighboring germinal cells. In considering these data, we offer the following rather fanciful set of speculations. Local proteolytic activity could be involved in the restructuring of the tubule which occurs when spermatocytes in early meiotic prophase are prepared for translocation into the adluminal compartment. These processes take place at stages VII and VIII of the cycle, the same stages at which spermiation occurs. The availability of plasminogen in the basal compartment would permit plasminogen activator to catalyze the formation of plasmin, a protease with a specificity much like that of trypsin. If the protease then degraded proteins in the extracullar matrix in the vicinity of boundary tissue immediately adjacent to Sertoli cells at stages VII and VIII, this would facilitate movement of Sertoli cell cytoplasmic extensions around the clone of spermatocytes in the basal compartment of stage VIII, and thereby initiate translocation of this nest of cells to the adluminal compartment.

From the combined data (Figs. 2-4), we conclude that the amount of plasminogen activator produced by Sertoli cells increases during development, and that this is dependent upon the stages of spermatogenesis, with greatest production by Sertoli cells in stages VII and VIII of the cycle of the seminiferous epithelium.

III. RATES OF GLUCOSE METABOLISM BY SERTOLI CELLS IN CULTURE

At phyisological levels of glucose in the medium (5.5 mM), glucose is utilized at the rate of approximately 500 nmol/mg protein/h by Sertoli cells in culture (Robinson and Fritz, 1981). Over 90% of the glucose utilized is converted to anions, most of which (77% to 90%) has been identified as lactate (Table 2). These rates are not influenced by added hormones of dbcAMP under experimental conditions thus far examined.

Recently, Jutte et al. (1981) have compared the effects of lactate and glucose on various metabolic processes in isolated rat pachytene spermatocytes and spermatids. Lactate addition increased rates of respiration, protein synthesis, and RNA synthesis by germ cells more than glucose did. Lactate was also more effective than glucose in preventing degeneration of spermatocytes and spermatids maintained in culture (Jutte et al., 1981). Nakamura et al. (1981) have shown that addition of lactate or pyruvate to preparations of spermatids increased the incorporation of labeled leucine into proteins by nearly 10-fold, whereas substitution of glucose of lactate or pyruvate stimulated protein synthesis by only two-fold. Lactate and pyruvate were also more effective substrates than glucose in maintaining ATP levels in spermatid preparations (Nakamura et al., 1981). It may therefore be concluded that lactate appears to be a preferred substrate by spermatocytes and spermatids.

These data suggest that the high rates of aerobic glycolysis by Sertoli cells (Table 2) may well prove to be physiologically important in supplying lactate and other substrates derived from glucose metabolism to germinal cells in the adluminal compartment. It has long been appreciated that in the follicle, lactate and other metabolites produced from glucose by granulosa cells are preferred substrates for oocyte metabolism and maturation (Biggers et al., 1967; Nilsson, 1974; Tsafriri et al., 1976; Ahrén et al., 1978; Hillensjö, this volume). In both ovary and testis, it therefore appears that somatic supporting cells provide intermediary products of glucose metabolism, which are the substrates preferred by germinal cells in the gonad.

IV. THE NATURE OF INTERACTIONS BETWEEN GERM CELLS AND SUPPORTING SOMATIC CELLS IN TESTIS AND IN OVARY

A. General

The control of spermatogenesis by FSH and androgens is most probably mediated via direct actions of hormones on testicular somatic cells, but not on germinal cells (for review, see Fritz, 1978). Germinal cell development is dependent upon the chemical microenvironment created by neighboring somatic cells and is thereby indirectly dependent on hormones which influence somatic cells.

Jost et al. (1973) presented data in support of the possibility that sexual differentiation, which occurs earlier in the gonad of the male fetus, is characterized by a close approximation of presumptive Sertoli cells to germinal cells in the gonad destined to become a testis, whereas presumptive granulosa cells in the gonad destined to become an ovary move away from nearby germinal cells. This is thought by Jost et al. (1973) to be associated in some manner with the inhibition of meiosis of germ cells in the testis by neighboring Sertoli cells, and the lack of inhibition of meiosis by less restricted granulosa cells in the ovary. The nature of putative meiosis inhibitory factor(s) from Sertoli cells is unknown. Its possible relationship to Müllerian duct inhibitory factor from Sertoli cells (Josso et al., 1977), or to the oocyte maturation inhibitory factor from granulosa cells (Channing et al., 1980; Tsafriri, this volume) remains totally speculative. Evidence demonstrating the existence of meiosis inducing and preventing factors in fetal testis and ovary cells has been presented by Byskov and Saxen (1976). Other interpretations are possible. For example, meiosis could be initiated in germ cells in both fetal ovary and testis, but the process in spermatocytes could be aborted more readily than atresia occurs during oocyte meiosis in the fetal gonad.

Supporting gonadal somatic cells could influence the development of neighboring germinal cells by secreting specific chemical agents. This possibility will be considered in detail. In addition direct cellular interactions could be involved. Several types of unique antigenic determinants have been observed on the surfaces of germinal cells more advanced than zygotene spermatocytes (Millette and Bellve, 1977; O'Rand and Romrell, 1977; Tung and Fritz, 1978). However, the possible functional significance of these surface components during germinal cell development is unknown. Similarly, the significance of specialized types of junctional complexes between somatic cells and germinal cells remains speculative (Ross, 1976; Russell, 1980). In the case of granulosa cells and oocytes, gap junctions are abundant (Anderson and Albertini, 1976), and metabolic exchange seems likely (Amsterdam et al., 1976; Beers, 1979; Gilula, 1977; Gilula et al., 1978; Moor et al., 1980).

Considerable evidence has accumulated concerning the nature of various substances produced and secreted by Sertoli cells and by granulosa cells, and the nature of hormonal modulation of some of these processes. Although mechanisms remain unknown, it appears increasingly likely that germinal cell development in ovary and in testis is dependent upon the specialized chemical composition in follicular or seminiferous tubule fluids. This unique microenvironment is created and maintained by neighboring somatic cells. In addition germ cell development could be influenced by the transfer of materials from surrounding somatic cells (Moor et al., 1980, 1981).

B. The Barrier between Lymph and Germinal Cells

The formation of a barrier by Sertoli cells results in the functional separation of germ cells from lymph surrounding the seminiferous tubule. Germ cells residing within the functional barrier created by gonadal somatic cells exist in a unique microchemical environment consisting of components in follicular or tubular fluid. The composition of fluid in the adluminal compartment of the seminiferous tubule is very different from that of the plasma (Setchell and Waites, 1975), whereas the composition of antral fluid in large follicles is much closer to that of the ultrafiltrate of plasma (Edwards, 1974).

The anatomical nature of the barrier separating oocytes from lymph is clearly different from that separating spermatocytes and spermatids from lymph. In early post-natal development, Sertoli cells have abundant gap junctions, but these do not contribute towards barrier formation. The number of tight junctional complexes (zona occludens) between neighboring Sertoli cells increases during gonadal development, eventually forming a functional circumferential barrier which has specific permeability properties and which exclude the passage of large molecules (Flickinger, 1967; Fawcett, 1975). The Sertoli cell is relatively huge, extending from basement membrane to lumen, a distance of 100 μ or more. In the mature testis, each Sertoli cell forms junctional complexes with each of its neighbors, and the total number of tight junctions per cell equals or exceeds the number of gap junctional complexes. Sertoli cells do not undergo mitotic division after sexual maturity is attained (Clermont and Perey, 1957; Steinberger and Steinberger, 1971; Griswold et al., 1977). However, the volume of Sertoli cells alters greatly during changes in functional status, dependent upon cell association patterns at different stages of the cycle of the seminiferous epithelium, and upon hormonal status. Functional compartments clearly exist within the seminiferous tubule, in which germinal cell mitosis and commitment to meiosis occurs (basal compartment); in which the lengthy prophase of meiosis carries spermatocytes to the zygotene stage (intermediary compartment) (Russell, 1978); and the adluminal compartment, in which meiosis is completed, all of spermiogenesis takes place, and spermatozoa are released into the lumen in the highly intricate process of spermiation (Russell and Clermont, 1977). A major point to be made in comparing the somatic cells of seminiferous tubules with those of follicles is that a single layer of Sertoli cells, of extraordinary sophistication and complexity, is responsible for compartmentation within the tubule. No physical membrane analogous to the zona pellucida exists around the adluminal compartment of the seminiferous tubule.

In contrast, a barrier analogous to that in the seminiferous tubule is not present in the follicle. Granulosa cells form abundant gap junctions with each other, but tight junctional complexes do not exist. Granulosa cells around the oocyte of a primary follicle are implicated in the synthesis of components required for formation of the zona pellucida, which surrounds and isolates the oocyte. The cumulus oophorus (corona radiata) cells send forth extensions which cross the zona pellucida and form intimate connections with oocyte organelles (Zamboni, 1970, 1974; Baker, 1972; Gilula et al., 1978). Various materials not taken up by isolated oocytes may be transferred to the oocyte from surrounding granulosa cells (Moor et al., 1980, 1981). In the mature follicle, just prior to ovulation, the connections between granulosa cells and oocyte are disrupted, and gap junctional complexes cease to exist (Anderson and Albertini, 1976; Gilula et al., 1978).

While ovulation and spermiation are formally analogous, it would be simplistic to expect identical mechanisms in the two sets of processes. Similarities between tubulobulbar complexes in Sertoli cell-spermatids prior to spermiation (Russell and Clermont, 1976), and the nature of the ultrastructure of connections between cumulus oophorus cells and oocytes prior to ovulation (Channing et al., 1980; Gilula et al., 1978; Zamboni, 1970, 1974), provide striking evidence of functional homology. It is not inconceivable that these homologous structures are associated with the remarkable stimulation of the formation of plasminogen activator which takes

Table 3. Inverse Relationships between Rates of Replication of Germ Cells and Supporting Somatic Cells in Ovary and Testis at Different Stages of Development

CELL TYPE	RELATIVE MITOTIC INDICES	
	Fetal and Perinatal Gonad	Sexually Mature Gonad
Granulosa cells	Low	High *
Oogonia	High	Zero
Sertoli cells	High	Zero
Spermatogonia	Low **	High

* The mitotic index of granulosa cells is highest in large pre-ovulatory follicles.

** During gonadogenesis, the mitotic index of early spermatogonia may be as high as that of early oogonia.

place in granulosa cells during ovulation (Strickland and Beers, 1976), and in Sertoli cells during spermiation (section II), respectively. However, the functions of plasminogen activator are different in the two cases. Rupture of the follicular wall required for ovulation is most likely dependent on plasmin formation from plasminogen in follicular fluid. In contrast, spermiation occurs in the apparent absence of plasminogen in tubular fluid of the adluminal compartment since plasminogen is probably restricted to the basal compartment.

Rates of germinal cell replication are inversely related to rates of division of gonadal supporting somatic cells, except during the earliest stages of gonadal development (Table 3). Relations schematically summarized may reflect differences in the nature of the diffusion barrier formed by granulosa and Sertoli cells. In testis, the mitotic index of Sertoli cells steadily decreases during post-natal sexual maturation. By 6 to 7 weeks of age in rats, the index approaches zero both in vivo (Nagy, 1972) and in vitro (Griswold et al., 1977). A functional barrier is formed by 3 to 4 weeks of age, at which time late stage pachytene spermatocytes have appeared. If meiosis does occur earlier, the process is aborted. Subsequently, steady rates of spermatogenesis are established and maintained until senescence.

Table 4. Summary of Biochemical Similarities between Granulosa and Sertoli Cells*

	Property	Granulosa Cells	Sertoli Cells
1.	Receptors to FSH	+	+
2.	Receptors to androgens	+	+
3.	Response to FSH		
	a) Adenylate cyclase	+	+
	b) Plasminogen activator	+	+
	c) Aromatase activity	+	+[i]
	d) Proteoglycan synthesis	+	+
4.	Inhibin Production	+	+
5.	High rates of aerobic glycolysis	+	+
6.	Receptors to H-Y antigen	+	+

i = immature

*See legend to Tables IV and V

LEGEND TO TABLES IV AND V

The sources of much of the information summarized are reviewed elsewhere (Dorrington et al., 1978; Ross and Lipsett, 1978; Fritz, 1978). Information on the production of inhibin by granulosa cells and Sertoli cells is reviewed by de Jong (this volume); the formation of "oocyte maturation inhibitor" by granulosa cells is reviewed by Tsafriri (this volume), and the utilization of various substrates by granulosa cells is reviewed by Hillensjo (this volume).

Evidence suggesting the release of H-Y antigen by Sertoli cells, and the existence of specific receptors for H-Y on granulosa and Sertoli cells, has been reviewed by Ohno (1979). Question marks indicate that information is not available, or that conflicting observations have not yet been resolved.

Sources of information not in the above reviews include the following: for proteoglycan synthesis by Sertoli cells, see Elkington and Fritz (1980); for inositol synthesis by Sertoli cells, see Robinson and Fritz (1979); for responses of granulosa cells to GnRH, and for the absence of responses of Sertoli cells, see Gore-Langton et al. (1981); and for the production of gonadocrinin by granulosa cells, see Ying et al. (1981).

Table 5. Summary of Biochemical Differences between Granulosa and Sertoli Cells*

Property	Granulosa Cells	Sertoli Cells
I. Only Sertoli Cells		
1. Anti-Müllerian hormone	-	$+^{f}$
2. Inositol production	-	+
3. H-Y antigen release	-	+
4. Androgen binding protein	?	+
II. Only Granulosa Cells		
1. Receptors and responses to LH or HCG	$+^{m}$	-
2. *De novo* steroidogenesis	$+^{m}$	-
3. Progesterone synthesis	$+^{m}$	-
4. Receptors and responses to prolactin	+	-
5. Receptors to estrogen	+	-
6. Receptors and responses to GnRH	+	?
7. Oocyte maturation inhibitor	+	?
8. Gonadocrinin	+	?

f = fetal m = immature

*See legend to Tables IV and V

It is well known that in mammals and birds all of oogenesis (except the final maturation of primary oocytes, arrested at diplotene) takes place during fetal and early perinatal life. In contrast, the mitotic index of granulosa cells depends upon the degree of maturity of the follicle in which the granulosa cells are located. Rapidly growing follicles, stimulated by estrogen and FSH, have highest rates of division of granulosa cells, whereas granulosa cells in small primary follicles have a low mitotic index (Peters, 1978; Richards, 1979). In fetal ovary the rate of formation of oocytes is high when the mitotic index of granulosa cells is low, and vice-versa in the maturing follicle. A reciprocal relation exists in testis: the mitotic index of Sertoli cells is high when the rate of formation of spermatocytes is low during fetal and perinatal stages of development, and vice-versa in the mature seminiferous tubule. This relatively obvious point has been belabored simply to emphasize that granulosa cells and Sertoli cells have differing rates of development prior to terminal differentiation. Immature Sertoli cells appear most closely homologous to immature granulosa cells. Most similarities and fewest biochemical differences are observed here (Table 4). As cyto-differentiation occurs, leading to terminal differentiation and specialized functions, other differences between mature granulosa and Sertoli cells become more evident (Table 5).

Tight junctional complexes between Sertoli cells are characterized by an extensive array of up to 50 rows of intramembrane particles on complementary grooves in highly oriented rows, evident in freeze-fractured membranes (Gilula et al., 1976). High trans-epithelial cell resistance is associated with membranes having very low non-specific permeability or "leakiness." A large potassium gradient exists across Sertoli cells (5 mM in plasma and approximately 50 mM in primary tubular fluid) (Setchell and Waites, 1975). This probably could not be maintained in the absence of the extraordinary tight junctional complexes among Sertoli cells. The absence of similar tight junctional complexes between granulosa cells would not favor the formation or maintenance of so high a concentration of potassium in follicular fluid.

V. DIFFERENCES BETWEEN INITIAL AND LATER WAVES OF SPERMATOGENESIS

A. General: A Consideration of the Nature of "Sterility" in Mice with Chromosomal Translocations

Male sterility occurs much more frequently than female sterility in mice having various chromosomal translocations (Searle, 1978; de Boer and Searle, 1980).

However, it is noteworthy that several of these strains can be propagated because testes from "sterile" males frequently produce viable spermatozoa during the initial waves of spermatogenesis. The arrest in later waves of spermatogeneis, frequently at some stage of primary spermatocyte meiosis, may be analogous to the considerable "atresia" which normally occurs during oogenesis. It is well appreciated that during fetal ovarian development, only a small percentage of oocytes from the initial potential population survives (Baker, 1972).

In mice bearing translocations leading to apparent sterility in the male but not in the female, it is possible that cells in the testis and in ovary suffer from the same genetic disability, but the effects of the translocation become manifest as total arrest of gametogenesis only when the gonad is challenged by the necessity for steady state maintenance of gametogenesis. Stated in an alternate manner, the initial waves of spermatogenesis may more closely resemble the single wave of oogenesis than subsequent waves of spermatogeneis do.

This supposition implies a difference in requirements for initiation of gametogenesis and maintenance of gametogenesis. It infers that the initial waves of spermatogenesis are in some manner different from those of subsequent steady state waves of spermatogenesis. While "atresia" may occur, it is incomplete, just as it is in the first and only wave of oogenesis.

B. Changes in Steroid Metabolism in the Testis during Puberty

Several sets of unique events are indeed associated with the successful initiaton of spermatogenesis following the onset of puberty. The most obvious are hormonal changes, in which levels of androgens are changing radically. During the first few days postnatally in rats, the number of Leydig cells diminishes (Christensen, 1975), and the initially high neonatal testosterone levels in plasma fall. As a new population of Leydig cells proliferates the amounts of 5α-reductase and 3α-hydroxysteroid dehydrogenase activity in Leydig cells are increased, resulting in the formation of large amounts of dihydrotestosterone and androstanediol relative to levels of testosterone produced by the testis. This is highest at about day 30 in the rat, and diminishes thereafter until normal adult levels of low testicular dihydrotestosterone and androstanediol are reached (Rivarola et al., 1972; 1975; van der Molen et al., 1975). High levels of androstanediols are being produced during the time when the seminiferous tubule barrier is becoming functionally competent (Fawcett, 1975; Tindall et al., 1975; Vitale et al., 1973). Another change in steroid metabolism taking place during this period involves the production of 17β-estradiol from testosterone. Highest rates of aromatase activity in Sertoli cells, dependent on FSH stimulation, occur in preparations from testes of young rats, and aromatase activity become minimal in cells from testes of 30 day old or older Wistar rats (Dorrington et al., 1978). In summary, during the initial wave of spermatogenesis, the rates of production of dihydrotestosterone and androstanediol by Leydig cells have reached their peak and begun to decline. At this stage of development, the rate of production of 17β-estradiol is also decreasing. In contrast, by 25 to 30 days of age in rats, testosterone levels are lowest, rising slowly thereafter until adult levels are reached (van der Molen et al., 1975).

C. Changes in Intratesticular Temperature at Puberty, and Changes in Temperature Sensitivity of Spermatogenesis

A second obvious change involves the descent of the testis into the scrotum at the time of the onset of puberty prior to the completion of the first complete wave of spermatogenesis. At this stage of development, the intratesticular temperature falls several degrees below that of the body core temperature. It is well appreciated that spermatogenesis in the cryptorchid testis of mammalian adults is arrested, and that exposure of the testis to elevated temperature results in the death of several classes of germinal cells, the most vulnerable of which in the rat are reported to be pachytene spermatocytes and young spermatids (for review, see van Demark and Free, 1970). However, spermatogenesis in the adult appears to be more temperature sensitive than spermatogenesis in immature rats. This has been shown in transplantation experiments in which tests from neonatal rats have been surgically transferred to an area adjacent to a cryptorchid testis of an adult rat. While the most advanced germinal cells in the adult testis of the host rat consisted primarily of spermatogonia and a few early spermatocytes, the germinal cells in the transplanted neonatal testis initially advanced almost all the way through meiosis. When this wave of pachytene spermatocytes died, subsequent waves of spermatogenesis were less successful (Chowdhury and Steinberger, 1972). More recently, Huckins and Cunningham (1981) have demonstrated that in tests of neonatal rats made unilaterally cryptorchid, the first waves of spermatogenesis appeared to be nearly identical in normal and cryptorchid testes up to 20 days of age. Even at 31 days of age, the abdominal testis contained modest numbers of spermatids. In contrast, by 45 days of age Sertoli cells and spermatogonia were the predominant cell types in most tubules of the cryptorchid testis, whereas advanced spermatids were present in the scrotal testis. Bergh (1981) has shown that the morphology of Sertoli cells is altered in germ cell depleted cryptorchid testes.

D. Progression of Spermatogonia to Pachytene Spermatocytes in Organ Cultures of Seminiferous Tuble Preparations from Neonatal Rats

An analogous observation relating to differences between the first and subsequent waves of spermatogenesis concerns the behavior of rat testis tubules in organ culture. In seminiferous tubule preparations from neonatal rats in organ culture, germ cells readily progress to pachytene spermatocytes, but they then abort. Subsequent waves do not progress so far. In seminiferous tubules from adult rats, germ cells rarely progress beyond the stage of leptotene spermatocytes (Steinberger and Steinberger, 1976). We have made similar observations with Sertoli cell-germ cell co-cultures (Solari and Fritz, unpublished observations).

Pachytene spermatocytes may initially appear in the seminiferous tubule at times prior to completion of a competent tubule barrier. Evidence for this can be found in the investment of pachytene spermatocytes from testes of 25 day old rats with materials which ordinarily do not penetrate the barrier, such as horse-radish peroxidase (Vitale et al., 1973). These data indicate that the initial waves of germinal cells can progress well into the prophase of meiosis without being entirely separated from components in the basal compartment. Thus, the unique microchemical environment in the adluminal compartment is not essential for the formation of pachytene spermatocytes during the first waves of spermatogenesis.

In contrast, all pachytene spermatocytes in tubules from testis preparations of adult rats appear to be located exclusively in the adluminal compartment, separated from components in the basal compartment by a functionally competent barrier (Fawcett, 1975; Gilula et al., 1976).

E. Differences in Responses to FSH by Sertoli Cells at Different Stages of Development

We have reviewed elsewhere the evidence that Sertoli cells prepared from testes of immature rats are more responsive to FSH than are cells from testes of older animals (Fritz, 1978). As previously stated, increased aromatase activity in response to FSH diminishes greatly in Sertoli cells prepared from testes of rats older than 30 days of age. This is not a consequence, however, only of decreased adenylate cyclase activity, since the response to dibutyryl cyclic AMP is also reduced in cells from mature animals (Dorrington et al., 1978). Addition of FSH increases the incorporation of [^{3}H]-thymidine into DNA of Sertoli cells prepared from testes of rats younger than 30-day-old, but not from testes of rats older than 40-days-old (Griswold et al., 1977).

While FSH is clearly required for one or more stages of initiation of spermatogenesis, it appears that androgens alone are sufficient to maintain spermatogenesis in adult hypophysectomized rodents (for review, see Fritz, 1978).

F. Possible Effects of the Absence of Advanced Germinal Cells on the Behavior of Sertoli Cells during Initial Waves of Spermatogeneis

As indicated above (Section II), plasminogen activator levels are considerably lower in testes during the first wave of spermatogenesis than in the sexually mature gonads. These low levels may render some aspect of the initial wave of spermatogenesis different from that in subsequent waves. The rapid increase in levels of testicular plasminogen activator activity during development between 40 and 60 days of age (Fig. 2) probably reflects an influence of advanced spermatids on rates of production of plasminogen activator by Sertoli cells. This interpretation is favored by the higher rates of production of plasminogen activator by cells in stages VII and VIII of the cycle of the seminiferous epithelium than by cells in other stages (Fig. 4). In essence, we postulate that Sertoli cell metabolism is altered by the presence of advanced germinal cells.

G. Are Sertoli Cells Different during Initial and Subsequent Waves of Spermatogenesis?

In brief recapitulation, Sertoli cells in the fetal gonad have unique functions, including a possible role in the organization of the indifferent gonad (Jost, 1973), in inhibiting meiosis of germinal cells, and in producing anti-Mullerian hormone. The capacity to produce anti-Mullerian hormone is lost shortly after birth (Josso et al., 1977). Aromatase activity in Sertoli cell preparations from testis of rats is not normally expressed except in tests from immature rats (Dorrington et al., 1978). During early post-natal life, Sertoli cells alter their morphology, and these changes are maintained by cell preparations in culture (Solari and Fritz, 1978). The formation of tight junctional complexes continues until a functionally competent barrier is completed (Flickinger, 1967).

Other changes also are occurring during the cytodifferentiation and maturation of Sertoli cells. For example, Sertoli cells at different stages of the cycle of the seminiferous epithelium have a characteristic cytology, and produce different products (Parvinen et al., 1980). These differences obviously are not detectable prior to completion of the first wave of spermatogenesis. During puberty, Sertoli cells may become temperature sensitive, concomitant with the descent of testes into the scrotum. For example, the production of androgen binding protein, one of the products of Sertoli cells (Fritz et al., 1974; 1976), is diminished in cryptorchid testes (Hagenas and Ritzen, 1976). Finally, several responses of Sertoli cells to FSH are most pronounced in preparations from testes of immature animals, and least in preparation from normal testes of adults.

VI. COMMENTS ON THE DIFFERENTIATION OF SERTOLI CELLS AND GRANULOSA CELLS

A. Speculations on the Comparison of Initial Waves of Spermatogenesis with Oogenesis

It is possible that biochemical and functional changes in Sertoli cells during their maturation and cytodifferentiation are causally related to differences between initial vs subsequent waves of spermatogenesis.

If this supposition has merit, immature Sertoli cells may share additional properties with immature granulosa cells which have not been characterized biochemically. For example, granulosa cells at all stages of differentiation are adapted to survive at 37°. Sertoli cells prior to descent of the testis into the scrotum must have the same capacity, but apparently become temperature sensitive as they differentiate following gonadal maturation at lower intratesticular temperatures. During the transition phase before becoming temperature sensitive, Sertoli cells may support the development of germinal cells more effectively, thereby permitting initial waves of spermatogenesis to proceed through meiosis, even at 37° or in organ culture.

Information reviewed suggests that initial waves of spermatogenesis, up to the formation of spermatocytes in late stages of meiosis, more closely resemble the single wave of oogenesis than do subsequent waves of spermatogenesis involved in steady state rates of production of spermatozoa.

B. Speculations on Comparisons between Somatic Cell-Germ Cell Relations in Testis and in Ovary

It is apparent that immature Sertoli cells share many functional and biochemical properties with immature granulosa cells (Table 4). As indicated above, another property should be added, namely the capacity to function normally at body core temperature, and to support the first waves of gametogenesis at 37°. In addition, both cell types in culture show the same kinds of morphological changes in response to FSH (Tung et al., 1975; Lawrence et al., 1979). In spite of an alleged common embryological origin, however, there are still obvious differences between the two types of cells. Most obviously, genetic influences are imposed by the Y chromosome, including the presence of H-Y antigen(s) on all male somatic cells and the putative absence of this antigen from all female somatic cells. However, a specific receptor for the H-Y antigen is thought to be present on granulosa cells (Ohno, 1979). The capacity of immature Sertoli cells but not granulosa cells to produce and secrete anti-Mullerian hormone (Josso et al., 1977) must somehow reflect this genetic heritage. In addition, the presence of estrogen receptors and GnRH receptors in granulosa cells but not in Sertoli cells is also implicated.

On the other hand, as Sertoli cells and granulosa cells terminally differentiate, the differences become more profound (Table 5). As granulosa cells mature they develop receptors to prolactin and LH, and they simultaneously develop the capacity to produce progestertone. The number of gap junctions between granulosa cells in different layers of the follicle prior to ovulation greatly exceeds the number of gap junctions between Sertoli cells in the seminiferous tubule. Granulosa cells do not have tight junctions, whereas tight junctional complexes between Sertoli cells are essential for seminiferous tubule function and the maintenance of spermatogenesis.

In spite of these profound differences, mature granulosa cells and Sertoli cells continue to share several interesting properties. As discussed above (Section II), these include the production of plasminogen activator, the rates of which depend upon stimulation by FSH, and upon the stage of development of follicle or stage of the cycle of the seminiferous epithelium. In ovary and in testis, germ cells are thought to be "nourished" by surrounding somatic cells. The high rates of aerobic glycolysis by Sertoli and granulosa cells provide lactate, which may be a preferential substrate for testicular germinal cells and oocytes, respectively. Many happy hours remain before we understand mechanisms involved, and the basis for the evolution of these symbiotic relations.

VII. SUMMARY AND CONCLUSIONS

Data have been reviewed indicating that a considerable number of biochemical similarities exist between immature Sertoli cells and immature granulosa cells. More differences appear as Sertoli cells and granulosa cells differentiate during gonadal maturation. Even here, however, several similarities remain in these homologous cells, including the capacity to synthesize and release plasminogen activator in response to FSH at particular stages of development.

The concept has been presented that the initial waves of spermatogenesis share more properties with the single wave of oogenesis than do subsequent waves of spermatogenesis in the mature adult. Differences between initial and subsequent waves of spermatogenesis are described, with emphasis on changes in Sertoli cells which occur during testicular maturation. The nature of interactions between gonadal somatic cells and germinal cells in testis has been compared with that in ovary. To illustrate these points, the control of the synthesis and secretion of plasminogen activator by Sertoli cells in culture has been described, and compared with comparable processes in granulosa cells. In addition, we have considered the possible functions in germinal cell development of high rates of aerobic glycolysis which are characteristic of Sertoli cells and granulosa cells.

ACKNOWLEDGMENTS

It is a pleasure to express deep gratitude to colleagues in this department, especially to Dr. Jennifer Dorrington, for valuable discussions of various topics in reproductive and cell biology, and for friendly interactions and criticisms during our "chatter sessions." I also wish to thank Drs. Martial Lacroix and Ranga Robinson, who provided much of the data on the properties of Sertoli cells reviewed here, and to acknowledge the excellent assistance of Donna McCabe, who patiently and skillfully helped me to prepare this manuscript under pressure.

Work from this laboratory was supported by grants from the Canadian MRC and NICHD.

REFERENCES

Ahrĕn, K., Dekel, N., Hamberger, L., Hillensjö, T., Hultborn, R., Magnusson, C., and Tsafriri, A., 1978, Metabolism and morphological changes produced by gonadotrophins and cyclic AMP in the oocyte and cumulus oophorus of preovulatory rat follicle, Ann. Biol. Anim. Biochem. Biophys., 18, (2B):409.
Amsterdam, A., Josephs, R., Lieberman, M. E., and Lindner, H. R., 1976, Organization of intramembrane particles in freeze-cleaved gap junctions of rat graafian follicles: Optical diffraction analysis, J. Cell Sci., 21:93.
Anderson, E., and Albertini, D. F., 1976, Gap junctions between the oocyte and companion follicle cells in the mammalian ovary, J. Cell Biol., 71:680.
Baker, T. G., 1972, Oogenesis and ovulation, in: "Germ Cells and Fertilization," C. R. Austin and R. V. Short, eds., pp. 14-45, Cambridge University Press.
Beers, W. H., 1979, The transmission of hormonal stimulation by a cell contact dependent mechanisms: In vitro studies and implications for hormone action in vivo, in: "Hormones and Cell Culture," Cold Spring Harbor Conferences on Cell Proliferation, Vol. 6, Cold Spring Harbor Laboratory.
Beers, W., Strickland, S., and Reich, E., 1975, Ovarian plasminogen activator: relationship to ovulation and hormonal regulation, Cell, 6:387.
Bergh, A., 1981, Morphological signs of a direct effect of experimental cryptorchidism on the Sertoli cells in rats irradiated as fetuses, Biol. Reprod., 24:145.
Biggers, J. D., Whittingham, D. G., and Donahue, R. P., 1967, The pattern of energy metabolism in the mouse oocyte and zygote, Proc. Natl. Acad. Sci. (USA), 58:560.
Byskov, A. G., and Saxen, L., 1976, Induction of meiosis in fetal mouse testis in vitro, Develop. Biol., 52:193.
Channing, C. P., Schaerf, F. W., Anderson, L. D., and Tsafriri, A., 1980, Ovarian follicular and luteal physiology, Internat. Rev. Physiol., 22:117.
Chowdhury, A. K., and Steinberger, E., 1972, The influence of a cryptorchid milieu on the initiation of spermatogenesis in the rat, J. Reprod. Fert., 29:173.
Christensen, A. K., 1975, Leydig cells, in: "Handbook of Physiology: Male Reproduction," Vol. V of section 7, R. O. Greep and E. W. Hamilton, eds., pp. 57-94, Amer. Physiol. Society, Washington, D.C.
Clermont, Y., and Perey, B., 1957, Quantitative study of the cell population of the seminiferous tubules in immature rats, Am. J. Anat., 100:241.
de Boer, P., and Searle, A. G., 1980, Workshop on chromosomal aspects of male sterility in mammals: summary and synthesis, J. Reprod. Fertil., 60:259.

de Jong, F. H., and Sharpe, R. M., 1977, The onset and establishment of spermatogenesis in rats in relation to gonadotrophins and testosterone levels, J. Endocrinol., 75:197.
Dorrington, J. H., Fritz, I. B., and Armstrong, D. T., 1978, Control of testicular estrogen synthesis, Biol. Reprod., 18:55.
Edwards, R. G., 1974, Follicular fluid, J. Reprod. Fertil., 37:189.
Elkington, J. S. H., and Fritz, I. B., 1980, Regulation of sulfoprotein synthesis by rat Sertoli cells in culture, Endocrinology, 107:970.
Fawcett, D. W., 1975, The ultrastructure and functions of the Sertoli cell, in: "Handbook of Physiology: Male Reproduction," Vol. V of section 7, R. O. Greep and E. W. Hamilton, eds., pp. 21-55, Amer. Physiol. Soc., Washington, D.C.
Flickinger, C. J., 1967, The postnatal development of the Sertoli cell of the mouse, Z. Zellforsch. Mikroskop. Anat., 78:92.
Fritz, I. B., 1978, Sites of action of androgens and follicle stimulating hormone on cells of the seminiferous tubule, in: "Biochemical Actions of Hormones," Vol. 5, G. Litwack, ed., pp. 249-281, Academic Press, New York.
Fritz, I. B., Kopec, B., Lam, K., and Vernon, R. G., 1974, Effects of FSH on levels of androgen binding protein in the testis, in: "Hormone Binding and Target Cell Activation in the Testis," M. Dufau and A. R. Means, eds., pp. 311-327, Plenum Press, New York.
Fritz, I. B., Parvinen, M., Karmally, K., and Lacroix, M., 1981, Localizaton of tsticular plasminogen activator in discrete portions (Stages VII and VIII of the seminiferous tubule, in: "Cell Biology of Testis," New York Academy of Sciences, in press.
Fritz, I. B., Rommerts, F. G., Louis, B. G., and Dorrington, J. H., 1976, Regulation by follicle stimulating hormone and dibutyryl cyclic AMP of androgen binding protein formation by Sertoli cell-enriched cultures, J. Reprod. Fertil., 46:17.
Gilman, J., 1948, The development of the gonads in man, with a consideration of the role of fetal endorines and the histogenesis of ovarian tumors, Contrib. to Embryol. #210, Carnegie Inst., 32:81.
Gilula, N. B., 1977, Gap junctions and cell communication, in: "International Cell Biology," B. R. Brinkley and K. R. Porter, eds., pp. 61-69, Rockefeller University Press, New York.
Gilula, N. B., Epstein, M. L., and Beers, W. H., 1978, Cell-to-cell communication and ovulation, J. Cell Biol., 78:58.
Gilula, N. B., Fawcett, D. W., and Aoki, A., 1976, Ultastructural and experimental observations on the Sertoli cell junctions of the mammalian testis, Devel. Biol., 50:142.
Gondos, B., 1977, Testicular development, in: "The Testis, Vol. IV," A. D. Johnson and W. R. Gomes, eds., pp. 1-37, Academic Press, New York.

Gore-Langton, R. E., Lacroix, M., and Dorrington, J. H., 1981, Differential effects of luteinizing hormone-releasing hormone on follicle-stimulating hormone-dependent responses in rat granulosa cells and Sertoli cells in vitro, Endocrinology, 108:812.

Griswold, M. D., Solari, A., Tung, P. S., and Fritz, I. B., 1977, Stimulation by FSH of DNA synthesis and of mitosis in cultured Sertoli cells prepared from testes of immature rats, Molec. Cellular Endocrinol., 7:151.

Hagenas, L., and Ritzen, E. M., 1976, Impaired Sertoli cell function in experimental cryptorchidism in the rat, Molec. Cell Endocrinol., 4:25.

Huckins, C., and Cunningham, G. R., 1981, Morphological and immunological characteristics of unilaterally cryptorchid rat testes, in: "Cell Biology of Testis," New York Academy of Sciences, in press.

Josso, N., Picard, J.-Y., and Tran, D., 1977, The Anti-Mullerian hormone, Rec. Prog. Hormone Res., 33:117.

Jost, A., 1970, Hormonal factors in the sex differentiation of the mammalian foetus, Phil. Trans. Roy. Soc. Lond. B., 259:119.

Jost, A., Vigier, B., Prepin, J., and Perchellet, J. P., 1973, Studies on sex differentiation in mammals, Sex differentiation, Rec. Prog. Hormone Res., 29:1.

Jost, A., Magre, S., and Cressant, M., and Perlman, S., 1974, Sertoli cells and early testicular differenitation, in: "Male Fertility and Sterility," R. E. Mancini and L. Martini, eds., pp. 1-11, Academic Press, New York.

Jutte, N. H. P. M., Grootegoed, J. A., Rommerts, F. F. G., and van der Molen, H. J., 1981, Exogenous lactate is essential for metabolic activities in isolated spermatocytes and psermatids, J. Reprod. Fertil., 62:(2)399.

Lacroix, M., and Fritz, I. B., 1980, Characterization of a plasminogen activator secreted by rat Sertoli cells in culture, J. Cell Biol., 87: No. 2, Part 2, 152a.

Lacroix, M., Parvinen, M., and Fritz, I. B., 1981, Localization of testicular plasminogen activator in discrete portions (Stages VII and VIII) of the seminiferous tubule, Biol. Reprod., 25:143.

Lacroix, M., Smith, F. E., and Fritz, I. B., 1977, Secretion of plasminogen activator by Sertoli cell enriched cultures, Molec. Cell Endocrinol., 9:227.

Lacroix, M., Smith, F., and Fritz, I. B., 1979, The control of plasminogen activator secretion by Sertoli cells in culture, and its possible role in spermatogenesis, in: "Hormones and Cell Culture," Cold Spring Harbor Conferences on Cell Proliferation, Vol. 6, Cold Spring Harbor Laboratory, p. 239.

Lawrence, T. S., and Ginzberg, R., Gilula, N. B., and Beers, W. H., 1979, Hormonally induced cell shape changes in cultures rat ovarian granulosa cells, J. Cell Biology, 80:21.

Leblond, C. P., and Clermont, Y., 1952, Definition of the stages of the cycle of the seminiferous epithelium in the rat, Ann. New York Academy of Sciences, 55:548.

Merchant, H., 1975, Rat gonadal and ovarian organogenesis with and without germ cells. An ultrastructural study, Develop. Biol., 44:1.

Millette, C. F., and Bellve, A. R., 1977, Temporal expression of membrane antigens during mouse spermatogenesis, J. Cell Biol., 74:86.

Moor, R. M., Smith, M. W., and Dawson, R. M. C., 1980, Measurement of intercellular coupling between oocytes and cumulus cells using intercellular markers, Experimental Cell Res., 126:15.

Moor, R. M., Osborn, J. C., Dran, D. G., and Walters, D. E., 1981, Selective effect of gonadotrophins on cell coupling nuclear maturation and protein synthesis in mammalian oocytes, J. Embryol. Morph., 61:347.

Nagy, F., 1972, Cell division kinetics and DNA synthesis in the immature Sertoli cells of the rat testis, J. Reprod. Fertil., 28:389.

Nakamura, M., Hino, A., and Kato, J., 1981, Regulation of protein synthesis in spermatids from rat testes: Role of adenosine triphosphate (ATP), J. Biochem. (Tokyo), 89:1309.

Nilsson, L., 1974, Acute effects of gonadotrophins and prostaglandins on the metabolism of isolated ovarian follicle from PMSG-treated immature rats, Acta Endocr. (Kbh), 77:540.

Ohno, S., 1979, Major Sex Determining Genes, Springer Verlag, New York.

O'Rand, M. G., and Romrell, L. J., 1977, Appearance of cell surface auto and isoantigens during spermatogenesis in the rabbit, Develop. Biol., 55:347.

Parvinen, M., Marana, R., Robertson, D. M., Hansson, V., and Ritzen, E. M., 1980, Functional cycle of rat Sertoli cells: differential binding and action of FSH at various stages of the spermatogenic cycle, in: "Testicular Development Structure and and Function," A. Steinberger and E. Steinberger, eds., pp. 425-432, Raven Press, New York.

Parvinen, M., and Vanha-Perttula, T., 1972, Identification and enzyme quantitation of the stages of the seminiferous epithelial wave in the rate, Anat. Rec., 174:435.

Peters, H., 1978, Folliculogensis in mammals, in: "The Vertebrate Ovary: Comparative Biology and Evolution," R. Jones, ed., pp. 121-144, Plenum Press, New York.

Richards, J., 1979, Hormonal control of ovarian follicular development: A 1978 perspective, Rec. Prog. Hormone Res., 35:343.

Rivarola, M. A., Podesta, E. J., and Chemes, H. E., 1972, In vitro testosterone ^{14}C metabolism by rat seminiferous tubules at different stages of development: Formation of 5α-androstandiol at meiosis, Endocrinol., 91:537.

Rivarola, M. A., Podesta, E. J., Chemes, H. E., and Cigorraga, S., 1975, Androgen metabolism in the seminiferous tubule, Curr. Top. Molec. Endocrinol., 2:25.

Robinson, R., and Fritz, I. B., 1979, Myoinositol biosynthesis by Sertoli cells, and levels of myoinositol biosynthetic enzymes in testis and epididymis, Can. J. Biochem., 57:962.

Robinson, R., and Fritz, I. B., 1981, Metabolism of glucose by Sertoli cells in culture, Biol. Reprod., 24:1032.

Ross, M. H., 1976, The Sertoli cell junctional specialization during spermiogenensis and at spermiation, Anat. Rec., 196:79.

Ross, G. T., and Lipsett, M. B., 1978, Homologies of structure and function in mammalian testes and ovaries, Internat. J. Androl. Suppl., 2:39.

Russell, L., 1977, Movement of spermatocytes from the basal to the adluminal compartment of the rat testis, Amer. J. Anat., 148: 313.

Russell, L., 1978, Desmosome-like junctions between Sertoli cells and germ cells in rat testis., Amer. J. Anat., 148:201.

Russell, L. D., 1980, Sertoli-germ cell interrelations: A review, Gamete Research, 3:179.

Russell, L., and Clermont, Y., 1976, Anchoring device between Sertoli cells and late spermatids in rat seminiferous tubules, Anat. Rec., 185:259.

Searle, A. G., Beechey, C. V., and Evans, E. P., 1978, Meiotic effects in chromosomally derived male sterility of mice, Ann. Biol. Anim. Biochem. Biophys., 18:(2B).

Setchell, B., and Waites, G. M. H., 1975, The blood testis barrier, in: "Handbook of Physiology. Male Reproduction, Vol. V, Sec. 7, Endocrinology Section," R. O. Greep and E. W. Hamilton, eds., pp. 143-172, Amer. Physiol. Soc., Washington, D.C.

Solari, A., and Fritz, I. B., 1978, The ultrastructure of immature Sertoli cells maturation-like changes during culture, and the maintenance of mitotic potentiality, Biol. Reprod., 18:329.

Stickland, S., and Beers, W. H., 1976, Studies on the role of plasminogen activator in ovulation. In vitro response of granulosa cells to gonadotrophins, cyclic nucleotides and prostaglandins, J. Biol. Chem., 251:5694.

Steinberger, A., and Steinberger, E., 1971, Replication pattern of Sertoli cells in maturing rat testis in vivo and in organ culture, Biol. Reprod., 4:84.

Steinberger, A., and Steinberger, E., 1976, Action of gonadotropins in testis organ and cell culture, Prog. Reprod. Biol., 1:42.

Tindall, D. J., Vitale, R., and Means, A. R., 1975, Androgen binding protein as a biochemical marker of formation of the blood-testis barrier, Endocrinol., 97:636.

Tsafriri, A., Lieberman, M. E., Ahren, K., and Lindner, H. R., 1976, Dissociation between LH-induced aerobic glycolysis and oocyte maturation in cultured Graafian follicles of the rat, Acta Endocrinol. (Kbh), 83:151.

Tung, P. S., Dorrington, J. H., and Fritz, I. B., 1975, Structural changes induced by follicle stimulating hormone or dibutyryl cyclic AMP on presumptive Sertoli cells in culture, Proc. Natl. Acad. Sci. (U.S.A.), 72:1838.

Tung, P. S., and Fritz, I. B., 1978, Specific surface antigens on rat pachytene spermatocytes and successive classes of germinal cells, Developmental Biol., 64:297.

van Demark, N. L., and Free, M. J., 1970, Temperature effects, in: "The Testis, Vol. III," A. D. Johnson, W. R. Gomes, and N. L. van Demark, eds., pp. 233-312, Academic Press, New York.

van der Molen, H. J., Grootegoed, J. A., de Greef-Bijleveld, M. J., Rommerts, F. F. G., and van der Vusse, G. J., 1975, Distribution of steroids, steroid production and steroid metabolizing enzymes in rat testis, Curr. Topics Molecular Endocrinol., 2:3.

Vitale, R., Fawcett, D. W., and Dym, M., 1973, The normal development of the blood-testis barrier and the effects of clomiphene and estrogen treatment, Anat. Rec., 176:333.

Ying, S.-Y., Ling, N., Bohlen, P., and Guillemin, R., 1981, Gonadocrinins: Peptides in ovarian follicular fluid stimulating the secretion of pituitary gonadotropins, Endocrinology, 108:1206.

Zamboni, L., 1970, Comparative studies on the ultrastructure of mammalian oocytes, in: "Oogenesis," J. D. Biggers and A. W. Schuetz, eds., pp. 5-45, University Park Press, Baltimore.

Zamboni, L., 1974, Fine morphology of the follicle wall and follicle cell-oocyte association, Biol. Reprod., 10:125.

Top Row:

Leo Reichert, Alexander Tsafriri, Brian Rigby, Aaron Hsueh, Darrel Ward, Robert Ryan, Lars Bjersing, Harry Lipner, Sidney Strickland, Irving Fritz, Torbjörn Hillensjö, Fernand Labrie, Jerome Strauss III, Frank de Jong.

Bottom Row:

Larry Anderson, Eimei Sato, M. R. Sairam, Neena Schwartz, Cornelia P. Channing, S. Y. Ying, Lakshmi Kumari, Sheldon Segal, James Hammond.

Bellagio, 1981

Conference on Intraovarian Control Mechanisms

LIST OF CONTRIBUTORS

Larry D. Anderson - Department of Anatomy, University of Maryland School of Medicine, 660 West Redwood Street, Baltimore, Maryland 21201.

Lars Bjersing - Institute of Pathology, University of Umeå, S-901 87 Umeå, Sweden.

Cornelia P. Channing - Department of Physiology, University of Maryland School of Medicine, 660 West Redwood Street, Blatimore, Maryland 21201.

Irving B. Fritz - Department of Medical Research, University of Toronto, Toronto, Ontario M5G 1L6.

James M. Hammond - Division of Endocrinology, Milton S. Hershey Medical Center, Pennsylvania State University, Hershey, Pennsylvania 17022.

Torbjörn Hillensjö, Department of Physiology, University of Göteborg, P. O. Box 33031, S-400 33 Göteborg, Sweden.

Aaron J. W. Hsueh - Department of Reproductive Medicine, University of California, San Diego, La Jolla, California 92093.

F. H. de Jong - Department of Biochemistry, Erasmus University Rotterdam, Rotterdam, The Netherlands.

G. Lakshmi Kumari, National Institute of Health & Family Welfare, Munirka, New Delhi 110067, India.

Fernand Labrie - Department of Molecular Endocrinology, Le Centre Hospitalier de l'Université Laval, 2705 Laurier Boulevard, Quebec G1V 4G2, Canda.

Florence Ledwitz-Rigby - Department of Biological Sciences, Northern Illinois University, DeKalb, Illinois 60115.

Harry Lipner - Department of Biological Sciences, Florida State University, Tallahassee, Florida 32313.

Leo E. Reichert, Jr. - Department of Biochemistry, Albany Medical College, Albany, New York 12208.

Robert J. Ryan - Department of Cell Biology, Mayo Clinic, Rochester, Minnesota 55901.

M. R. Sairam - Reproduction Research Laboratory, Clinical Research Institute of Montreal, 110 ave. des Pins ouest, Montreal, Quebec H2W 1R7, Canada.

Eimei Sato - Department of Animal Science, College of Agriculture, Kyoto University, Kyoto, Japan 606.

Neena B. Schwart - Department of Biological Sciences, Northwestern University, Evanston, Illinois 60201.

Jerome F. Strauss, III - Department of Obstetrics and Gynecology, University of Pennsylvania School of Medicine, Philadelphia, Pennsylvania 19104.

Sidney Strickland - The Rockefeller University, 1230 York Avenue, New York, New York 10021.

A. Tsafriri - Department of Hormone Research, Weizman Institute of Science, P. O. Box 26, Rehovot, Israel.

Darrell N. Ward - Division of Biochemistry, University of Texas System Cancer Center, M. D. Anderson Hospital and Tumor Institute, 6723 Bertner Avenue, Houston, Texas 77030.

Shao-Yao Ying - Laboratories for Neuroendocrinology, The Salk Institute, P. O. Box 85800, San Diego, California 92138.

Sheldon J. Segal - The Rockefeller Foundation, 1133 Avenue of the Americas, New York, New York 10036.

AUTHOR INDEX

SUBJECT INDEX

MIX
Papier aus verantwortungsvollen Quellen
Paper from responsible sources
FSC® C105338

If you have any concerns about our products,
you can contact us on
ProductSafety@springernature.com

In case Publisher is established outside the EU,
the EU authorized representative is:
Springer Nature Customer Service Center GmbH
Europaplatz 3, 69115 Heidelberg, Germany

Printed by Libri Plureos GmbH
in Hamburg, Germany